Climate Change and Cultural Transition in Europe

Climate and Culture

Series Editors

Carmen Meinert
Claus Leggewie

VOLUME 4

The titles published in this series are listed at *brill.com/clac*

Climate Change and Cultural Transition in Europe

Edited by

Claus Leggewie
Franz Mauelshagen

BRILL

LEIDEN | BOSTON

Cover illustration: The Rhone Glacier around 1900 and in 2008. COPYRIGHT: JÜRG ALEAN, EGLISAU, SWITZERLAND.

Library of Congress Cataloging-in-Publication Data
Names: Leggewie, Claus, 1950- editor. | Mauelshagen, Franz, 1967- editor.
Title: Climate change and cultural transition in Europe / edited by Claus Leggewie, Franz Mauelshagen.
Description: Leiden ; Boston : Brill, 2018. | Series: Climate and culture, ISSN 2213-0519 ; 4 | Includes bibliographical references and index.
Identifiers: LCCN 2017042288 (print) | LCCN 2017049973 (ebook) | ISBN 9789004356825 (E-book) | ISBN 9789004356429 (hardback : alk. paper)
Subjects: LCSH: Climate change mitigation--Social aspects--Europe. | Climate change mitigation--Europe.
Classification: LCC TD171.75 (ebook) | LCC TD171.75 .C55 2017 (print) | DDC 304.2/5094--dc23
LC record available at https://lccn.loc.gov/2017042288

Typeface for the Latin, Greek, and Cyrillic scripts: "Brill". See and download: brill.com/brill-typeface.

ISSN 2213-0519
ISBN 978-90-04-35642-9 (hardback)
ISBN 978-90-04-35682-5 (e-book)

This book is printed on acid-free paper and produced in a sustainable manner.

Contents

PART 3
Memoirs

PART 4
Present

PART 5
Prospects

Foreword

Man influences the environment and climate and the consequences are now felt around the globe. National or regional efforts to restrict or at least contain the damage can only be insufficient: in principle environmental and climate protection needs a global concept.

Paradoxically, the way we *perceive* environmental and climate change and handle damage is closely linked to local or regional patterns of perception. It is these particularistic perceptions that often lead to different, in many instances opposing reactions to preventive and curative environmental and climate protection measures.

This local view is grounded not only in different ways of socio-economic development in different regions of the world, but also in differences in *cultural patterns*. Think, for example, of the strongly varying notions of the actual problems at hand, or of different policy styles and politico-social environments. Also, the disturbance of the environment and climate causes relatively rapid social changes, in which the interpretation of symbols for the relationship between man and nature plays an important part. The history of climate and culture, patterns of perception of environmental and climate change and an informed assessment of the future direction of environmental and climate policy in various parts of the world have to be taken into account in order to get to grips with the problem.

From a variety of angles, such as the history of ideas, historiography, the study of civilisation, and the political sciences, the monographs and edited volumes in *Climate and Culture* will all deal with the following questions:

- How do local and regional cultures perceive changes in the environment and climate in past and present?
- How did and do they adjust to them?
- How do their various representatives and spokesmen introduce their respective views to the global debate and into emerging international negotiating systems?

The following titles will be included in the series:

Volume 1: *Nature, the Environment and Climate Change in East Asia*, edited by Carmen Meinert, 2013.

Volume 2: *Climate Change in South and Southeast Asia*, edited by Barbara Schuler, 2014.
Volume 3: *Cultural Dynamics of Climate Change and the Environment in Northern America*, edited by Bernd Sommer, 2015.
Volume 4: *Climate Change and Cultural Transition in Europe*, edited by Claus Leggewie and Franz Mauelshagen 2017.

Carmen Meinert
Claus Leggewie

Acknowledgements

The publication of this book was only possible through the support of numerous people and institutions. First of all, we have to thank the co-editor of the book series *Climate and Culture*, Carmen Meinert, for her patience.

Our deep appreciation goes also to the contributors of this volume. We are very grateful for their scholarly work but also their long patience. Editing a volume of more than four hundred pages with eleven contributions by fourteen authors with a broad spectrum of academic expertise was a very laborious process, which took much more time than we—and most contributors—thought.

The anonymous reviewers of the work also deserve our gratitude! Due to their profound critique and suggestions the anthology as a whole as well as many individual contributions have gained a lot.

Furthermore, I am most obliged to Nele Fabian, Thorben Pelzer, Sebastian Sponheuer, and Ben Müller. They accompany the editing process of the whole Climate and Culture series at the KWI. Without their invaluable knowledge and practical assistance, their enduring commitment and routine, and their patience this volume could not have been published. Iain Sinclair has done a tremendous job in proofreading the whole volume. Patricia Radder from Brill exemplarily accompanied the production process and gave valuable assistance, too. I owe my gratitude to all of them.

Special thanks go to Jürg Alean from Eglisau, Switzerland for his permission to use his photographs of the Rhône Glacier on the front cover of this book.

Claus Leggwie and Franz Mauelshagen

List of Figures and Tables

Figures

Tables

Abbreviations

AD	Anno Domini (in the year of the Lord)
ADP	Platform for Enhanced Action
AIDS	Acquired Immunodeficiency Syndrome
am	ante meridiem
AOSIS	Alliance of Small Island States
Art.	Artikel (English: article)
AWG-LCA	Ad Hoc Working Group on Long-term Co-operative Action
BASIC	Brazil, South Africa, India and China
BCE	before the Common Era/before Christian
BNA	Federal Network Agency (Bundesnetzagentur, Germany)
BMU	Federal Ministry for the Environment, Nature Conservation and Nuclear Safety (Bundesministerium für Umwelt) Naturschutz und Reaktorsicherheit, Germany)
BRICS	Brazil, Russia, India, China, South Africa
CDU	Christian Democrat Union (Germany)
CE	Common/Current/Common Era
Cf.	confer
ch.	Chapter
CHF	Swiss Franc
CHP	Combined Heat and Power
CLIWOC	Climatological Database for the World's Oceans
CO_2	Carbon dioxide
COP	Conference of the Parties
ed.	edited by
EEG	Renewable Energy Sources Act (Erneuerbare-Energien-Gesetz, Germany)
ENSO	*El Niño*—Southern Oscillation
ESC	Economies Sociétés Civilisations
et al.	et alii/aliae/alia
ETS	Emissions Trading Scheme
EU	European Union
esp.	especially
FDP	Free Democratic Party (Germany)
Frankfurt a.M.	Frankfurt am Main
Freiburg i.Br.	Freiburg im Breisgau
G7	Group of Seven
G8	Group of Eight

G20	Group of Twenty
GDP	Gross Domestic Product
GHG	Greenhouse Gas
Gr.	Greek
HIV	Human Immunodeficiency Virus
HWWI	Hamburg Institute of International Economics (Hamburgisches WeltWirtschaftsInstitut, Germany)
Ibid.	Ibidem
i.e.	id est
IEA	International Energy Agency
IIASA	International Institute for Applied Systems Analysis
IMC	International Meteorological Committee
IMF	International Monetary Fund
INDCs	Intended Nationally Determined Contributions
IPCC	Intergovernmental Panel on Climate Change
KfW	German Investment and Development Bank (Kreditanstalt für Wiederaufbau, Germany)
kWh	Kilowatt Hour
Lat.	Latin
LCO	Little Climatic Optimum
LDCs	least developed countries
LIA	Little Ice Age
Loc. cit.	loco citato
LSCE	Climate and Environment Sciences Laboratory (Laboratoire des Sciences du Climat et de l'Environnement, France)
LULUCF	Land Use, Land-Use Change and Forestry
MDG	Millennium Development Goals
MXD	Maximum Latewood Density
MW	Megawatt
NAO	North Atlantic Oscillation
NDCs	Nationally Determined Contributions
OcCC	Organe consultatif sur les changements climatiques/Beratendes Organ für Fragen der Klimaforschung
OECD	Organisation for Economic Co-operation and Development
pm	post meridiem
PV	photovoltaic
PVV	Party for Freedom (Partij voor de Vrijheid, The Netherlands)
RECLIDO	*REd para la reconstrucción del CLIma a partir de fuentes DOcumentales*
REDD+	Reducing Emissions from Deforestation and Forest Degradation
RPC	Representative Concentration Pathways

RWS	*Rijkswaterstaat*
SDGS	Sustainable Development Goals
UN	United Nations
UK	United Kingdom
UNFCCC	United Nations Framework Convention on Climate Change
US	United States (of America)
USA	United States of America
vol.	Volume
VW	Volkswagen
WMO	World Meteorological Organisation
WWF	World Wide Fund for Nature

Notes on Contributors

Brandi, Clara

is an economist and political scientist. She is Senior Researcher and Project Leader in the Department at the German Development Institute/Deutsches Institut für Entwicklungspolitik (DIE) and specializes in. She studied economics at the University of Wisconsin, Madison, and at the Albert-Ludwigs-Universität, Freiburg, where her Master thesis won the Friedrich-August-von-Hayek-Award. As Michael-Wills-Scholar, she completed the MPhil in Politics at the University of Oxford. She received her PhD from the European University Institute, Florence. Her recent publications include: Brandi, Clara, Sander Chan, and Steffen Bauer. "Aligning Transnational Climate Action with International Climate Governance: The Road from Paris." *Review of European Community & International Environmental Law* (RECIEL) 25.2 (2016): 238-247; and: Brandi, Clara. "Sustainability Standards and Sustainable Development – Synergies and Trade-Offs of Transnational Governance." *Sustainable Development* 25 (2017): 25-34.

Glaser, Rüdiger

is a Geographer. He is full Professor of Physical Geography at the University of Freiburg and specialises in Historical Climatology, Global Change Studies and IT based concepts. His recent publications include: Camenisch, Chantal, Kathrin M. Keller, Melanie Salvisberg, Benjamin Amann, Martin Bauch, Sandro Blumer, Rudolf Brázdil, Stefan Brönnimann, Ulf Büntgen, Bruce M.S. Campbell, Laura Fernández-Donado, Dominik Fleitmann, Rüdiger Glaser, Fidel González-Rouco, Martin Grosjean, Richard C. Hoffmann, Heli Huhtamaa, Fortunat Joos, Andrea Kiss, Oldřich Kotyza, Flavio Lehner, Jürg Luterbacher, Nicolas Maughan, Raphael Neukom, Theresa Novy, Kathleen Pribyl, Christoph C. Raible, Dirk Riemann, Maximilian Schuh, Philip Slavin, Johannes P. Werner, Oliver Wetter. "The Early Spörer Minimum. A Period of Extraordinary Climate and Socio-economic Changes in Western and Central Europe." *Climate of the Past Discussions* (2016). Accessed April 4, 2017. <http://www.clim-past-discuss.net/cp-2016-7/cp-2016-7.pdf>; and: Riemann, Dirk, Rüdiger Glaser, Michael Kahle, and Steffen Vogt. "The CRE tambora.org – New Data and Tools for Collaborative Research in Climate and Environmental History." *Geoscience Data Journal* (2016). <http://dx.doi.org/10.1002/gdj3.30>.

Himmelsbach, Iso
is a sociologist and historian and earned a PhD in Historical Climatology. He is a research associate in the department of Physical Geography at the University of Freiburg and specialized in flood and drought research. His recent publications include: Himmelsbach, Iso, Rüdiger Glaser, Johannes Schönbein, Dirk Riemann, and Brice Martin. "Reconstruction of Flood Events Based on Documentary Data and Transnational Flood Risk Analysis of the Upper Rhine and Its French and German Tributaries since AD 1480." *Hydrology and Earth System Sciences* 19 (2015): 4149-4164; and Martin, Brice, Axel Drescher, Marie Fournier, Ouarda Guerrouah, Florie Giacona, Rüdiger Glaser, Iso Himmelsbach, Nicolas Holleville, Dirk Riemann, Johannes Schönbein, Marie-Claire Vitoux, and Lauriane With. "Les évènements extrêmes dans le fossé rhénan entre 1480 et 2012. Quels apports pour la prévention des inondations?" *La Houille Blanche* 2 (2015): 82-93.

Kemfert, Claudia
has been Head of the Department of Energy, Transportation, Environment at the German Institute of Economic Research (DIW Berlin) since April 2004, and Professor of Energy, Economics, and Sustainability at Berlin's Hertie School of Governance since 2009. She studied economics at Oldenburg, Bielefeld, and Stanford University. Her research activities concentrate on the evaluation of climate and energy policy strategies. In 2016 she was appointed as a member of the German Advisory Council on the Environment. Her recent publications include Kemfert, Claudia, Friedrich Kunz, and Juan Rosellón. "A Welfare Analysis of Electricity Transmission Planning in Germany." *Energy Policy* 94 (2016): 446-452; and Kemfert, Claudia. *The Battle about Electricity: Myths, Power and Monopolies*. Hamburg: Murmann, 2013.

Le Roy Ladurie, Emmanuel
is a historian of pre-modern history with a broad spectrum of expertise ranging from agrarian history, the history of the inquisition, to climate history. He is Professor emeritus at the Collège de France, where he held the Chair of History of Modern Civilization from 1973-1999. He is one of the founding fathers of historical climatology. His publications in this field include: Le Roy Ladurie, Emmanuel. *Histoire du climat depuis l'an mil*. Paris: Flammarion, 1967 (engl. translation 1973); Le Roy Ladurie, Emmanuel. *Histoire humaine et comparée du climat*. Paris: Fayard, 2004; and: Le Roy Ladurie, Emmanuel, and Daniel Rousseau. *Naissance de l'histoire du climat*. Paris: Hermann, 2013.

Leggewie, Claus
is a political scientist and historian and specialises in interdisciplinary cultural studies. He served as the director of the Institute for Advanced Studies in the Humanities (Kulturwissenschaftliches Institut) and as a member of the Global Change Advisory Council of the Federal Republic of Germany unteil 2017. At Giessen University he holds the Ludwig Börne Chair and is a Senior Fellow at the Institute for Advanced Sustainability Studies in Potsdam. His recent publications include: Leggewie, Claus. *Anti-Europäer. Breivik, Dugin, al Suri & Co.* Suhrkamp: Berlin, 2016; Leggewie, Claus. *Europe first! Eine Unabhängigkeitserklärung*. Ullstein: Berlin, 2017. With Erik Meyer he is the editor of *Global Pop. Das Buch zur Weltmusik.* Metzler: Stuttgart, 2017.

Mauelshagen, Franz
is an environmental historian and historian of science. He is a Senior Fellow at the Institute for Advanced Sustainability Studies e.V., Potsdam, and specialises in the History of Climate, Climatology and Climate Science. His current work is mainly on the Anthropocene and the interference between historical and geological timescales. His recent publications include: Sam White, Christian Pfister, and Franz Mauelshagen. *Palgrave Handbook of Climate History.* Basingstoke: Palgrave Macmillan, 2017; and Mauelshagen, Franz. "Bridging the Great Divide—The Anthropocene as a Challenge to the Social Sciences and Humanities." In *Paul J. Crutzen: The Anthropocene—A New Phase in Earth History*, edited by Benner, Susanne, Gregor Lax, Paul J. Crutzen, Jos Lelieveld, and Ulrich Pöschl. London et al.: 2017.

Pfister, Christian
is an Environmental Historian. He is emeritus Professor of Economic, Social, and Environmental History at the University of Bern, currently Senior Researcher at the Oeschger Centre for Climatic Change Research at the same university. He specialises in Environmental History with a research focus on Climate and Disaster History. His recent publications include: Pfister, Christian, Christian Rohr, and Jover Antoine. "Euro-Climhist: Eine Datenplattform der Universität Bern zur Witterungs-, Klima- und Katastrophengeschichte." *Wasser Energie Luft* 109 (2017): 45-48; Pfister, Christian. "When Europe Was Burning. The Multi-Seasonal Drought of 1540 and the Arsonist Paranoia." In *Historical Disaster Experiences. Towards a Comparative and Transcultural History of Disasters Across Asia and Europe, edited by Gerrit Jasper Schenk,* 155-185. Heidelberg: Springer, 2017.

Parker, Geoffrey
is *Andreas Dorpalen* Professor of European History and an Associate of the Mershon Center at The Ohio State University. He specializes in Reformation, European history and military history and has written and co-written numerous prizewinning books, including Parker, Geoffrey. *The Grand Strategy of Philip II*. New Haven, CT: Yale University Press, 1998; and Parker, Geoffrey. *Imprudent King. A new biography of Philip II*. New Haven, CT: Yale University Press, 2014.

Riemann, Dirk
currently works in a postdoc position at Freiburg University. His research interests include climate reconstructions and the impact on societies. Beside this his interests lie in the development of Collaborative Research Environments as tool for inter- and trans- disciplinary scientific collaboration. Furthermore, he is doing research on the use of sustainable energy sources and necessary transformation processes. His recent publications include: Camenisch, Chantal, Kathrin M. Keller, Melanie Salvisberg, Benjamin Amann, Martin Bauch, Sandro Blumer, Rudolf Brázdil, Stefan Brönnimann, Ulf Büntgen, Bruce M.S. Campbell, Laura Fernández-Donado, Dominik Fleitmann, Rüdiger Glaser, Fidel González-Rouco, Martin Grosjean, Richard C. Hoffmann, Heli Huhtamaa, Fortunat Joos, Andrea Kiss, Oldřich Kotyza, Flavio Lehner, Jürg Luterbacher, Nicolas Maughan, Raphael Neukom, Theresa Novy, Kathleen Pribyl, Christoph C. Raible, Dirk Riemann, Maximilian Schuh, Philip Slavin, Johannes P. Werner, Oliver Wetter. "The Early Spörer Minimum. A Period of Extraordinary Climate and Socio-economic Changes in Western and Central Europe." *Climate of the Past Discussions* (2016). Accessed April 4, 2017. <http://www.clim-past-discuss.net/cp-2016-7/cp-2016-7.pdf>; and Riemann, Dirk, Rüdiger Glaser, Michael Kahle, and Steffen Vogt. "The CRE tambora.org – New Data and Tools for Collaborative Research in Climate and Environmental History." *Geoscience Data Journal* (2016). <http://dx.doi.org/10.1002/gdj3.30>.

Sieglerschmidt, Jörn
is a historian of Early Modern History and was lecturer at the universities of Konstanz, Karlsruhe and Mannheim. Until 2011 he was senior curator at the State Museum of Labour and Technology at Mannheim (now Technoseum) and at the Library Service Centre Baden-Württemberg. Since then he is retired. Past and current fields of interest and research are historical geography, economic, social, environmental history, and the semiotics of history. Selected recent publications are: Siglerschmidt, Jörn, and Bernd Herrmann. *Umweltgeschichte im Überblick* [*Environmental History—An Overview*]. Wiesbaden:

Springer Fachmedien, 2016; and: Herrmann, Bernd, and Jörn Sieglerschmidt. *Umweltgeschichte in Beispielen* [*Examples and Illustrations from Environmental History*]. Wiesbaden: Springer Fachmedien, 2017.

Schmitt, Lea
is a cultural scientist. She was Fellow at the Institute for Advanced Study in the Humanities in Essen from 2009 to 2016 and specialises in cultural dimensions of climate change and the implementation of energy transition by citizens' participation. Since 2016 she is consultant at the project development company MWSP which tracks the task to develop civilian use of former military areas in Mannheim. Her recent publications are Schmitt, Lea. *Klima, Raum und Zeit im Wandel. Eine ethnographische Untersuchung von Adaptationen und Konflikten auf der westfriesischen Insel Ameland.* München: oekom, 2015; and Leggewie, Claus, Christa Reicher, and Lea Schmitt, ed. *Geschichten einer Region. AgentInnen des Wandels für ein nachhaltiges Ruhrgebiet*, Dortmund: Kettler, 2016.

Vogt, Markus
is professor of Christian social ethics at the Ludwig-Maximilians University of Munich. He studied Theology and Philosophy in Munich, Jerusalem, and Lucerne. In the year 2015 he was appointed as dean of the Catholic-theological faculty at the LMU. Since 1995 he is acting as an advisor for the ecological research group of the German Bishop Conference. Since 2011/2012 he is research professor and Fellow at the Rachel Carson Center for Environment and Society (RCC) In the year 2016 he became spokesman of the bioecology committee of experts of the Bavarian state government. His recent publications include: Vogt, Marcus and Wilhelm Korff, ed. *Gliederungssysteme angewandter Ethik.* Freiburg i.Br.: Herder, 2016; and Haber, Wolfgang, Martin Held, and Markus Vogt, ed. *Die Welt im Anthropozän: Erkundungen im Spannungsfeld zwischen Ökologie und Humanität.* Munich: oekom, 2016.

Vogt, Steffen
is a Geographer. He was a lecturer at the University of Freiburg specialised in Historical Climatology and IT based concepts. He is now working in the private sector with svGeosolutions GmbH. His recent publications include: Riemann, Dirk, Rüdiger Glaser, Michael Kahle, and Steffen Vogt. "The CRE tambora.org – New Data and Tools for Collaborative Research in Climate and Environmental History." *Geoscience Data Journal* (2016). Accessed April 4, 2017. <http://dx.doi.org/10.1002/gdj3.30>; Vogt, Steffan, Rüdiger Glaser, Jürg Luterbacher, Dirk Riemann, Ghazi. Al Dyab, Johannes Schönbein, Elena Garcia-Bustamente.

"Assessing the Medieval Climate Anomaly in the Middle East: The Potential of Arabic Documentary Sources." *PAGES News* 19/1 (2011): 28-29.

INTRODUCTION

Tracing and Replacing Europe's Carbon Culture

Claus Leggewie and Franz Mauelshagen

Europe is the second smallest continent, covering only about two percent of the Earth's overall surface and about 6.8 percent of its land area. It is part of the Eurasian landmass, which is why some people question whether it is a continent at all. The editors of the *Climate & Culture* book series have nevertheless decided that Europe deserves a volume of its own, particularly with regard to its role in the making of anthropogenic climate change. There is little doubt that greenhouse gas emissions, which have triggered global warming, first reached significant quantities in Europe as a consequence of industrialisation. Moreover, as a political unit, the supranational European Union (EU in the following) has come to represent Europe more than any other political organisation or any geographic definition does, particularly after the EU's expansion following the breakdown of the socialist Eastern bloc and the political order of the Cold War. As a consequence, European climate politics is as much determined by the EU and its representatives as it is by any of the member states or non-EU nation states. Clearly, the EU has determined European climate politics to a great extent, at least for several decades up to the present. That said, the EU's future suddenly seems at risk after the Brexit vote and the election of Donald Trump as 45th president of the United States in 2016. Political cooperation in the EU has suffered already for several years following the success of neo-nationalist parties and governments in a number of member states. It remains to be seen how much this will affect EU climate politics in the near and not so near future.

Following the structure of this book and the *Climate & Culture* book series, we distinguish between three temporal dimensions in Europe's relation to climate change: the past, the present, and the future. The past has established a causal link between the social, economic and cultural transformation of Europe between 1750 and 2000 and the making of climate change (see part 2 of this introduction). But before we describe it in greater detail, we introduce the term 'carbon culture' (part 1) to create a conceptual framework for our account. With regard to the present, we will look at the Paris agreement and the efforts made by European societies and institutions to implement effective climate protection policies (part 3). The European Union has positioned itself as an advocate of global climate policies, despite sometimes contradictory positions

 | DOI 10.1163/9789004356825_002

among its member states as to how urgently the climate challenges need to be addressed and which new energy technologies are to be implemented. Finally, our introduction discusses narratives of climate change and alternative futures (part 4), starting from a brief account of climate science and its future scenarios, which involve a variety of uncertain futures of socio-economic development. Focusing on narratives in fiction (in various formats, such as literature and film) and the humanities we intend to emphasize the challenge that climate change still presents for the arts and the humanities, as both seem to struggle with including 'nature' or the non-human. We conclude this introduction with a brief outline of this volume's contents (part 5).

1 Carbon Culture

Political borders have constantly changed throughout Europe's history. The same applies to geographic and cultural borders. Just think about religion, which many consider a key element of culture: Europe was never simply or 'purely' Christian. Schisms between Christian faiths have cut through the political geography of Europe almost since the emergence of Christianity. The Jewish diaspora has always had a strong presence. And Islam has its fair share in European history as well. Before the Reconquista, Islam had gained a foothold in Spain and played a major role in the renewal of Antique philosophical traditions during the Renaissance. Periods of peaceful co-existence with Christianity alternated with periods of conflict. The Ottoman Empire was a cause for constant concern on the Balkan as well as for the Holy Roman Empire. Nevertheless, the history of relations between 'Turkey' and the rest of Europe cannot be reduced to continuous conflict, as has sometimes been done, even by historians. Following the religious turmoils of the Reformation, the Ottomans became an essential element in the balance-of-power structure of Europe, not only as a threat, but also as an ally for various parties. Today, Muslims are embedded in all European nations through citizenship and immigration.

These considerations suffice to declare diversity the most significant feature of European culture. It is also the most obvious. But is it more than a mere triviality, which applies almost universally? How does the cultural diversity that characterises Europe differ from that of other places? Is it possible to go beyond an abstract understanding of diversity?

Some historians have pointed to other, more specific characteristics such as Europe's 'eccentricity' and its 'exterritoriality'.[1] It is eccentric in that its foundations were laid in the Middle East and the Mediterranean, from where much of its cultural heritage diffused into (other parts of) Europe. Its extraterritoriality emerged from the Columbian exchange, colonisation, and the (often unintended) diffusion of all kinds of things 'beyond the pillars of Hercules': knowledge and scientific traditions; technologies and cultural practices; capitalism; but also plants and animal species; and of course people and their genes.[2] Europe's expansion and the spread of (some of) its cultural heritage add to the complications involved with defining what is specifically European about European culture. Aspects that once used to be special have 'settled' elsewhere, often crossing the World Oceans on their way. Languages such as Spanish, Portuguese, French and English are some examples. In all the history of human colonisation, language communities have formed permanent transoceanic diasporas never before the Columbian Exchange.[3] Modern science is another example. As it has followed the paths to modernisation everywhere, Europe can claim exclusive rights to it no more.

Any attempt at defining 'European culture' based on geography, religion, or history obviously requires a lot of caution. Mainstream cultural theories today have pointed to the danger of essentialism and advised on avoiding it by defining culture as 'practice' or 'a system of practices'. However, their preference for emphasizing diversity in all matters cultural creates its own dilemma—that of lack of concreteness and specificity—as well as the consequential danger of making the very concept of culture almost too flexible and fluid to remain useful.

Here, we propose an alternative for the purpose of this introduction: the concept of *carbon culture*. This is a concrete proposal responding to the demand to connect culture and cultural change with climate and climate change, which is the intention of this volume and the entire *Climate & Culture* book series. By 'carbon culture' we mean the totality of all cultural practices, either underlying the burning of fossil fuels or enabled by it, and, thus, emerging from it.[4] The existence of a carbon culture obviously requires the knowledge,

---

1 Brague, Remi, *Europa, seine Kultur, seine Barbarei. Exzentrische Identität und römische Sekundarität* (Wiesbaden: Springer, 2012).

2 The concept was invented by Crosby, Alfred W., *The Columbian Exchange: Biological and Cultural Consequences of 1492* (Westport, Conn.: Greenwood Pub. Co., 1972).

3 Ostler, Nicholas, *Empires of the Word: A Language History of the World* (London: HarperCollins, 2005), 325-521.

4 The paramount importance of energy regimes and their transformation in the history of material cultures, the emergence of civilisations, and complex modern societies has long been

infrastructures, and technologies of industrialism. Emerging cultural patterns are clearly the more tricky part, as one can easily find out by way of a simple thought experiment: try to imagine, which of your daily activities do not, in one way or another, involve the burning of fossil fuels. It is much more difficult to answer this question than the opposite one. In modern socitети all activities related to mobility or transport are based on the burning of fossil fuels. Using your car to go to work in the morning is an obvious case: burning fossil fuel is what your engine does. But public transport does not solve the problem, because the energy needed to run the railway or buses also comes from the burning of fossil fuels. Only in some places is this beginning to change.

Mobility is by no means the only large sector involved with most people's everyday activities. Construction is another one. And just like mobility it is energy intensive and, hence, releases a lot of carbon into the atmosphere. But that is not even the end of the story. The built infrastructure that emerges from construction uses heat and electricity. That is what we all do at home or in our workplace. And, once again, that energy comes from the burning of fossil fuels, at least for the most part and in a vast majority of industrial countries.

Mobility and transport, construction and electricity—however meaningful these sectors of economic activity are in modern societies, they only provide precisely the most obvious framework for our argument. It is equally important, though less trivial, to discover the traces of fossil fuels in production chains and the material world we live in. Industrial technologies have produced an enormous increase in material variety, to which anthropogenic materials such as plastic have made an important contribution. As a consequence, mining has dramatically expanded, providing new materials for industrial mass production or unprecedented amounts of materials already

emphasized. White, Leslie A., "Energy and the Evolution of Culture," *American Anthropologist* 45.3 (1943) was the pioneering article. More recent publications on the same theme are Goudsblom, Johan, *Fire and Civilisation* (London: Penguin Books, 1992); Sieferle, Rolf P., *The Subterranean Forest: Energy Systems and the Industrial Revolution* (Cambridge: White Horse Press, 2001) (first German edition 1982); Burke, Edmund, "The Big Story: Human History, Energy Regimes and the Environment," in *The Environment and World History*, ed. Burke, Edmund, and Kenneth Pommeranz (Berkeley: 2009); Smil, Vaclav, *Energy in Nature and Society: General Energetics of Complex Systems* (Cambridge MA: The MIT Press, 2008), and Stiner, Mary C. et al., "Energy and Ecosystems," in *Deep History: The Architecture of Past and Present*, ed. Andrew Shryock et al. (Berkeley: University of California Press, 2011), to name but a few. Despite a growing literature, mainstream cultural theory and social science still underrate energy regimes or even ignore them entirely.

used.[5] The *material culture of industrialisation* is markedly distinct from anything we find in pre-industrial worlds.[6] It is the signature of carbon culture. The so-called 'material turn' in the humanities and social sciences is only beginning to explore this, and it still seems a long way away from systematically considering 'materiality' in context with climate, the environment, or the earth system.[7] It is often simply overlooked that the enormous expansion of material flows in all industrial societies is showing the signs of environmental disruption on a planetary scale.

Of course, neither carbon culture nor its signature material culture is entirely homogeneous across industiralised countries. Using the singular 'carbon culture' is not to deny variety, but to emphasize distinction from the varieties produced by pre-industrial material cultures and the relative energy scarcity they had to deal with. We also want to emphasize resemblance between material cultures building on the extensive burning of fossil fuels today. That resemblance can be illustrated by material flows of affluent economies and their evolution in time:[8]

1. In all affluent economies, consumption of nonrenewable organics increases in time; fossil fuels dominate in this category, and their increase is indicative of a rise in energy demand in a society;

---

5 The importance of mining and its history has been stressed by Bardi, Ugo, *Extracted: How the Quest for Mineral Wealth Is Plundering the Planet: A Report to the Club of Rome* (White River Junction, Vermont: Chelsea Green Publishing, 2014), chapter 2.

6 See Smil, Vaclav, *Making the Modern World: Materials and Dematerialization* (Chichester: Wiley, 2014).

7 Miller, Peter N., *Cultural Histories of the Material World* (Ann Arbor: The University of Michigan Press, 2013) is an excellent introduction to the material turn and its scope across disciplines. See also Bennett, Jane, *Vibrant Matter: A Political Ecology of Things* (Durham, N.C.: Duke University Press, 2010). Both these seminal publications also demonstrate the limits of the dominant concept of 'materiality' and Latour's Actor Network Theory (ANT) with its focus on 'things'. Energy, raw materials and their (potential) use as resources cannot easily be grasped with these categories. Nevertheless, the approaches just mentioned also guided most of the contribitions to two workshops held at the Max Planck Institute for the History of Science in 2016 and 2017, the first on 'Resources and Economies of Knowledge in the Anthropocene' (September 12-13, 2016; coorganised by Nadine Hee, Helge Wendt and Franz Mauelshagen); the second on 'Energy Transformations: Perspectives from the Humanities' (January 17-18, 2017; organised by Jürgen Renn and Robert Schlögl).

8 In the following we paraphrase Smil, *Making the Modern World: Materials and Dematerialization*, 93. Compared with Smil's own summary, we have added numbering, modified the order slightly, and added an important aspect (No. 4) as well as some explanations.

2. absolute consumption goes up in every material category, enabled by an expanding fossil energy regime; not only is the extraction of material resources (mostly from the earth's crust) energy-intensive, but also their transformation into products and their distribution over long distances by means of transportation;
3. the importance of biomaterials declines substantially relative to other materials;
4. raw materials are transformed into new types of anthropogenic materials (glass, aluminium, steel, plastic etc.) by chemical, often energy-intensive processes;
5. the importance of recycling rises, which follows indirectly from the previous point; biomaterials are renewable and mostly decompose, when released as waste into the environment; thus, expanded use of non-renewable materials increases the importance of recycling to avoid scarcity;
6. the total mass of bulk construction materials dominates the entire spectrum of material flows in other words: bulk construction is the single most material-intensive human activity;
7. demand for metals expands;
8. shifts in per capita consumption are material-specific; demand rises slowly or quickly; availability declines steadily and slowly, but sometimes also very suddenly and quickly, depending on a variety of circumstances.

In sum, the concept of 'carbon culture' highlights the emergence of a historically new pattern of material flows—both, within the social world, and between the social world and the environment—based on cheap and abundant energy from the burning of fossil fuels. Carbon cultures extract unprecedented amounts and varieties of material resources from the environment, transform them into all kinds of products or even new anthropogenic materials, and release them back into the environment as emissions or waste. Byproducts such as CO_2 and other greenhouse gases (GHGs) change the chemical composition of the atmosphere and, thus, lead to climate change. That is how carbon cultures are causally linked with climate change. The point is that the carbon cultures of industrialism represent a new pattern of material culture and a connected set of human behaviors, unprecedented in human history. In other words, controlling GHG emissions is likely to affect everything that belongs to carbon culture, which indicates the complexity of the problems posed by climate change today.

2 Europe's Share in the Making of Global Warming

In narratives of the making of global warming, Europe has featured prominently among those parts of the world where the natural greenhouse effect has been enhanced due to a massive release of CO_2 and other GHGs into the atmosphere. Carbon culture emerged from the Industrial Revolution in England. It quickly spread across Europe and over the Atlantic. Since industrialisation, many other nations outside Europe and the United States have bailed into it as well. That applies to Russia and the former Soviet Union and, more recently, to the emerging economies of the BRICS countries. Across the entire spectrum of old industrial and industrialising nations today, the positive correlation between GDP, energy consumption, and CO_2 emissions is evident. [9]

The wealth of industrial nations around the world has been built on relatively cheap fossil fuels. Among them, the European Union is the third-largest emitter of GHGs today and, therefore, a leading force behind changes in the chemical composition of the atmosphere. In his book *The Unbound Prometheus*, David Landes has described the economic history of the western hemisphere in terms of acceleration and extreme dynamism, of inherent innovative capacities, the generation of collective and individual wealth, and a new entrepreneurial mentality. These were the forces of industrial modernity in Europe and the United States. In the most recent globalisation, this pattern spread around the globe, creating an international division of labour, global trade and investment.

The foundation of this dynamic development was coal first, then crude oil and natural gas. All these fossil resources were welcomed as a blessing and appreciated as a source of wealth and prosperity. Coal mining and processing structured Europe's industrial culture; it influenced personal and collective identities, social movements, and industrial relations; and these in turn were patterns for ideological battles, political parties, trade unions, associations et cetera. Recent decades of industrial history have proven, in countless battles over the closing of coalmines and the subsequent structural changes in the old centres of industrialisation, how difficult coal dependency is to overcome. While carbon culture has created new spheres of conflict over public health (threatened by emissions), political participation of the working class, state welfare, and the distribution of wealth, at the same time, and beyond these

9 See: Figure 1. Progress towards Meeting Europe 2020 and Kyoto Targets, in *Report from the Commission to the European Parliament and the* Counci, ed. European Commissionl (Brussels: European Commission, 2015), 4, accessed June 6, 2017. <http://eur-lex.europa.eu/resource.html?uri=cellar:adf850de-8de4-11e5-b8b7-01aa75ed71a1.0007.02/DOC_1&format=PDF>.

spheres of conflict, it has also established a community of interest, which includes nearly everbody. In the 'affluent societies' that emerged with the shift from coal to oil after World War II, this community of interest seems even more obvious.[10] It has established some kind of complicity in the making of climate change involving everyone living the lifestyle of affluent societies.[11]

Since the 1970s, when the era of cheap oil came to an end during the oil crisis, the future of industrialisation has become a battlefield on how to transform the energy regime into a more stable state. [12] It had been stated for decades that the burning of oil would not sustain for long, even by the standards of human history.[13] During the 1970s, global warming entered that same battlefield and developed into a major cause of distress for industrial mentalities.[14] Since the 1980s, this effect has been enhanced by the emergence of China and India as major emitters of GHGs.

Europe's exterritoriality—in other words: imperialism and its uneven chronology of colonisation and decolonisation—has played a key role in the proliferation of carbon culture around the world. The United States and Canada developed their own carbon cultures more or less at the same speed as England, France and Germany. In Latin America and the Caribbean the carbonisation of the energy regime was delayed by several decades, thus resembling similar trajectories in Spain or Portugal. However, the Latin

10 We are using the term 'affluent society' for the post-WWII style of consumerism here. Recent histories of consumption and consumerism no longer use the term 'consumer society' as a synonym, as they prefer different, more long-term chronologies. See, for example, Trentmann, Frank, ed., *The Oxford Handbook of the History of Consumption* (Oxford, New York: Oxford University Press, 2012), which goes back to Antiquity, and Trentmann, Frank, *Empire of Things : How We Became a World of Consumers, from the Fifteenth Century to the Twenty-First* (London: Allen Lane, 2016), here in particular 9-10.

11 'Complicity' is a term used by Ghosh, Amitav, *The Great Derangement: Fiction, History and Politics in the Age of Global Warming* (Chicago, IL: Chicago University Press, 2016), in particular 9-10.

12 Pfister, Christian, "The '1950s Syndrome' and the Transition from a Slow-Going to a Rapid Loss of Global Sustainability," in *The Turning Points in Environmental History*, ed. Frank Ükötter et al., (Pittsburgh: University of Pittsburgh Press, 2010) has argued that the boom that followed WWII was based on cheap oil and marks the transition to unsustainability.

13 The standard reference for this is Hubbert, Marion K., "Energy from Fossil Fuels," *Science* 109.2823 (1949).

14 Weart, Spencer R., *The Discovery of Global Warming* (Cambridge: Harvard University Press, 2008); see also Schellnhuber, Hans J., *Selbstverbrennung: Die fatale Dreiecksbeziehung zwischen Klima, Mensch und Kohlenstoff* (München: C. Bertelsmann, 2015).

American and Caribbean cases are complicated by dissimilar political histories of decolonisation and independence.[15]

The political map of imperialism looks different again in Asia. In his Berlin Family Lectures, delivered at the University of Chicago in fall 2015, Amitav Ghosh argued that India's and China's late joining of the club of fossil economies was due to imperial power politics hampering the modernisation of both economies. Under different circumstances they might have started the transition to a fossil energy regime decades earlier, which would then almost certainly have accelerated climate change.[16] Thus, one could argue that the role of imperialism is ambivalent in that it denied Asians their fair share of ecnomic prosperity, while delaying the Great Acceleration.[17] However, gaining time on global warming was merely an uninteded side-effect of European imperialism in Asia, which was made up for by the globalization of western developmental patterns. This may have prevented the emergence of a greater variety in economic modernization, based on different energy regimes and material flows.[18]

3 The Paris Agreement and European Climate Policy

Setting a two-degree target for global warming has been a consensual starting point for a vast majority of climate scientists to calculate emission (reduction) scenarios and inform climate politics. An IIASA Working Paper from 1975 already stated "If there were global temperatures more than 2°C of 3°C above the current average temperature, this would take the climate outside of the range of observations which have been made over the last several hundred

15 For estimates of energy consumption in Latin America and the Caribbean at an early stage of developing a carbon culture see Rubio, Mar et al., "Energy as an Indicator of Modernization in Latin America, 1890-1925." *The Economic History Review* 63.3 (2010).

16 Ghosh, *Great Derangement*, part II: history, in particular p. 108-111.

17 The Great Acceleration refers to rapid economic growth after WWII and the resulting effects it has had on the earth system. The term is particularly prominent in current debates about the Anthropocene. See Steffen, Will et al., "The Anthropocene: Conceptual and Historical Perspectives." *Philosophical Transactions of the Royal Society of London: A Mathematical, Physical and Engneering Sciences* 369.1938 (2011) and Steffen, Will et al., "The Trajectory of the Anthropocene: The Great Acceleration." *The Anthropocene Review* 2.1 (2015).

18 Mauelshagen, Franz, "Bridging the Great Divide—the Anthropocene as a Challenge to the Social Sciences and Humanities," in *Religion and the Anthropocene*, ed. Celia Deane-Drummond et al., (Eugene OR: 2017), 93-94.

thousand years."[19] Human experience became the standard argument for a 2°C climate target—that is, 2°C above pre-industrial levels. The target influenced the Kyoto protocol and was also adopted by the IPCC. European countries in particular pushed for the recognition and implementation of this upper limit. In 1996, the European Council of Ministers of the Environment became the first political body to formally support the idea.[20] Signatories to that statement included personalities now operating at the forefront of international climate politics—John Gummer (now chair of the UK's Committee on Climate Change) and Germany's then minister of the environment, Angela Merkel, who is currently spearheading Germany's transition to renewables.

After a period of stagnation and depression following COP15 in Copenhagen 2009, the Paris Agreement in 2015 was a breakthrough for climate protection policies. It set a 1.5-2°C target for global warming and adopted the UN's Sustainable Development Goals (SDGs). These ambitious goals are a challenge to European societies, national governments, and supranational bodies of the EU. The EU has projected its climate policy development defining a number of achievements to be reached in 2020, 2030 and 2050. The 2020 climate and energy package was put together in 2007 and included three central targets: cut GHG emissions by twenty percent (based on 1990 levels), a twenty percent share of renewables in energy supply, and an improvement in energy efficiency, also by twenty percent.[21] The 2030 climate and energy framework adopted in 2014 develops these targets further in terms of forty percent cuts in GHGs, a 27 percent share of renewables, and a 27 percent improvement in energy efficiency.[22] The roadmap to a low-carbon economy reached in 2050 includes a domestic emissions cut by eighty percent (below 1990 levels).[23]

While the EU is on track in reaching its 2020 goals,[24] climate policies need to improve substantially to address the 2030 targets. Accordingly, even more

19 Nordhaus, William D., "Can We Control Carbon Dioxide?" *IIASA Working Paper* (1975), accessed July 12, 2016. <http://pure.iiasa.ac.at/365/1/WP-75-063.pdf>.

20 European Commission, "2020 Climate & Energy Package," last modified july 8, 2011, accessed July 11, 2016. <http://ec.europa.eu/clima/policies/strategies/2020/index_en.htm>.

21 European Commission, "2020 Climate & Energy Package," last modified July 8, 2011, accessed July 11, 2016. <http://ec.europa.eu/clima/policies/strategies/2020/index_en.htm>.

22 European Commission, "2030 Climate & Energy Framework," last modified July 8, 2011, accessed July 11, 2016. <http://ec.europa.eu/clima/policies/strategies/2030/index_en.htm>.

23 European Commission, "2050 Low-Carbon Economy," last modified July 8, 2011, accessed July 11, 2016. <http://ec.europa.eu/clima/policies/strategies/2050/index_en.htm>.

24 See: Figure 1. Progress towards Meeting Europe 2020 and Kyoto Targets. <http://eur-lex.europa.eu/resource.html?uri=cellar:adf850de-8de4-11e5-b8b7-01aa75ed71a1.0007.02/

ambitious action is needed to reach the 2050 goals. Emissions need to be cut in a wide range of sectors, as the figure published by the European Commission shows.[25] Post-industralism, first taken to be a burden on European development, can now be perceived as an opportunity for carbon-free reinvention.

Of all sectors, transportation has been most resistant to cutting emissions. Actually, emissions have increased every year since 1990. Among other reasons, the construction of hybrid and electric cars continues to lag. The consumer response to their prototypes shows too little acceptance. Moreover, patterns of mobility behaviour have not changed at all. Travelling mid- to long-distance by car or air is as popular as ever, both in private and in business travel.[26] Using more sustainable means of public transport clearly needs to be more broadly internalised by European societies.

Far-reaching emission cuts across all economic sectors require appropriate governance architectures, especially when touching on such sensitive issues as individual mobility. The 2030 framework, especially recognised that such further reductions could not only be based on a more or less effective emissions trading scheme and investments in innovation or national action, but would require a new mode of governance. Consequentially, in 2015, the concept of an Energy Union was presented, to emerge as a new governance framework. A familiar idea as it seems, recognising that the European Union was initially founded as an economic framework. However, the actual frame for the Energy Union has yet to be fleshed out. Right now, it is more a combination of existing policies without a transformative mode of governance fostering integration with respect to the Anthropocene era.

Furthermore, the EU has to invest its soft power in G7 and G20 summits that could be used to develop a road map to implement the Paris Agreement.[27] Many other crises and developments are capable of pushing the climate agenda to the sidelines, such as migration, instabilities in the Middle East,

---

DOC_1&format=PDF>.

25 European Commission, "2050 Low Carbon Economy," last modified June 16, 2017, accessed June 17, 2017. <https://ec.europa.eu/clima/policies/strategies/2050_en>.

26 European Commission, "Reducing Emissions from Rransport," last modified June 16, 2017, accessed June 17, 2017. <https://ec.europa.eu/clima/policies/transport_en>. Notes: (*) Excluding LULUCF (Land Use, Land–Use Change, and Forestry) emissions and International Bunkers (**) Excluding International Bunkers (international traffic departing from the EU) (***) Emissions from Manufacturing and Construction and Industrial Processes (****) Emissions from Fuel Combustion in Agriculture/Forestry/Fisheries, Other (Not elsewhere specified), Fugitive Emissions from Fuels, Solvent and Other Product Use, Waste, Other.

27 Leggewie, Claus et al. *WGBU 2016, Policy Paper G20*, forthcoming.

Brexit, or the election of new nationalist governments that challenge European institutions. However, in an alternative scenario, the issue of climate change might as well foster cooperation in a way that also helps bring the above-mentioned crises closer to a soution. Consequently, breakthroughs in sustainability thinking in academia, at the normative level, regarding the institutionalisation of sustainable practices and empirical evidence for societal groups to gain self-efficacy, as well as substantially changed values are also essential.[28]

It is therefore scientifically imperative to test and develop new and different kinds of scenarios and narratives. These scenarios have to include a wide range of possible futures in terms of sustainable production, mobility, and lifestyles in the context of the overall emissions allowed for the 2020, 2030, and 2050 reduction targets. Will Europe continue to be a frontrunner and be transformed, together with its societies, into a climate neutral region fostering intense cooperation? Or will new nationalisms arise that might result in a few individual green countries, but not allow a comprehensive European answer to the challenge posed across the planetary boundaries? At the moment, the answer to these questions is still open and this volume is an attempt to understand and inform such climate- and culture-related developments.

4 Narratives of Climate Change and the Future

The (anthropogenic) greenhouse effect has been discussed for almost two hundred years now. In the 19th century, science was a force driving technological development and the expansion of natural resource extraction. But at the same time it was an autonomous system of thought, which also developed into a bastion of awareness of the destructive forces unleashed by technological progress. In the 1820s, the French mathematician and physicist Joseph Fourier assumed that certain gases trap some of the solar radiation reflected from the surface of the earth—like the glass of a greenhouse. Consequently John Tyndall invented the term 'greenhouse effect'. Writing in the 1860s, Tyndall suggested that slight changes in the chemical composition of the atmosphere were capable of changing the earth's heat balance.[29]

28 Szulecki, Kacper et al. "Giving Shape to the Energy Union: Evolution, National Expectations and Implications for EU Energy and Climate Governance," (17 June 2015). Available at SSRN: <http://ssrn.com/abstract=2623892> or <http://dx.doi.org/10.2139/ssrn.2623892>.

29 A now classic account of the science of climate change is Weart, Spencer R., *The Discovery of Global Warming* (Cambridge MA, London: Harvard University Press, 2003), see chapter 2 for an account of scientific discovery on how greenhouse gasses work. Some of the

In the 1890s, the Swedish chemist Svante Arrhenius experimented with burning coal and estimated that doubling the CO_2 concentration in the planet's atmosphere would increase its temperature by 2.5°C to 4.0°C. After a pause, Guy Stewart Callendar, an amateur meteorologist, brought the greenhouse effect back on the agenda in the 1930s.[30] However, further into the 20th century the question of CO_2 accumulation in the atmosphere as well as that of warming remained surrounded by doubts. The rate of absorption of CO_2 by the oceans was among the most relevant uncertainties. How CO_2 concentration in the atmosphere changed, if at all, as a consequence of the burning of fossil fuels, continued to be an open question until Charles Keeling developed a reliable method to measure it and began applying it at Mauna Loa on Hawaii starting in 1958.[31] After only a few years, there were no more doubts about atmospheric CO_2 rising. Only seven years into Keeling's measurments, U.S. President Lyndon B. Johnson issued a "special message to the Congress on Conservation and Restoration of Natural Beauty" stating (among other things) that "this generation has altered the composition of the atmosphere on a global scale through […] a steady increase in carbon dioxide from the burning of fossil fuels."[32]

---

original publications, or extracts from them, have been collected and republished in Archer, David , and Raymond T. Pierrehumbert., ed. *The Warming Papers: The Scientific Foundation for the Climate Change Forecast* (Hoboken: Wiley-Blackwell, 2011).

30 See Callendar, Guy S., "The Artificial Production of Carbon Dioxide and Its Influence on Temperature," *Quarterly Journal of the Royal Meteorological Society* 64.275 (1938). On Callendar: Fleming, James R., *The Callendar Effect. The Life and Work of Guy Stewart Callendar (1898-1964), the Scientist Who Established the Carbon Dioxide Theory of Climate Change* (Boston, Mass.: American Meteorological Society, 2007). Callendar's work has been revisited and evaluated in a recent study, see Hawkins, Ed et al., "On Increasing Global Temperatures: 75 Years after Callendar." *Quarterly Journal of the Royal Meteorological Society* 139.677 (2013).

31 Two recent books on climate science describe the backgrond of Keeling's work emphasising the meaning of the military sector for the emergence of earth system science in general after World War II: Hamblin, Jacob D., *Arming Mother Nature: The Birth of Catastrophic Environmentalism* (New York: Oxford University Press, 2013), and Howe, Joshua P., *Behind the Curve: Science and the Politics of Global Warming* (Seattle: University of Washington Press, 2014).

32 The full text of this message from February 8, 1965 has been made available by The American Presidency Project at <http://www.presidency.ucsb.edu/ws/?pid=27285>, accessed December 8, 2016. See also Lavelle, Marianne, "A 50th Anniversary Few Remeber: LBJ's Warning on Carbon Dioxode," *The Daily Climate*, February 2, 2015, accessed July 12, 2016. <http://www.dailyclimate.org/tdc-newsroom/2015/02/president-johnson-carbon-climate-warning>.

In the 1960s, it was still a long way to go from traditional, descriptive climatology to climate science, which has since developed an understanding of the causes of climate change. Modelling past, present and future climates is an essential part of climate science today.[33] Climate models are based on quantitative methods and computations. Integrated assessment models go beyond simple climate models, in that they not only take account of other non-natural variables such as demographic and economic changes, but also anticipate human behavioural patterns (e.g. changes in lifestyle) and institutional reactions (not least in the form of environmental and climate policies), and calculate possible consequences into the model.

This involves a variety of scenarios, which, based on different starting points (from the IPCC's RCP 2.6 scenario to the RCP 8.5 scenario), show the spectrum of different possible futures from the business-as-usual case of the energy markets. If the latter scenario would continue, it might result in global average temperatures eight degrees above pre-industrial levels. In contrast, the best-case scenario of a timely withdrawal from fossil energy sources on a global scale would allow limiting the temperature rise to about two degrees. Nevertheless, even with these scenarios we are only beginning to understand the importance of social, normative and cultural changes and their impact on the climate. Social and cultural scientists need to discuss with their modelling colleagues which variables could be used as indicators for value changes. The important role of change agents has been stressed recently, but the discussion needs to be extended based on more profound research cooperation and trans-disciplinary outreach.

In any case, the humanities are invited to take notice of these models and scenarios, understand them and turn them into 'narratives', thereby creating equally plausible scenarios for a socio-economical and socio-psychological transformation of ways of producing and living, perception patterns, and visions of the future. In doing so, they will capitalise on their expertise in understanding and analysing plots, frames, stories, histories, and discourses. In short: Linguistic resources also used by societal and political agents, or which they employ as 'meta-narratives' for bringing the systemic and action-oriented spheres together.

---

33 Except for the books written by Weart, Hamblinm and Howe mentioned in previous footnotes, we refer to Edwards, Paul N., "History of Climate Modeling," *Wiley Interdisciplinary Reviews: Climate Change* 2.1 (2011) here. Also, for a brief account of the history of climate science see Heymann, Matthias and Dania Achermann, "From Climatology to Climate Science in the 20th Century," in *Palgrave Handbook of Climate History*, ed. White, Sam, Christian Pfister and Franz Mauelshagen (Basingstoke: 2017).

One important resource in this respect is literary fiction. Yet, authors of 'serious fiction' have made little effort to tell the story of climate change. In his already mentioned Berlin Family Lectures Amitav Ghosh, a renowned author of fiction himself, provided a stunning explanation for this. He assessed that 'serious literature' commented, reflected, criticised, or supported every political movement in the twentieth century, but has almost entirely failed to do anything like this with regard to climate change. In Ghosh's interpretation, literary modernism, though pushed forward time and again by the idea of avant-garde, got trapped in a bourgeois uniformitarian perspective on the non-human world. Thus, the socio-political realm of the human has dominated novel writing. This has left little space, if any, for climate change, the catastrophes that come along with it, or even 'nature', which has been deprived of any independent agency in the eyes of modernity. Literature itself has become part of carbon culture, or in Ghosh's own words: "Global Warming's resistance to the arts begins deep underground"[34] — in the geological layers where fossil fuels were shaped millions of years past in the earth's history. It seems that those fuels have blinded fiction for global warming, as in the age of oil the literary imagination "became ever more radically centred on the human".[35]

There are, of course, exceptions from the rule: Abdel Rahman Munif's five-part series of novels *Cities of Salt* (1984-89), for which Ghosh invented the term 'petrofiction'[36]; Liz Jensen's eco-thriller *Rapture* (2009); Barbara Kingsolver's *Flight Behaviour* (2012). Some novels have indeed successfully dealt with the material dimensions of carbon culture and its environmental consequences.[37] But the fact that these examples are indeed exceptions confirms the rule. Authors of fiction and artists seem reluctant to build scenarios and narratives of a *successful* transformation to a sustainable future into a literary form or into visual media. Fascinating as the discovery of climate change and the struggle for the recognition of the two-degrees-Celsius guardrail may be, they shy away from the danger that narratives of successful change might turn out as kitsch— or simply naïve.

It seems a much easier task to take on the messengers of dangerous climate change than the message. In *Far North* (2009) by Marcel Theroux, the world is largely uninhabitable, not due to climate change but because of actions

34 Ghosh, *Great Derangement*, 73.

35 Ibid., 66.

36 On petrofiction see: "The Oil Encounter and the Novel (1990)", reprinted in Ghosh, Amitav, *Incendiary Circumstances: A Chronicle of the Turmoil of Our Times* (Boston: Houghton Mifflin, 2005), 138-151.

37 Ghosh, *Great Derangement*, 73-81.

combating global warming that paradoxically altered the climate. Michael Crichton's techno-thriller *State of Fear* (2004) exposes a group of eco-terrorists attempting to create 'natural' disasters to convince the public of the dangers of global warming. The popularity of conspiracy narratives easily assists general criticism of the dominating scientific consensus on anthropogenic global warming and accuses its proponents of spreading fear in modern societies. The same applies to the award-winning novel *Fallen Angels* (1991) by Larry Niven, Jerry Pournelle, and Michael Flynn, in which left and right wing Luddites and radical environmentalists oppose technological progress and endanger the welfare of modern societies. In *Solar* (2010), Ian McEwan tells the story of a physicist in the process of fighting climate change through artificial photosynthesis.

Even in the genre of science fiction, climate change seems to establish rather slowly.[38] Climate fiction (or 'cli-fi') usually results in dystopian books, plays, or movies representing the "future as [a] catastrophe".[39] That alone is no reason to deny science fiction a serious discussion. In fact, catastrophism has gained ground elsewhereand for a reason including in the sciences. Science fiction resists climate change for other, more principled reasons, as Ghosh has argued: "it is precisely not an imagined 'other' world apart from ours; nor is it located in another 'time' or another 'dimension.' By no means are the events of the era of global warming akin to the stuff of wonder tales; yet it is also true that in relation to what we think of as normal now, they are in many ways uncanny; and they have indeed opened a doorway into what we might call a 'spirit world'—a universe animated by nonhuman voices."[40]

38 Emmerich's *The Day after Tomorrow* is now almost a classic. A more recent movie is *Tomorrow*, dir. Melanie Laurent, Cyril Dion (France, 2015). For an analysis of *The Day After Tomorrow* and the climate scenario used as part of the plot see Mauelshagen, Franz, "Climate Catastrophism: The History of the Future of Climate Change," in *Historical Disasters in Context: Science, Religion, and Politics*, ed. Andrea Janku et al. (London, New York: 2012).

39 Classic dystopian authors are the British J.G. Ballard with *The Wind from Nowhere* (1961), *The Drowned World* (1962) and *The Burning World* (1964), as well as John Barnes with *Mother of Storms* (1994) and Saci Lloyd with *The Carbon Diaries: 2015*. However, the greatest output is now in 'disaster movies', comic books, and video games. Horn, Eva, *Zukunft Als Katastrophe* (Frankfurt a.M.: S. Fischer, 2014) looks at both, literary fiction and movies, and dedicates some of her analysis to climate (change) fiction, most of which indeed turns out to be a variant of science fiction. By using the phrase 'future as (a) catastrophe' we alluded to the title of her book.

40 Ghosh, *Great Derangement*, 72-73.

5 Outline of This Book

This volume on Europe shares the same structure with the other volumes in the *Climate & Culture* book series. Thus, following a section on *ideas*, we look into the *past*, then discuss the *present*, and finish with future *prospects*. Obviously, different levels of cultural variety connected with geographic focuses require flexibility in how the editors of each volume approach these sections. North America holds a much smaller number of different nation states and cultures than Asia or Africa. Culturally diverse regions virtually exclude the option of surveying the complex relationship between climate and culture nation by nation. That also applies to Europe. Trying to cover 'everything' geographically or culturally would have turned this publication into a handbook, which was not our purpose. Therefore, neither of the sections provides comprehensive surveys on specific nations or cultures. Instead, we chose thematic focuses specific for each section, which we considered most interesting and relevant. We are fully aware that the European map as we draw it here has many gaps compared with a handbook approach.

Opening the *Ideas* section, Jörn Sieglerschmidt surveys explanatory theories of weather and weather patterns. Linking antique, medieval, and modern approaches, Sieglerschmidt discerns an epoch-spanning search for the inherent truth of weather dynamics, and the attempt to establish norms for weather forecasting as well as investigations of probability factors regarding weather phenomena.

The second chapter by Markus Vogt draws our attention to the potential of Christian ethics in assisting rational approaches towards coping with climate change. Ceding intrinsic values to nature, Christians must rethink and reinterpret the dominion mandate, and have to take a position of responsibility, for example by acknowledging climate change as a "sign of the times". Vogt's survey highlights the need for religious institutions to generate an emotional reaction of society to environmental crises.

The section on the *Past* focuses on climate history and historical climatology, which is unique in the entire book series. As a consequence of this thematic focus, all the contributions to this section deal with early modern Europe, prior to the emergence of carbon culture and the onset of anthropogenic climate change. Agriculture was still the dominant economic sector in that period, highlighting the vulnerability of societies to climate variability during the Little Ice Age. Studying the responses of societies in that period is relevant to our understanding of adaptive processes.

All of the authors in the *Past* section have made important contributions to European historical climate research in the past. To begin with, when he started working on climate history in the 1950s following his now famous dissertation on the *Peasants of the Languedoc*, Emmanuel Le Roy Ladurie became the pioneer of historical climatology among historians. His *Histoire du climat depuis l'an mil* (1967) was most influential and inspired other historians as well as geographers to pick up the thread and push historical climate research to new limits. The book marked a first peak in Le Roy Ladurie's own study of climate. He returned to the subject again thirty years later, when he wrote his four volumes on *Histoire humaine et comparée du climat* (2004). His contribution to this volume is a personal account of the historical development of climate history.[41]

In his contribution, Geoffrey Parker expands on his award-winning book *Global Crisis: War, Climate Change and Catastrophe in the Seventeenth Century*, published in 2013. Here he investigates the 1590s—already a decade of severe crises and very much the beginning of a long seventeenth century in terms of both, crisis *and* climate. Parker points out the extraordinary, non-cyclical nature of this decade within the Little Ice Age. Even crises not directly provoked by climate should be considered indirect consequences of unusually cold European summers, reduced solar activity, El Niño events, and volcanic eruptions.

After Le Roy Ladurie's ground-breaking early engagement in climate history, Christian Pfister soon emerged as his most influential successor. His name represents progress in climate reconstructions from documentary evidence as well as in the historical study of the economic and social consequences of climatic fluctuations and extreme events. In his contribution, he presents results of his already much-debated study of the 1540 drought in Europe, in which he combines once more both aspects of historical climate research, i.e. reconstruction and impacts. In Pfister's assessment the 1540 drought was at least as warm, longer lasting, and dryer than the record-breaking spring and summer of 2003. Here, he describes the evidence of his reconstructions and elaborates on the economic and social disruptions following the extremes in 1540; and he finishes with a scenario of what a drought of that same scale would mean today.

The section finishes with a chapter by Dirk Riemann, Rüdiger Glaser, Johannes Schönbein, Steffen Vogt, and Iso Himmelsbach explaining methods

41 This account was first presented in French at the conference "Climate History—Past and Future / Histoire de Climat – Passé et Future", held at the German Historical Institute in Paris, 5-6 September 2011. Two years later, an extended French version was published under the same title: Le Roy Ladurie, Emmanuel et al. *Naissance De L'histoire Du Climat* (Paris: Hermann, 2013).

of climate reconstruction based on data from societal and natural archives for the last millennium. By combining conclusions from various sources (chronicles, weather journals, annals, and natural climate indicators such as tree rings) they provide a holistic understanding of climatic fluctuations. In a second step, they explore the connection between climate varibability, social discourse and food supply based on selected German cases.

The section on the *Present* collects case studies on carbon culture and its transformation, which also involves measures of adaptation to climate change. Carla Brandi investigates the European Union's leadership in the aftermath of the 2009 UNCCC in Copenhagen (COP 15). With the EU under pressure, she highlights three areas causing the loss of faith in the EU's position of eminence: preference cohesion, credibility and opportunities of power. Recognising a renewed trust in the Union's ability to lead, Brandi explores negotiations, struggles and concerted practices during subsequent climate summits.

In the chapter that follows, Claudia Kemfert takes an economic point of view and examines the opportunities that could result from a rapid energy transition in Germany. Kemfert considers Germany exceptionally well positioned for generating economic growth from switching to renewables. Despite this position, she also points out insufficiencies in the use of energy sources, as well as in investments in the energy supply.

Careful political handling of climate change requires consistent and permanent communication with all relevant agents, if only to avoid a general feeling of reluctance in some parts of society due to economic concerns. In her case study, Lea Schmitt inquires into conflict-ridden confrontations on the Frisian island of Ameland. To use them to their full potential, adaption measures to climate change must be integrated into local power relations.

Finally, the section on *Prospects* has two contributions by the editors of this volume. In his article on 'futurisation', Claus Leggewie argues that after decades of a rather bright and optimistic futurology, in our days the 'future' is mostly evaluated in much more sceptical and pessimistic terms. This applies to science-based scenarios (e.g. about the consequences of climate change, the degradation of biodiversity or the carrying capacity of Planet Earth in general) as well as to fiction (and science fiction). The reaction of political decision makers is what is called 'futurisation' in this article; the concept of boundaries (to economic growth or greenhouse gas emissions or the amount of public and private debt) almost naturally introduces a political approach focusing on the chances of future generations. Climate change has brought about a cultural reorientation regarding the dominant time consciousness of post-modern societies, as is shown mostly with reference to the German case.

Last but not least, Franz Mauelshagen argues that modern societies have fundamentally changed their relationship with the future and the conditions under which to plan for it by imposing their own dynamic on the environment and climate. As a consequence, uncertainties are undermining traditional prognostics based information about the past. In traditional risk calculations, 'nature' was static scenery in front of which the drama of history unfolded. With climate and the environment changing more rapidly, this is can no longer be taken for granted. The history of scenario planning after World War II indicates that the age of uncertainty began some time around 1970. This context helps to see the broader picture behind the new challenges climate change poses to the insurance industry in Europe and elsewhere. Expanding insurance globally has been recommended as a crucial strategy for adapting to climate change. However, the European case shows that changing frequencies and severities of meteorological and climatological hazards pose unprecedented adaptive challenges to the insurance industry itself.

PART 1

Ideas

CHAPTER 1

Complexion and Climate: An Attempt at an Outline of Weather Outlooks in Europe from the Beginnings Until Today*

Jörn Sieglerschmidt

Abstract

The explanations for the historically changing dynamics of weather conditions vary considerably, depending on when and where they were formulated. These differing views shall be examined and explained in terms of how they affected individuals and groups, how they reflect fundamental problems of philosophical understanding in their respective times, and how they contributed to an understanding of daily or long term weather phenomena. The history of explanations shows a play between the endeavour of catching the truth and getting a reliable weather forecast on the one hand and the probability typical for all weather phenomena on the other.

La Météorologie est un refoulé de l'histoire. De la grande et des petites, histoires des sciences et de la philosophie.[1]

1 Introduction

Our perception of nature is largely being affected by natural science as a new kind of faith. Differing, alternative creeds are deemed unscientific and

* The Latin word *complexio* can be translated as "physical constitution or temperament". In modern times the sense of the word is more expansive, including the terms of natural philosophy described in chapter 3.

1 Serres, Michel, *La naissance de la physique dans le texte de Lucrèce. Fleuves et turbulences.* (Paris: Editions de Minuit, 1977), 85 [The Birth of Physics. Manchester: Clinamen, 2000].

 | DOI 10.1163/9789004356825_003

are therefore incapable of winning major appeal. In the following article, it shall be shown that scientific experience is more than experiment and quantification (ch. 2); that through the 19th century, a systemic-structural way of looking at nature, and a holistic means of explaining it, has prevailed in natural philosophy (ch. 3); that this way of thinking can be witnessed in literary, musical, and other artistic expressions, which rebind natural phenomena to everyday experience (ch. 4); that only a natural science based on experiment and quantification has given up upon this means of explanation, or is incapable of honouring such a means due to their methods, as natural science can only explain what is controllable within their technical models (ch. 5). Therefore, it is time to readjust the one-sided positions of natural science with natural philosophical reflections. It is the de-solidarisation which occurred by separating humankind from nature that needs to be revoked.[2]

Throughout the whole of human history, the artefacts that humans left behind show that they have taken the surrounding natural world and, thus, climatic conditions into account on a daily basis. It can certainly be assumed that talk of the weather has always been important to everyday conversation.[3] Humans were always conscious that the weather had an influence on their fate. However, the explanations for the historically changing dynamics of weather conditions vary considerably, depending on when and where they were formulated. In the following, these differing views shall be examined and explained in terms of how they affected individuals and groups. A distinction shall be made between short- and medium-term climatic developments, whereby climate indicates long-term development and is, as a generic term, sometimes synonymous with the weather as perceived over an extended period.

The concept of nature will not be further elaborated upon,[4] but interpreted in the usual manner as our own nature as well as the nature that surrounds us. Alongside human nature, there is a second nature acquired during a lifetime, which includes not only the fruits of upbringing and education but also the world of artefacts that humans create, that is, what in general is known as culture or civilisation. Like the concept of nature, the concept of culture shall not

2 Descola, Philippe, *Par-delà nature et culture* (Paris: Gallimard, 2005), 665-690; Wulf, Andrea, *The Invention of Nature. The Adventures of Alexander von Humboldt—The Last Hero of Science* (London: John Murray, 2015).

3 Harley, Trevor A., "Nice Weather for the Time of the Year: The British Obsession with the Weather," in *Weather, Climate, Culture*, ed. Sarah Strauss, and Benjamin S. Orlove (Oxford: Berg, 2003), 103-118.

4 Sieglerschmidt, Jörn, and Birgit Biehler, "Natur," in *Enzyklopädie der Neuzeit*, vol. 8, ed. Friedrich Jaeger (Stuttgart: Metzler, 2008), 1133-1159, with further supporting documents.

be given a thorough definition,[5] but rather a rough understanding shall be evoked. During the modern era, the understanding of culture shifted from an involvement of humankind with nature that was being shared with other life-forms to an autonomous understanding of culture which included symbolic creations of culture by humankind. In this way, a constriction to elitist forms of culture occured. In this article, all perceptions and experiences are interpreted as cultures of nature.

The nature that surrounds humans will not be distinguished according to whether it remains uninfluenced by humans and is to a certain extent wild, on the one hand, or whether is domesticated by humans—which was in many cases understood primarily as culture up until 1800—on the other. This would neither make sense nor be at all possible, since we are concerned here with geophysical processes, which the human influence has in the interim sometimes played a role in massively changing. This surrounding nature is understood as the entirety of physical, chemical, and biological processes that take place in the atmosphere, in, and on the surface of, the ground and water. Thus, the natural environment will only be mentioned where direct reference is made to the concept of environment that Jakob von Uexküll coined.[6]

Weather phenomena are not only significant from a meteorological but also from a metaphorological point of view. Thus, the metaphor that plays on atmospheric or other aspects of geophysics is usually an allegory and not a background metaphor in the sense that Blumenberg intended.[7] Nonetheless: our language is interspersed with metaphors that play on geophysics. Our image worlds play with forms and varieties of weather. The seasons are a favoured subject of the arts, especially of painting and music. From early high cultures until today nothing has changed in this respect. The history of the *Geisteswissenschaften* and therefore of science too can illuminate this predicament on the basis of many examples and show that it is not simply a case of linear development that leads from pre-scientific concepts to the clarity—or, to paraphrase Blumenberg: to the truth and to the light—of systems founded on the natural sciences. If then relations can be established, it is only in certain narratives that such directness can be recognised.

5 Jaeger, Friedrich, "Kultur," in *Enzyklopädie der Neuzeit*, vol. 7, ed. Friedrich Jaeger (Stuttgart: Metzler, 2008), 253-281.

6 Herrmann, Bernd, *Umweltgeschichte: Eine Einführung in Grundbegriffe* (Berlin, Heidelberg: Springer, 2013), 28-31.

7 Haverkamp, Anselm, *Metapher: Die Ästhetik in der Rhetorik. Bilanz eines exemplarischen Begriffes* (München: Fink, 2007), 19-20; Blumenberg, Hans, *Paradigms for a Metaphorology* (Ithaca: Cornell University Press, 2010).

2 Experience

Any science of weather begins with the experience, that is, with the perception of weather events and their ordering with the intention of establishing rules that facilitate a weather forecast. The degree to which such a statement is bound up with all manner of presuppositions then becomes known when it is asked what exactly should be understood in terms of experience. Both Hippocrates and Aristotle—separated by almost a century—contributed decisively, in the context of the following two millennia, to the development of reflections concerning weather phenomena. For both, experience is indeed the basis of every insight, although it alone does not suffice to obtain such insight. The problem touched upon here concerns the relation between the specific and the general. For Aristotle, *empeiria* (experience), the empirical, referred to knowledge of the specific, while *technë* (art, craftmanship) or *epistëmë* (understanding, science) concerned the general. Both the empirical *knowledge that* and the scientific *knowledge as to why* are reliant on one another,[8] whereby the memory organises individual sense-data relating to experience, a view that is later summarised thus: *Experientia fit ex multis memoriis*.[9] For Aristotle the empirical leads to art, while ignorance, or lack of experience, generates chance, whereby experience falls back on concept formation. The Aristotelian distinction between *empeiria* (experience) and *epistëmë* (understanding, science), between historical and philosophical disciplines, was sustained in philosophical discourse until well into the nineteenth century, in so far as the empirical was equated with history, that is, the knowledge of singular facts, and *epistëmë* (understanding, science) with philosophy or (natural) science.[10]

The Aristotelian concept of the empirical is first put into question in empiricism with the attempt to see pure sense-data, that is sense-data uninfluenced

---

8 Kambartel, Friedrich, *Erfahrung und Struktur: Bausteine zu einer Kritik des Empirismus und Formalismus* (Frankfurt a.M.: Suhrkamp, 1968), 50-52.

9 Thomas Aquinas, *Summa theological*, vol. 2 (Parma: Petrus Fiaccadorus, 1852), 404 (Pars I, Questio 54, art. 2), accessed March 27, 2017. <http://www.mdz-nbn-resolving.de/urn/resolver.pl?urn=urn:nbn:de:bvb:12-bsb10744125-5>. Along with further antique literature, this edition is available at <https://www.deutsche-digitale-bibliothek.de/> for research purposes and is viewable/downloadable online free of charge as a digital copy from the copyright holder. All literature published before 1900 that is cited in this chapter has been retrieved through <https://www.deutsche-digitale-bibliothek.de/>, some through <http://www.europeana.eu> or Early English Books Online (EEBO) and The Making of the Modern World (Gale Cengage Learning).

10 Kambartel, *Erfahrung und Struktur:* [...] *Empirismus und Formalismus*, 58-60.

by antecedent experience or any kind of concept, as the basis for further knowledge.[11] Thus Francis Bacon had already achieved an important advancement of Aristotelian reflection as regards the natural sciences. It was Bacon who took the concept of experience, the *experientia* in the *experimenta*, and broadened it through the controlled investigation of nature. He distinguished casual (Lat. *vaga*) and directed (*ordinata*) experience. The former cannot lead to knowledge; it is only the latter that can impart scientific and philosophical insights. These are of an axiomatic nature and lead in turn to proposals for new experiments.[12] Bacon and Kant agree on much. Kant, however, goes further than Bacon as far as he clearly formulates that experience without the operation of reason would not be at all possible.[13] The famous dictum at the beginning of the introduction to *The Critique of Pure Reason* makes clear the agenda on which Kant bases the work of autonomous reason that creates for itself a world of its own so that it must not allow itself to be led by chance nature.

One of the goals of reason is experience as the synthesis of empirical knowledge, which is also referred to as understanding. Reason issues from itself the principles and designs that nature requires in order to answer questions that are guided by reason. Otherwise, accidental insights unconstrained by laws can be expected; but reason requires insights that follow certain laws.[14] In the process, the understanding provides rules, which are the products of the understanding and may not be related back to other rules:[15]

> Experience in the sense of empirical *knowledge*, that is, the reliability of nature and, as its precondition, the necessary connection between events, does not result of itself from nature but must be worked into nature. Experience is thus a task of reason that must be achieved in the service of the autonomy of human actions.[16]

11 Kambartel, *Erfahrung und Struktur:* [...] *Empirismus und Formalismus*, 19-48, 59.

12 Ibid., 77-78; Bacon de Verulamio, Franciscus, *Novum Organum Scientiarum*, Editio Secunda (Amstelædami: Joannes Ravesteinius, 1660), 27 (Lib. 1, Aph. 1), 84-85 (Lib. 1, Aph. 82), 154-158 (Lib. II, Aph. 6-10), accessed March 27, 2017. <http://www.mdz-nbn-resolving.de/urn/resolver.pl?urn=urn:nbn:de:bvb:12-bsb10042656-8>.

13 Kant, Immanuel, "Kritik der reinen Vernunft," in *Werke in sechs Bänden*, vol. 2, ed. Wilhelm Weischedel (Darmstadt: Wissenschaftliche Buchgesellschaft, 1966), 45; Kambartel, *Erfahrung und Struktur:* [...] *Empirismus und Formalismus*, 96-97, 100.

14 Kambartel, *Erfahrung und Struktur:* [...] *Empirismus und Formalismus*, 108-111.

15 Ibid., 129-131.

16 Ibid., 111; 93-94.

Friedrich Kambartel formulates here the close connection between scientific (as in natural science) knowledge and human freedom, an insight that Friedrich Schiller explained in more detail with the Kantian difference between reason (*Vernunft*) and mind (*Verstand*).[17]

Despite this new perspective there remains a continuity between the antique and modern, not to mention today's view of experience. In respect of which it is hardly surprising that scientific meteorology begins where rules established as a result of experience make possible a prognosis of future natural occurrences. Mathematics is thus for Kant, just as it is for Bacon, necessarily of assistance.[18] It should also be remembered that with Bacon and Kant, a narrowing of the concept of science comes into play, such that science is intended to be limited to that which clearly demonstrates experimentally controlled and mathematically formulated causal relationships. In this way, the natural sciences hit their limits in the 20th century at the latest, and they know it, too. At the same time, the new conditions are reflected upon in philosophical discussion.[19]

Humans have had experiences of the kind just discussed ever since they begun to reflect on nature. From the early civilisations onward, certain reoccurring weather phenomena were either related to the acts of gods or identified directly with a deity. Since the early Stone Age (6000-4000 BCE), it was already possible for astrologers and astronomers to draw up precise calendars indicating years with both equinoxes and solstices. These calendars were also linked to the passing of the seasons.[20] Hunters and gatherers too required experience of weather dynamics, in order to be able to hunt successfully and harvest fruits at the right point in time. With the Neolithic revolution, long- and short-term weather forecasts became essential for assessing the likely success of the harvest. Babylonian and Chaldean priests were, as astronomers and astrologers, responsible for predicting natural catastrophes such as floods, droughts, storms, earthquakes. At this point in time there were already written documents that show how the observation of appearances in the sky of, for

---

17 Schiller, Friedrich, "Über das Erhabene," in *Schillers Werke: Nationalausgabe*, vol. 21. (Weimar: Böhlau, 1963), 38-54, esp.48, 38-49, 6.

18 Bacon, *Novum Organum*, 157-158 (Lib.2, Aph.8); Kambartel, *Erfahrung und Struktur*: [...] *Empirismus und Formalismus*, 220-221.

19 Kambartel, *Erfahrung und Struktur*: [...] *Empirismus und Formalismus*, 206, 214-220; Cartwright, Nancy, *The Dappled World: A Study of the Boundaries of Science* (Cambridge: University Press, 1999); Hacking, Ian. *The Taming of Chance* (Cambridge: University Press, 1990), 1-5, 142-149, 200-215.

20 Körber, Hans-Günther, *Vom Wetteraberglauben zur Wetterforschung* (Leipzig: Edition, 1989), 10-11.

example, halos around the moon or sun were associated with agricultural predictions.[21]

The introduction of written records constitutes one important turning point in the observation both of immediate weather events and of the weather over an extended period. The other takes place in Greece through the attempt to explain weather events scientifically based on empirical findings, without making the gods responsible for these events. Writers such as Hesiod, Homer, Aristophanes, and Herodotus also provide physical explanations for the weather. But it is Hippocrates and, in particular, Aristotle who first systematised available knowledge and categorised it according to geophysical qualities.[22] Bacon still recognised this as progress.[23]

Written records are decisive for the advancement of knowledge of simple rules pertaining to weather that were previously preserved in oral traditions. Empirical observations in written form mean that the results of these observations remain available for further analysis. Oral traditions had contributed to the passing on of knowledge of many millennia.[24] So it has been observed that some of the Hippocratic texts were made for oral presentation.[25] However, much is also lost in this way or reduced to convenient formulas[26] that hardly allow for systematic development. A long time ago, Jack Goody and Ian Watt attributed to oral cultures a homeostatic organisation of knowledge[27] that was unable to develop dogmatics at all. Instead, it adapted to changing circumstances of a social or natural world without perceiving any such discontinuity in the tradition. It was only once writing reified experience, that is, made it into

---

21 Fierro, Alfred, *Histoire de la météorologie* (Paris: Denoël, 1991), 13-16.

22 Körber, *Vom Wetteraberglauben zur Wetterforschung*, 46-60; Fierro, *Histoire de la météorologie*, 20-26.

23 Bacon, *Novum Organum*, 92 (Lib. 1, Aph.89).

24 Henshaw, Anne, "Climate and Culture in the North: The Interface of Archaeology, Paleoenvironmental Science, and Oral History," in *Weather, Climate, Culture*, ed. Sarah Strauss, and Benjamin S. Orlove (Oxford: Berg 2003), 217-231; Ogilvie, Astrid E.J., and Gísli Pálsson, "Mood, Magic, and Metaphor: Allusions to Weather and Climate in the Sagas of Icelanders," in *Weather, Climate, Culture*, ed. Sarah Strauss, and Benjamin S. Orlove, (Oxford: Berg, 2003), 251-274.

25 Jouanna, Jacques, "Rhetoric and Medicine in the Hippocratic Corpus," in *Rhetoric and Medicine in the Hippocratic Corpus: Greek Medicine from Hippocrates to Galen*, ed. Jacques Jouanna, 39-53 (Leiden: Brill, 2012).

26 Propp, Vladimir Yakovlevich, *Morfologija skazki Морфология сказки* [Morphology of the Folk Tale], *Voprosy poėtiki* [Problems of Poetics] 21 (Leningrad: Akademija, 1928).

27 Goody, Jack, and Ian Watt, "The Consequences of Literacy," in *Comparative Studies in Society and History* 5 (1962/1963), 304-345; of particular relevance here: 340.

a thing, that knowledge could accumulate in such a way as to be capable of further dogmatic development.

The folklore of weather is reflected in the numerous rules of the weather that were passed down orally in all European cultures. People were not only existentially dependent on the weather in agriculture but also with regard to seafaring and attempted to use experience gained through extended observation in order to make predictions. Such rules are sometimes contradictory.[28] But many of these rules, and particularly those that tend to be limited to short-term predictions, show a high degree of probability when measured against long-term weather dynamics. Observers of past times were also able—as is documented in this way—to accurately assess a range of phenomena, from the overall weather situation in central Europe to the maritime or continental climate and from the effects of polar or subtropical air to the significance of certain wind currents. However, it should be taken into consideration that the regional specificity of many rules cannot be known and that one rule may do justice to the large regional differences in England but be completely misleading in Italy.

Taken together, these weather rules attest to a good facility for observation on the part of our predecessors.[29] In addition to temperature, moisture and the appearance of the sky (in terms of, for example, rainbows, halos around the moon and sun, northern lights) the response of plants and animals to specific types of weather was also observed. Since antiquity, frogs and toads have been the best-known weather prophets. It is certain that many domesticated animals such as cats, dogs, chickens, ducks, horses, donkeys, pigs react far earlier to certain types of weather than humans do and, therefore, facilitate short-term predictions. Wild animals such as spiders or swallows are also still accepted today as indicators of weather. The sensitivity of the sick to the weather was also known about, for example, those with a scar: *Héros, vous ressentez vos blessures? Annoncez-le, la pluie est sûre.*

Various stages in the growth of plants determined by certain weather patterns have been also observed over the ages. Such phenological observations can also be found in the written records of the modern period. Phenological data remains relevant today. It is well known that rain can be expected if field

28 Fierro, *Histoire de la météorologie*, 31-42; Strauss, Sarah: "Weather Wise: Speaking Folklore to Science in Leukerbad," in *Weather, Climate, Culture*, ed. Sarah Strauss, and Benjamin S. Orlove, (Oxford: Berg, 2003), 48-51.

29 Malberg, Horst, *Bauernregeln: Aus meteorologischer Sicht* (Berlin: Springer, 2003), 9-72; Chassany, Jean-Philippe, *Les couleurs du vent: La météo d'hier et d'aujourd'hui* (Paris: Maisonneuve et Larose, 1986).

bindweed, thistles, pimpinella or dandelions close their flowers or the stalks of clover straighten up. Owing to a proverb, the hygrometric properties of pine-cones are well known in Walloon, Belgium: "Quand il fait beau, j'ouvre mes fenêtres. Quand il fait laid, je les referme. Qui suis-je?"[30]

Seasonal and, to some extent, monthly rules pertaining to weather and to agriculture were widespread in antiquity. Whether in Virgil's *Georgics* or Pliny the Elder's *Naturalis Historia* or the works of Lucius Iunius Moderatus Columella, there were clues everywhere that related to calendars and related back to Greek precursors, the *parapëgmata*.[31] During the Renaissance, this tradition—expanded by later authors such as Rutilius Taurus Aemilianus Palladius, Konrad von Megenberg, Thomas von Cantimpré, Petrus de Crescentiis, to mention but a few—was passed down along with the antique authors in printed form to the modern age.[32]

Another kind of text that was widespread in Europe from the sixteenth century was the so-called *Hausväterliteratur*.[33] Here two authors who were particularly influential in central Europe should be mentioned: a priest from Parchim named Johann Coler[34] and a country nobleman from Silesian Lower Austria called Wolf Helmhardt von Hohberg.[35] Literature on husbandry

30 Fierro, *Histoire de la météorologie*, 37-42; Malberg, *Bauernregeln*, 121-130.

31 Körber, *Vom Wetteraberglauben zur Wetterforschung*, 67-68; Taub, Liba, *Ancient Meteorology* (London, New York: Routledge, 2003), 20-37.

32 Vergilius Maro, Publius, *Bucolica* [...], *Georgica* [...] (Florenz : 1487/1488), accessed March 27, 2017. <http://nbn-resolving.de/urn:nbn:de:bvb:12-bsb00065268-3>; Plinius Secundus, Gaius, *Naturalis historiae libri XXXVII* (Lutetia: Chaudière, 1516), accessed March 27, 2017. <http://nbn-resolving.de/urn:nbn:de:bvb:12-bsb00065268-3>; Columella, Iunius Moderatus, *De re rustica libri XII* (Lugduni: Seb. Gryphius, 1548), accessed March 27, 2017. <http://www.mdz-nbn-resolving.de/urn/resolver.pl?urn=urn:nbn:de:bvb:12-bsb10171885-3>; Palladius, Rutilius Taurus Aemilianus, *De re rustica Libri XIIII* (Paris: Robertus Stephanus, 1543), accessed March 27, 2017. <http://reader.digitale-sammlungen.de/resolve/display/bsb10172095.html>.

33 Humpert, Magdalene, *Bibliographie der Kameralwissenschaften* (Köln: Schroeder, 1937); for France see: Serres, Olivier de [Seigneur de Pradel]. Le Theatre d'agriculture et mesnage des champs. Seconde edition reueuë et augmentee par l'auteur. Paris: Abr[aham] Savgrain, 1603, accessed March 27, 2017. <http://www.mdz-nbn-resolving.de/urn/resolver.pl?urn=urn:nbn:de:bvb:12-bsb10229138-5>.

34 Coler, Johann, *Calendarium perpetuum et libri oeconomici* [...] (Wittemberg: Paul Helwig, 1603), accessed March 27, 2017. <http://www.mdz-nbn-resolving.de/urn/resolver.pl?urn=urn:nbn:de:bvb:12-bsb10228551-6>.

35 Hohberg, Wolf Helmhardt von, *Georgica Curiosa oder Adeliches Land=Leben* [...], 2 Vols. (Nürnberg: Endter, 1695), accessed March 27, 2017. <http://digital.slub-dresden.de/werkansicht/dlf/8112/1/>.

includes many topics regarding the maintenance of a noble country household, for instance; it provides guidelines for the construction of a country estate, (moral) rules for married life, the upbringing of children etc. Many aspects of agriculture, forestry, and housekeeping as well as the administration of estates were treated in such works. As for the building of houses, the situation of a building played a role—as Hippocrates had already described—due to the various influences of climate.

Much use was made of calendars that not only showed the course of the sun through the day but also provided recommendations as to tasks that should be carried out on the land, in the forest or in the house according to seasonal weather patterns. Also included under household activities were certain medicinal procedures. As such, astro-meteorological knowledge had an extensive role to play. Nonetheless, many authors attempted to incorporate location-specific circumstances and give practical tips that did not hail from natural philosophy alone. As such, Johann Coler referred emphatically to local experience.[36] In England, Anthony Fitzherbert, jurist and landowner, preferred to rely on the experience of farmers rather than science.[37]

Thus, the calendars and publications offering general practical and location-specific advice distinguished themselves from the practices, prognostics, and almanacs that were becoming more widespread at the time. In addition to seasonally established holy days and other days considered as significant for future weather trends—so-called *Lostage* (fateful days, weeks or months determining the weather of future seasons)—during the year, these publications drew upon astro-meteorological forecasting of the weather. Today, Ice Saints (12-15 May), Sheep Chill (10-20 June) or Seven Sleepers Day (27 June) and other calendric rules are known as *Lostage*.[38]

So-called farmers' practices appear from the end of the fifteenth century in numerous editions, among which Leonhard Reynman's slim volume on weather of 1510 was reprinted particularly frequently. Fourteen chapters contain numerous natural weather indicators (optical manifestations and clouds) and astro-meteorological rules that, to a certain extent, had already emerged

36 Coler, Johann, *Calendarium perpetuum* [...]. Das ist: Ein stetswehrender Calender/ Auch Zwantzig nothwendige vnd gantz nützliche Haußbücher: [...] (Wittemberg: Johann Wilhelm Fincelius, Paul Helwig, 1632), 327.

37 Fitzherber[t], [Anthony], *The Booke of Husbandry: Very profitable and necessarye for all maner of persons* [...] (London: Iohn Awdely, 1573); Skeat, Walter W., ed., *Fitzherberts Book of Husbandry 1534* (London: Trübner, 1882), 91.

38 Körber, *Vom Wetteraberglauben zur Wetterforschung*, 77-79; Malberg, *Bauernregeln*, 18-19; Fierro, *Histoire de la météorologie*, 49-51; Hellmann, Gustav, *Beiträge zur Geschichte der Meteorologie*, vol. 1 (Berlin: Preußisches Meteorologisches Institut, 1914), 5-102.

in writings from antiquity and from Arab lands.[39] These works were just as unsuited to meteorological prognoses as the hundred-year calendar.[40] First published by Moritz Knauer, this calendar linked astrologically derived weather prognoses to information on agriculture and associated seasonal tasks. The calendar was repeatedly updated and reprinted.[41] Such calendars are still available today, even though the basis for our knowledge as concerns the weather has completely changed. It is not therefore surprising that the Prussian Academy of Sciences, which had a monopoly on a particular astronomical calendar that was much in demand, reinstated the astro-meteorological prognoses in the calendar that had been removed in 1779 after having been proven scientifically inaccurate. The Academy took the decision to do so after the returns on the calendar dramatically sank as a result of the missing material. After all, the Academy still had to finance itself at least in part from its own revenues.[42]

However, there were already critical voices in the early sixteenth century, when Sebastian Brant satirised astro-meteorological prognoses in *A Ship of Fools* and thus took up a motif that was already conventional in antiquity and that was current in the Renaissance too, alongside a permanent and firm belief in astrology.[43] Rabelaisian and similarly satirical in style, Johann Fischart's *Aller Practick Großmůtter* [The Grandmother of All Practices] appeared in

39 Reynman, Leonhard, [Wetterbüchlein.] *Von warer erkantnuß des wetters.* Also das ain ÿeder er seÿ gelert oder ungelert durch alle natürliche anzaigung die änndrung des wetters aÿgentlich vnd augenscheinlich erkennen mag etc. (München: Hannss Schobsser, 1510) [Neudrucke von Schriften und Karten über Meteorologie und Erdmagnetismus, vol. 1 (Berlin: Asher 1893)], accessed March 27, 2017. <http://nbn-resolving.de/urn:nbn:de:bvb:12-bsb00004843-3.>; Hellmann, Gustav, *Meteorologische Volksbücher, Ein Beitrag zur Geschichte der Meteorologie und zur Kulturgeschichte. Sammlung populärer Schriften der Gesellschaft Urania zu Berlin*, vol. 8 (Berlin: Paetel, 1895), 21-30; Hellmann, Gustav, *Versuch einer Geschichte der Wettervorhersage im 16. Jahrhundert, Abhandlungen der Preussischen Akademie der Wissenschaften Jg. 1924 Phys.-Math. Klasse 1* (Berlin: Akademie der Wissenschaften, 1924).

40 Malberg, *Bauernregeln*, 149-164.

41 Körber, *Vom Wetteraberglauben zur Wetterforschung*, 79-80, 97; Hellmann, *Meteorologische Volksbücher*, 46-62; Knauer, Mauritius, *Calendarium Oeconomicum Practicum Perpetuum. Das ist: Immerwährender Curieuser Hauß=Calender* [...], (Nürnberg: Johann Georg Lochner, 1741), accessed march 27, 2017. <http://www.mdz-nbn-resolving.de/urn/resolver.pl?urn=urn:nbn:de:bvb:12-bsb10300775-3>, accessed March 27, 2017.

42 Körber, *Vom Wetteraberglauben zur Wetterforschung*, 97-98.

43 Brant, Sebastian, *Das Narrenschyff* (Basel: Jo[hann] B[ergmann], 1494), 82^{v}–83^{r}, accessed March 27, 2017. <http://nbn-resolving.de/urn:nbn:de:bvb:12-bsb00036978-3>; Barclay, Alexander ed. [Sebastian Brant] *The Ship of Fools*. 2 vols. (Edinburgh, London: William

1572.[44] Others too, such as Johann Leonhard Rost, writer and astronomist, dealt critically with this kind of weather forecasting.[45]

The critique was then backed up as longer series of observations were compared with actual weather dynamics. Hermann IV, landgrave of Hessen-Kassel (1607-1658), was educated in the natural sciences and among the leading meteorologists of the early period of weather observation. He began observing the weather in 1623 and continued until 1646, and presented the results of his observations in print in order to test the reliability of astro-meteorological forecasts.[46] Despite his insistence on attributing a certain value to these prognoses,[47] he came to a negative conclusion regarding the weather rules.[48]

Weather observations recorded in written form that had to do above all with extraordinary weather events are to be found in historical narratives, calandars and other texts since the early modern period.[49] People's fears are present in the narratives in the form of the *ekpyrosis* (Greek for: world fire) and the *kataklysmos* (Greek for: great flood, diluvium).[50] Abnormal weather and atmospheric events always played an important role later too, since besides the possible forecast of such events, the cause of such visitations also remained a key concern. Until long into the modern period, it was above all divine figures of authority that in this way displayed to humans their power in either an encouraging or punitive sense.[51]

Paterson/Henry Sotheran, 1874), vol. 2, 18-21; Sieglerschmidt, Jörn, "Zodiacus," in *Enzyklopädie der Neuzeit*, vol. 15, ed. Friedrich Jäger (Stuttgart: Metzler, 2012), 539-544.

44 Kurz, Heinrich ed. *Johann Fischart's sämmtliche Dichtungen, 3 Theile.* (Leipzig: Weber, 1867), 33-39, accessed March 27, 2017. <http://www.mdz-nbn-resolving.de/urn/resolver.pl?urn=urn:nbn:de:bvb:12-bsb10105532-1>.

45 Körber, *Vom Wetteraberglauben zur Wetterforschung*, 97-98.

46 Cyriandrus [Greek for Herrmann], Uranophilus [Greek for: friend of the heaven], *Meteorologica:* [...] (Cassel: Salomon Schadewitz, Sebald Köhler, 1551 [1651]), accessed March 27, 2017. accessed March 27, 2017. <http://www.mdz-nbn-resolving.de/urn/resolver.pl?urn=urn:nbn:de:bvb:12-bsb10058302-0>.

47 Ibid., *Ander Theil*, 1.

48 Ibid., *Dritter Theil*, 59.

49 Weikinn, Curt, Michael Börngen, and Gerd Tetzlaff, ed. *Quellentexte zur Witterungsgeschichte Europas von der Zeitenwende bis zum Jahre 1850: Hydrographie*, 6 Vols. (Berlin: Akademie, 1958-1963; 2000/2002).

50 Böhme, Gernot, and Hartmut Böhme, *Feuer, Wasser, Erde, Luft: Eine Kulturgeschichte der Elemente* (München: Beck, 1996), 53.

51 Hellmann, Gustav, *Die Meteorologie in den deutschen Flugschriften und Flugblättern des 16. Jahrhunderts: Ein Beitrag zur Geschichte der Meteorologie, Abhandlungen der Preußischen Akademie der Wissenschaften Jg. 1921, Phys.-Math. Klasse 1* (Berlin: Akademie der Wissenschaften, 1921).

Evidence in Europe of long-term chronicles, so-called weather journals, dates back to the fourteenth century.[52] Sailing handbooks, log books and directions concerning shipping routes have been handed down ever since the fifteenth century.[53] Such (monthly, weekly, daily) weather observations were often prompted in the Middle Ages and the modern period by the practical need of forecasts for the purposes of agriculture and sea travel. Scientific curiosity also played a role, as in the cases of the Hessian landgrave or, some decades earlier, of Renward Cysat. Thus, Cysat's education in the natural sciences as a chemist came in useful when he was, from time to time, noting down in great detail observations of a phenological kind, in addition to those concerning temperature and precipitation. He also found it fitting to include observations concerning animals: for example, the advance of wild animals into urban areas during particularly hard winter months. Such observations seldom changed the framework of received knowledge systems. Certainly, scepticism was expressed from time to time, but theological and scientific doctrine remained for the most part unaffected. It was only when instruments such as the thermometer, barometer or hygrometer were invented in the seventeenth century and generated measurements that a new perspective on the weather emerged in Europe.

Throughout history, extraordinary climatic conditions and weather patterns often prompted resources to be provided for associated measures aimed at avoiding the worst effects. In some southerly areas of Europe, sufficient water was not always available throughout the year. Complex irrigation systems but also mill-like scoop wheels, like those that had already been developed, for example, in the early civilisations of western Asia, served during antiquity to channel water into lands used for agriculture as well as for human

52 Hellmann, Gustav ed., *Meteorologische Beobachtungen vom 14. bis 17. Jahrhundert, Neudrucke von Schriften und Karten über Meteorologie und Erdmagnetismus*, vol. 13 (Berlin: Asher, 1901).

53 Körber, Vom *Wetteraberglauben zur Wetterforschung*, 100-102; Pfister, Christian, *Wetternachhersage: Fünfhundert Jahre Klimavariationen und Naturkatastrophen* (Bern: Haupt, 1999), 22-25; Pfister, Christian, *Klimageschichte der Schweiz 1525-1860: Das Klima der Schweiz von 1525-1860 und seine Bedeutung in der Geschichte von Bevölkerung und Landwirtschaft* (Bern: Haupt, 1988), 25-31; Glaser, Rüdiger, *Klimageschichte Mitteleuropas: Tausend Jahre Wetter, Klima, Katastrophen* (Darmstadt: Wissenschaftliche Buchgesellschaft, 2001), 15-18; Pfister, Christian et al., "Daily Weather Observations in Sixteenth-Century Europe," *Climatic Change* 43 (1999): 111-150; Pfister, Christian, "Renward Cysat – ein 'interdisziplinärer' Pionier der Klimaforschung im Alpenraum," *Der Geschichtsfreund* 166 (2014): 187-208.

consumption.[54] Technical facilities such as water wheels required the controlled channelling of water from freely flowing sources and water stores in times of drought in order to be able to maintain supplies. Even more input was required for the construction of dikes on the North Sea, of which there is evidence dating back to the thirteenth century. These also required far-reaching legal regulation—just as the use of water for mills did—and showed too that something had been learnt from previous catastrophes or deficiencies.[55] Even the building of streets in Roman antiquity and the European modern era is to be seen not only as a means of increasing and accelerating transport capacities but also as protection against important military and trade routes being rendered impassable by the effects of climate. In addition to clothing and housing, in their everyday lives humans have always had the need of measures to protect them from the influence of certain weather patterns. Among societal practices aimed at overcoming extraordinary weather events, ranging from the ringing of church bells to ward off bad weather to supplicatory processions, numerous measures have been taken as a result of weather observations, and still are, in order to provide effective protection. The forms of memory culture consist not only in oral and written evidence but also in numerous historical objects—like dikes, scoop wheels or marks on the wall of a house indicating the the year and the level of historical floods—pertaining to continuous human analysis of the weather.[56]

3 Cosmos and Complexion

In many myths concerning the origin of the world, the world is described after the act of creation as an ordered, beautiful cosmos in which all things stand in meaningful relation to one another. The maxim *omnia in omnibus*, much discussed by Athanasius Kircher, still makes it clear in the seventeenth century[57] that macrocosm and microcosm draw on one another in many, in

54 Radkau, Joachim, *Natur und Macht: Eine Weltgeschichte der Umwelt* (München: Beck, 2000), 112-159.

55 Pfister, Christian, *"The Monster Swallows You:" Disaster Memory and Risk Culture in Western Europe, 1500-2000*. Rachel Carson Center Perspectives 2011, accessed May 5, 2017. <http://www.environmentandsociety.org/sites/default/files/web_edition_color.pdf>.

56 Glaser, Rüdiger, *Klimageschichte Mitteleuropas: Tausend Jahre Wetter, Klima, Katastrophen* (Darmstadt: Wissenschaftliche Buchgesellschaft, 2001), 20-21.

57 Leinkauf, Thomas, *Mundus combinatus: Studien zur Struktur der barocken Universalwissenschaft am Beispiel Athanasius Kirchers SJ (1602-1680)* (Berlin: Akademie, 2009), 89-91, 269-275.

inexhaustibly numerous ways. This is denoted as *complexio*. Besides the natural cosmos there is therefore a semiotic cosmos that links the various signs together to create a meaningful whole. The *book of nature* possesses a sign structure in the proper sense:[58] everything has its place in an unmistakable kind of combination with other elements in the system, everything has its own signature.

The baroque system of knowledge incorporates the elements (water, earth, air, and fire) into an encyclopedic view that encompasses all branches of human knowledge.[59] There are fluid transitions between animate and inanimate nature on the one hand and the social world of humans on the other. The animation—in the proper sense (*animus*, Latin for "soul")—of things transforms the world into an animated, an animist universe, where every thing, every process, every motion, every appearance and utterance of a living being has a certain semiotic meaning.[60]

From antiquity until the 18th century, nature is subdivided in accordance with a tetrahedral schema, first discussed by Empedocles and canonised via Hippocrates as well as Aristotle for use in medicine or natural philosophy. The four elements—earth, water, air and fire—coincide with four qualities—dry, wet, hot, cold. In the Hippocratic documents,[61] these qualities correspond to the bodily fluids or humours—blood, phlegm, yellow and black bile.

Subsequently, a taxonomy developed that incorporated not only geophysical aspects such as the seasons, wind directions, and phases of the day

58 Carnap, Rudolf, *Der logische Aufbau der Welt* (Berlin: Weltkreis 1928), 13-21 (Abschn. 11-16), esp. 17-18; Carnap, Rudolf, *The Logical Structure of the World. Pseudoproblems in Philosophy* (London: Routledge & Kegan, 1967), 21-30, esp. 25-27.

59 Meetz, Karen S., "Tempora triumphant: Ikonographische Studien zur Rezeption des antiken Themas der Jahreszeitenprozession im 16. und 17. Jahrhundert und zu seinen naturphilosophischen, astronomischen und bildlichen Voraussetzungen" (PhD diss., University of Bonn, 2003), 251, accessed March 27, 2017. <http://hss.ulb.uni-bonn.de/2003/0126/0126.pdf>; Horstmanshoff, Herman F.J., Antonie M. Luyendijk-Elshout, and F.G. Schlesinger, ed., *The Four Seasons of Human Life. Four Anonymous Engravings from the Trent Collection. Nieuwe Nederlandse bijdragen tot de geschiedenis der geneeskunde en der natuurwetenschappen*, vol. 60; *Pantaleon Reeks*, vol. 32 (Rotterdem, Durham: Erasmus Publishing, Trent Collection Duke University, 2002), 65-77.

60 Kohl, Karl-Heinz, *Die Macht der Dinge: Geschichte und Theorie sakraler Objekte* (München: Beck, 2003), 202; Porter, Martin, *Windows of the Soul: Physiognomy in European Culture 1470-1780* (Oxford: Clarendon, 2005), 21-28.

61 On authorship in *Corpus hippocraticum*, see: Jouanna, Jacques, *Hippocrate: Pour une archéologie de l'École de Cnide, Collection d'Études Anciennes*, vol. 141 (Paris: Belles Lettres, 2009); Jouanna, Jacques, *Hippocrate* (Paris: Fayard, 1992) [Am.: Hippocrates. Baltimore: Johns Hopkins University Press, 1999]; Böhme, and Böhme, *Feuer, Wasser, Erde, Luft*, 164.

within a tetrahedral form, but also numerous medicinal and bodily as well as biblical references. Certain weather conditions were thus not perceived simply as weather conditions *per se*, but open up complex semiotic references. Not all such references were consistent and identical but could be thoroughly contradictory.

A complex explanation of weather phenomena in the Aristotelian tradition was found in the *Meteorologia* of Robert Fludd on a copperplate engraving published in 1626 by the *Officina Bryana*, the printing house of the family de Bry, led at that time by Mathäus Merian. This copperplate engraving was also published as frontispiece of the *Metereologica* of Hermann IV, landgrave of Hessen-Kassel.[62] The *Short Explanation and Clarification of the Mirror and Account of this Meteorographia* (Brevis speculi seu catoptri huiusce nostri meteorographici explicatio et elucidatio) decribes initially God (Jahwe) with his helpers, the angels, as the source of all things that masters the geophysical processes. The copperplate engraving and the accompanying text offer a closed, rationally constructed intellectual universe, whose natural and supernatural forces influence and explain the meteorological processes. The kabbalistic and hermetic influences on Robert Fludd are shown by the reference to *Ensoph seu divinitatis infinitudo* (Ensoph and the infinity of God). The cosmological view is geocentric. In the sublunar world there are fiery meteors (the Greek term *meteōros* describes objects in the sublunary sphere) like comets, or watery meteors like clouds, rain, snow, fog and other phenomena. The geophysical processes are embedded in the forces of the planets, astrological forces.[63] Even Hermann IV is sceptical regarding astrometeorology, but nevertheless concludes that the astrometeorological forcasts are not always, but still generally reliable.

The peoples of the world are incorporated into the tetrahedral division of the world. Therein lies the link to antique geography and the theory of climatic zones. Climate (*klima* in Greek) means, among other things, the flattening of the earth towards the poles, later and more generally the area of the skies or of the earth, or weather conditions observed over time and determined by location.[64] It was already clear, before the discovery of the so-called third world,

62 Fludd, Robert, *Philosophia sacra & vere Christiana Seu Meteorologia Cosmica.* (Francofurti: Officina Bryana, 1626), 144-145, accessed March 27, 2017. digital.slub-dresden.de/werkansicht/dlf/16704/1/; Cyriandrus, *Meteorologica*; the copy consulted is from Niedersächsische Staats- und Universitätbibliothek Göttingen, sign. 4 Phil I 1012:6.

63 Rösche, Johannes, *Robert Fludd. Der Versuch einer hermetischen Alternative zur neuzeitlichen Naturwissenschaft* (Göttingen: V&R unipress, 2008).

64 Taub, *Ancient Meteorology*, 29.

that the position of the earth in relation to the sun was not only influenced by the seasons but also by the form and intensity of the sun's rays. In general, a temperate climate was considered particularly suitable for humans, in accordance with an antique notion of the moderate and therefore positive midpoint between the extremes. The cold and therefore barren polar regions as well as the hot equatorial regions were therefore considered unsuitable. The theory of climatic zones is dismissive of such extremes and distinguishes between preferred and disadvantaged areas within the temperate zones for the most part according to the theory's place of origin. As such, there was always already a game underway involving the identification of one's own and the foreign.

Greek literature provides numerous examples of characterisations of peoples. These were carried over as topoi and stereotypes via Latin literature into the modern period. A few, such as Herodotus, draw on their own experience or reports of direct experience, which connect particular ways of life with climatic conditions in the sense of an ethnography, an ethnological description.[65] Besides literary documents, the treatise "On Airs, Waters, Places" from the *Corpus hippocraticum* continued to be valued into the 19th century as an authority on how the natural surroundings influenced human health.[66] In the first part, the effects of the geographical situation are described with reference to, for example, a city, the type of water available, the air quality, and the effects that these have on the illnesses of women and men. Winds are judged according to the direction in which they blow. Thus, the cold north wind is considered purifying. Water is categorised according to its clarity, hardness and salt content or freshwater qualities. Hard and cold water is, for example, not supportive of good health. Rainwater is praised as particularly good. Rain is formed by clouds, which are driven against each other by opposing winds and compressed.[67] The effect that the seasons and associated trends in weather conditions have on human health are also discussed. Finally, the influences of climate and altitude are discussed as regards Asian and European peoples as well as their political institutions and temperaments (belligerent, soft, etc.).[68]

65 Weiler, Ingomar, "Ethnographische Typisierungen im antiken und mittelalterlichen Vorfeld der 'Völkertafel'," in *Europäischer Völkerspiegel: Imagologisch-ethnographische Studien zu den Völkertafeln des frühen 18. Jahrhunderts*, ed. Franz K. Stanzel (Heidelberg: Winter, 1999), 97-118.

66 Diller, Hans, ed. *Hippokrates: Über die Umwelt* (Corpus medicorum Graecorum Part 1.1,2) (Berlin: Akademie, 1999); Jouanna, Jacques, ed. *Hippocrate: Airs, eaux, lieux* (Paris: Les Belles Lettres, 1996).

67 *Hippocrates on Airs, Waters, and Places* (London: Wyman, 1881), 35, 37 (ch. 3), 48-49.

68 Ibid., 46-105 (Ch. 4-6); Glacken, Clarence J., *Traces on the Rhodian Shore: Nature and Culture in Western Thought from Ancient Times to the End of the Eighteenth Century* (Berkeley:

Hippocratic views on how climate affects human qualities and the bodily fluids were passed down via Galen and through the Middle Ages to the modern period without any major revisions. In many cases, the attributes are ethnocentrically motivated, that is, the self-attributes in the descriptions of e. g. Greeks, Romans, Jews tend to imply that the areas in which the respective people live are, in comparison to others, the most suitable. In the modern period, confessional differences emerge too. Thus Jean Bodin considered in *Methodus ad facilem historiarum cognitionem* (1566) and in *Six livres de la république* (1586) the northern climate zone in which the (protestant) English lived as being inauspicious, a view that did not go undisputed in England. Germany is also perceived to be included in the northern climate zone, something that Christoph August Heumann found problematic in *Acta Philosophorum* (1715), since cold air was supposed to lead to sluggish blood and passive *ingenia* (Latin, "innate character").[69]

Besides antique teachings on climate, Jean Bodin also considered the morphology of a region (mountain ranges, waterways, air currents and humidity) to be of significance for the character of a people. In weighing up geographical influences and the opportunities that upbringing in the form of education afforded, Bodin gives the latter the edge since it seemed to him that upbringing could compensate to some extent for the disadvantages of location, as the example of the Germans showed.[70] However, the determinative relation between location, peoples and their legal norms remained valid. At the same time, Montaigne's stance in "An Apology for Raymond Sebond" is critical of human superciliousness. But he considers the climate's determination of human nature as a given and delivers a treatment of the topic on which there were to be many variations during the course of the modern period.[71] Bartolomé de las Casas (1484/85-1566) deals critically with the theory of climate zones in his *Apologética historia sumaria*, where he considers the prevailing climate in South America, normally perceived as damaging to humans and human culture, to be very advantageous. In so doing, he propagates at the same time the

---

University of California, 1967), 82-88.

69 Zacharasiewicz, Waldemar, "Klimatheorie und Nationalcharakter auf der 'Völkertafel'," in *Europäischer Völkerspiegel: Imagologisch-ethnographische Studien zu den Völkertafeln des frühen 18. Jahrhunderts*, ed. Franz K. Stanzel (Heidelberg: Winter, 1999), 122-123, 131-132.

70 Zacharasiewicz, Waldemar, *Die Klimatheorie in der englischen Literatur und Literaturkritik von der Mitte des 16. Jahrhunderts bis zum frühen 18. Jahrhundert, Wiener Beiträge zur englischen Philologie*, vol. 77 (Stuttgart: Braumüller, 1977), 89-91.

71 Montaigne, Michel de, *Essais* (Frankfurt a.M.: Eichborn 1999), 287; Zacharasiewicz, *Die Klimatheorie in der englischen Literatur*, 101-104.

picture of the good and happy primitive human.[72] By contrast, Jean Baptiste Dubos (1670-1742) in *Réflexions critiques* (1719) takes detailed assumptions from climate theory and links them to the achievements, each of which is valued variously, of European peoples. In the second volume, he examines the effect of the passage of time on morality and art with a view to making physical causes, and particularly the air, responsible for the differences between peoples and their artistic production. However, in the process, he reflects that changes throughout time relativise the air's influence on the customs of peoples. For the air in a given country must also be considered subject to change. He makes a comparison with wine, the *terroir* of which is decisive even if the year of the grape plays an equally decisive role. In principle, he considers all changes to hinge upon the *vicissitudes du monde*, that is, the ongoing process of transition inherent in the passage of time and the course of world events.[73] The *Examen de Ingenios* (1575) by the doctor Juan Huarte de San Juan ran to many editions and was translated throughout Europe. In 1752, Gotthold Ephraim Lessing translated the work into German.[74] It attributed a good memory (damp brains) to the Germans as northerners, and pronounced powers of imagination to southerners (dry, warm brains). John Barcley too experienced a similar level of success in Europe with *Icon animorum* (1614).[75] He proposed that differences between people within a relatively confined space could also be explained with reference to climate.[76] In the so-called *Völkertafeln* (charts depicting various peoples), these properties were then represented in word and image for the broad public. The *Laconicum Europe Speculum* (Abbreviated Mirror of Europe), created by Johann Wolfgang Baumgartner and first printed in 1736/1737, is particularly striking.[77]

---

72 Delgado, Mariano, "Verliebter Blick: Die Natur der Neuen Welt in der Apologética historia sumaria des Bartolomé de las Casas," in *Begriff und Darstellung der Natur in der spanischen Literatur der frühen Neuzeit, Hispanistisches Kolloquium*, vol. 4, ed. Wolfgang Matzat, and Gerhard Poppenberg (München: Fink, 2012), 291-304.

73 Dubos, Jean Baptiste, *Réflexions critiques sur la poésie et sur la peinture, deuxième partie* (Paris: Jean Mariette, 1719), 136-301, accessed March 27, 2017. <http://www.mdz-nbn-resolving.de/urn/resolver.pl?urn=urn:nbn:de:bvb:12-bsb10573734-5>.

74 Huarte de San Juan, Juan, *Johann Huarts Prüfung der Köpfe zu den Wissenschaften:* [...] (Zerbst: Zimmermann, 1752), accessed March 27, 2017. <http://www.mdz-nbn-resolving.de/urn/resolver.pl?urn=urn:nbn:de:bvb:12-bsb10913288-8>.

75 Barcley, John, *Spiegel Menschlicher Gemüths Neigungen* (Frankfurt a.M.: Erhard Berger, 1660), accessed March 27, 2017. <http://www.mdz-nbn-resolving.de/urn/resolver.pl?urn=urn:nbn:de:bvb:12-bsb10913288-8>.

76 Zacharasiewicz, "Klimatheorie und Nationalcharakter," 128-129.

77 *Laconicum Europae Speculum* [...] (Augsburg: Martin Engelbrecht, 1736/1737); Rupnow, Dirk, "Laconicum Europae Speculum – Stereotype ohne Schimpf und Vorurteil?" in

After Daniel Defoe poked fun at the assertions of climate theories with reference to his fellow countrymen (*The climate makes them terrible and bold, And English beef their courage does uphold*),[78] Johann Gottfried Herder argued against Montesquieu (1689-1755) and his climate determinism. In 1748, Montesqieu had published *The Spirit of the Laws*, in which slavery is linked to the oppressive heat that robbed people of their strength and courage.[79] By contrast, Herder emphasised in *Outlines of a Philosophy of the History of Man* [*Ideen zur Philosophie der Geschichte der Menschheit*], 1784-1794, the neutral cultural diversity of peoples,[80]and in an astonishingly prophetic manner formulated the following:

> There is no question, but, as climate is a compound of powers and influences, to which both plants and animals contribute, and which every thing that has breath promotes in its reciprocating mutations, so man is placed in it as a sovereign of the Earth, to alter it by art. Since he stole fire from Heaven, and rendered steel obedient to this hand [Prometheus], since he has made not only beasts, but his fellow men also, subservient to his will, and trained both them and plants to his purposes; he has contributed to the alteration of climate in various ways. Once Europe was a dank forest; and other regions, at present well cultivated, were the same. They are now exposed to the rays of the Sun; and the inhabitants themselves have changed with the climate. The face of Egypt would have been nothing more then the slime of the Nile, but for the art and policy [domestic politics] of man. He has gained it from the flood; and both there, and in farther Asia, the living creation has adapted itself to the artificial climate. We may consider mankind, therefore, as a band of bold though diminutive giants, gradually descending from the mountains, to subjugate the earth, and change climates with their feeble arms. How far they are capable of going in this respect, futurity will show.[81]

Europäischer Völkerspiegel: Imagologisch-ethnographische Studien zu den Völkertafeln des frühen 18. Jahrhunderts, ed. Franz K. Stanzel (Heidelberg: Winter, 1999), 75-95.

78 Zacharasiewicz, *Die Klimatheorie in der englischen Literatur*, 508-509.

79 Weigand, Kurt, ed., *Montesquieu* [Charles-Louis de Secondat Baron de Brède et de], *Vom Geist der Gesetze* (Stuttgart: Reclam, 1965), 255-282; Glacken, *Traces on the Rhodian Shore*, 568-575; Vasak, Anouchka, Météorologies: *Discours sur le ciel et le climat, des Lumières au romantisme, Les dix-huitièmes siècles*, vol. 112 (Paris: Champion, 2007), 193-259.

80 Sieglerschmidt, Jörn, "Spielt die Natur mit dem Menschen? Zum Einfluß der Natur auf die soziale Umwelt des Menschen," in *Ad historiam humanam: Aufsätze für Hans-Christoph Rublack*, ed. Thomas Max Safley (Epfendorf: Bibliotheca Academica, 2005), 200.

81 Herder, Johann Gottfried, *Outlines of a Philosophy of the History of Man* (New York: Bergman, 1966), 176; Glacken, *Traces on the Rhodian Shore*, 569-570.

4 The Immoderate and the Poetic

Following on from the teachings of antique natural philosophy and its reception in the modern period, weather events are treated in all artistic media (text, tone, image and moving image) as reflections of human sensitivities, examples and metaphors of the relation between humans and nature. Cheerfulness is still related to good weather today, just as sadness is to bad weather, love and the coming into being of life to spring, and transience and death to autumn as well as the storm outside to that within us—to the extent that timeless artistic themes can be spoken of.[82] Life—not only human life—is incorporated in a cyclical continuum caused by geophysical conditions: the passing of a day, the course of a year. The course that other processes take was also seen as cyclical until well into the 19th century, whether these be natural (a lifetime), social (wealth, poverty) or political (forms of government).[83]

Lyric, epic and dramatic poetry, painting, sculpture and music and, today, photography and film too, have always dealt with weather-related themes and the way in which weather conditions change during the year. The arts use the representation of nature as a metaphor-rich mirror image of human nature that invites allegoresis, behind which there persistently stands an integral conception of nature fortified during antiquity by natural philosophy and expansively laid out during the Renaissance and Baroque periods, in which weather conditions have a prominent place.

It is particularly in the Baroque and early classical periods that associated themes from the *complexion* are taken up musically in, *inter alia*, *Four Seasons* arrangements (Antonio Vivaldi, 1678-1741, Joseph Haydn, 1732-1809) or *The Creation* (Joseph Haydn). Besides animal sounds, those of the weather, such as a thunderstorm for example, are also imitated. As for theatre (or, later, radio plays), associated devices also serve to reproduce weather sounds and weather events, including the allegorical thunder or mist. This is manifested in the staging of clouds that Nicolò Sabbatini suggested in 1637, whereby clouds could be used as a heavenly form of transport for humans.[84]

---

82 Becker, Karin ed., *La pluie et le beau temps dans la littérature Française: Discours scientifique et transformations littéraires, du Moyen Âge à l'époque moderne* (Paris: Hermann, 2012), 39-49.

83 Sieglerschmidt, Jörn, "Zyklizität," in *Enzyklopädie der Neuzeit*, vol. 15, ed. Friedrich Jäger (Stuttgart: Metzler, 2012), 645-654.

84 Sabbatini, Nicola, *Pratica di fabricar scene, e machine ne' teatri* (Pesaro: Concordia 1637); Sabbatini, Nicola, *Pratica di fabricar scene, E machine ne'teatri* (Ravenna: Pietro de' Paoli, Giovanni Battista Giouannelli, 1638); Guldin, Rainer, "Windschiffe, Die Wolke als

As in all areas of art, the semiotic universe of the 17th and 18th centuries was lost during the 19th century at the latest—and thus, that of antiquity too.[85] When in 1881 Franz Liszt (*Nuages Gris*) or in 1900 Claude Debussy (*Nuages*) chose to take up the theme of clouds, they did so with the clear intention of departing from the musical certainties and clear structures of composition. Rainer Guldin speaks of a formlessness and an impressionism that is in no way concerned any longer with imitating nature. This applies all the more to György Ligeti's *Clocks and Clouds* of 1972, the intention of which is to apply a theoretical model of Karl Popper's (1902-1994) such that, between the extremes of chance and necessity, human behaviour is broached as a question of intonation and described as being equivalent to clouds.[86] Nonetheless, one can still hear in Liszt's and Debussy's works echoes of what Herder once attributed to music: a "natural language of feeling" and a "language of the heart and of passions".[87]

This complex of themes was taken up in landscape painting at the same time in the same way. However, this was not the first time that weather events had been depicted. It is the world of the gods in the visual arts that either provides the thundering, lightning-hurling power of weather events or symbolises these. This power, which stands in opposition to humans and which humans are at the mercy of, relates first and foremost to catastrophes brought about as a result of the weather: floods, lightning bolts and conflagrations, earthquakes and volcano eruptions, storms, geophysical events with their sometimes long-term and devastating consequences for human existence—and that of other creatures. At the same time, there is however an alternative model in the form of the *locus amoenus*, the pleasant place that does not equate to paradise as far as such a thing is conceivable on this earth, with earthly passions such as love. The weather conditions, whether in the Garden of Eden or paradise, accord with an idea of a Golden Age that is physically in complete harmony with all

---

Fahrzeug und Bühnenrequisit," in *Wolkenbilder: Die Erfindung des Himmels*, ed. Stephan Kunz, Johannes Stückelberger, and Beat Wismer (München: Hirmer, 2005), 40-41.

85 Borst, Arno, *Das Buch der Naturgeschichte: Plinius und seine Leser im Zeitalter des Pergaments. Abhandlungen der Heidelberger Akademie der Wissenschaften, Philosophisch-historische Klasse, Jg. 1994, 2.* (Heidelberg: Winter, 1994); Lepenies, Wolf, *Das Ende der Naturgeschichte: Wandel kultureller Selbstverständlichkeiten in den Wissenschaften des 18. und 19. Jahrhunderts* (München: Hanser, 1976).

86 Guldin, Rainer, *Die Sprache des Himmels: Eine Geschichte der Wolken* (Berlin: Kadmos, 2006), 205-220.

87 Scholtz, Gunter, "Musik," in *Historisches Wörterbuch der Philosophie*, vol. 6., ed. Joachim Ritter, and Karlfried Gründer, 249-250 (Darmstadt: Wissenschaftliche Buchgesellschaft, 1984).

creatures. With the expulsion from this paradise, weather events become the punishment of the gods, a sign that humans should repent, and contemplate and invoke a better life.[88]

The visual repertoire of the modern era illustrates such events, including weather events themselves, in the framework of the tetrahedral schemata. Allegorical series of pictures depicting the planets, temperaments, stages of life, months, times of day and seasons, as well as the elements, feature the world and the passage of time in a state of transition, that is to say, they depict the *vicissitudo omnium rerum* (the vicissitude of all things). They order these natural occurrences according to a cosmological idea of time. Thus weather conditions feature too in the numerous prints that are intended, through their inscriptions and subtitles, to provide moral guidance,[89] or the ceiling frescos of Giorgio Vasari, Cristofero Gherardi, paintings of Jacopo Tintoretto, of Pieter Bruegel in the sixteenth, and of Rembrandt van Rijn or David Teniers in the seventeenth century.[90] Up to well into the 19th century, visual repertoires of this kind continued to draw parallels between, for example, age and the time of day or the season. The pictures of Giuseppe Arcimboldo (1526-1593) relating to the elements and seasons remain somewhat outside of this common practice of allegory. Arcimboldo is primarily concerned with the seasonal flora and fauna and how their presence relates to certain elements.[91]

With the beginning of the broad movements in landscape painting (including so-called marine art) in the 16th century, weather events are also covered.[92] Yet common allegorical and metaphorical meanings that persisted then have in no way become obsolete today. Only at the turn of the 19th century did the image become a medium with which to convey philosophical argument, the self-assertion of the individual in relation to nature. Caspar David Friedrich (1774-1840) is not only a much commented upon witness of this tension, but at the same time one of the most distinctive representatives of romanticism. So much was confirmed by Gotthilf Heinrich von Schubert's critique of his 1803 season and life cycle series. Schubert referred to the work's treatment of earthly life as well as the world of religion. On these pages, there was no longer any reference made to the weather conditions that otherwise accompanied such

88 Behringer, Wolfgang, *Kulturgeschichte des Klimas: Von der Eiszeit zur globalen Erwärmung* (München: Beck, 2009), 165-170.

89 Kaulbach, Hans-Martin, and Reinhart Schleier: *"Der Welt Lauf:" Allegorische Graphikserien des Manierismus* (Ostfildern: Hatje 1997), 9-10, 75-164.

90 Meetz, Karen Sabine, "Tempora triumphant," 212-220, 248-252.

91 Ricci, Franco Maria, ed., *Arcimboldo* (Genf: Weber, 1978).

92 Wiemann, Elsbeth, ed., *Die Entdeckung der Landschaft: Meisterwerke der niederländischen Kunst des 16. und 17. Jahrhunderts* (Stuttgart: Staatsgalerie; Köln: Dumont, 2005).

compositions: the focus lay now far more on meanings relating to human life-spans. That Friedrich and other painters of his time used weather events emphatically in order to take a philosophical view of the world is shown in *Wanderer above the Sea of Fog*.[93]

During the period around 1800, the allegorical and metaphorical world of antiquity and the modern is not completely left behind, but the significance of nature and weather events lies in how they are used to highlight the alienation of humans from both their own nature and that surrounding them. Anouchka Vasak speaks in this regard of a *désolidarisation de l'homme et de l'univers*.[94] Incidentally, the new interpretation of weather events also made use of the insights of a meteorology oriented, at the time, toward the natural sciences. Johann Wolfgang Goethe and Carl Gustav Carus took note in 1818 and 1822 respectively of Luke Howard's classification of clouds from the year 1803, which they interpreted in a manner entirely in keeping with romantic natural philosophy. In art, clouds remain metaphors for the infinite and incomprehensible.[95]

Karen Gloy has drawn attention with reference to the handling of the concept of nature around 1800 to how, on the one hand, there is the standpoint of idealistic philosophy that—represented above all by Johann Gottlieb Fichte and Friedrich Wilhelm Joseph Schelling—tackles the conceptual impact that the relation between nature and mind has on speculation.[96] Besides which, there are variants present in philosophy and poetry that alight upon feeling, mood, intuition and imagination and have Novalis (1772-1801) as their most prominent representative—these being a way of thinking that we still associate today above all with romanticism. Friedrich Hölderlin too is concerned with an aesthetic variant that hinges on the lost unity with nature, such that he

93 Schulze Altcappenberg, Hein-Thomas, "An der Wiege der Romantik: Die Jahreszeiten Caspar David Friedrichs von 1803," in *An der Wiege der Romantik: Caspar David Friedrichs* Jahreszeiten *von 1803*, ed. Klaus-Dieter Lehmann (Berlin: Staatliche Museen, 2006), 15-24; Böhme, Hartmut, "'Was birgt die Wolke?' Zur Kultur- und Kunstgeschichte von Wolken und Wetter," in *Wolkenbilder: Die Erfindung des Himmels*, ed. Stephan Kunz, Johannes Stückelberger, and Beat Wismer (München: Hirmer, 2005), 11-21.

94 Vasak, *Météorologies*, 12.

95 Howard, Luke, *On the Modification of Clouds, and on the Principles of their Production, Suspension, and Destruction:* [...]. (London: J. Taylor, 1803); Goethe, Johann Wolfgang von, *Poetische und prosaische Werke in zwei Bänden*, vol. 1 (Stuttgart: Cotta, 1836), 107-108; Stückelberger, Johannes, *Wolkenbilder: Deutungen des Himmels in der Moderne* (München: Fink, 2010), 41-51.

96 Gloy, Karen, *Das Verständnis der Natur. Zweiter Band: Die Geschichte des ganzheitlichen Denkens* (München: Beck, 1996), 71-103.

perceives weather events like lightning, thunder, torrential rain, clouds, and dew as valid signifiers of this unity. The task of art is for Novalis to perfect nature.[97] Distinctive, overarching or holistic ideas of nature and of weather can be found right up until today in heterodox strains of thought such as esotericism. Rudolf Steiner's (1861-1925) anthroposophy offers a prominent example that does not only draw on the natural philosophy of Goethe but also the ancient knowledge of nature that became widespread as a result of the Renaissance and that Steiner interpreted as a secret science. The cosmic references in anthroposophy are also important in agriculture.[98]

Allegorical and metaphorical references to weather have occurred in literary texts since antiquity in the same way as in the other arts. In the European literature of the Middle Ages, the weather, and particularly catastrophes, and numerous associated metaphorical references were a sign of the fragility of human existence, sometimes of the end of the world.[99]

In view of the ubiquitous presence of this tradition of pictorial language, it should come as no surprise if, in the 19th century, there was an attempt to be rid of some of the associated burden. Either there was a denial, as in the case of Charles Baudelaire (1821-1867), that nature had any influence whatsoever on imagination[100] or straightforward disappointment in the customarily sympathetic intentions behind, for example, a poem about the spring. Heinrich Heine (1797-1865), who otherwise wrote spring lyrics of a thoroughly traditional sentiment, came to the following conclusion in *Neuer Frühling* (1833):

> Ich kann nicht singen und springen
> Ich liege krank im Gras;
> Ich höre fernes Klingen,
> Mir träumt, ich weiß nicht was.[101]
> [I can't sing and skip,
> I lie sick in the grass;

97 Gloy, Karen, *Die Geschichte des ganzheitlichen Denkens*, 104-151.

98 Zander, Helmut, *Anthroposophie in Deutschland: Theosophische Weltanschauung und gesellschaftliche Praxis 1884-1945*, (Göttingen: Vandenhoeck & Ruprecht, 2007/2008), 651-674, 1586-1590; Gloy, Das Verständnis der Natur, 159-160.

99 Ducot, Joëlle, and Claude Thomasset, ed., *Le temps qu'il fait au Moyen-Âge: Phénomènes atmosphériques dans la littérature, la pensée scientifique et religieuse*, (Paris: Presses de l'Université, 1998).

100 Erlebach, Peter, *Natur und Spiritualität in der englischen Literatur- und Geistesgeschichte* (Heidelberg: Winter, 2011), 152-153.

101 Cited after: Schusky, Renate, "Klänge durch das Jahr: Die vier Jahreszeiten in der Musik," in *Das Reich der Jahreszeiten*, ed. Rainer Gruenter (Zürich: Offizin, 1989), 137-138.

I hear a distant music
And dream I don't know what.]

Aleksandr Puškin (1799-1837) speaks similarly in 1831 of spring as a season that he does not love since it heralds illness, mental and emotional suffering and unsettling dreams.[102] Finally, at the beginning of the 20th century, in the famous opening passage of *The Man Without Qualities*, Robert Musil (1880-1942) creates a sophisticated picture of the weather only to then smash it with irony: "A barometric low hung over the Atlantic. It moved eastwards toward a high-pressure area over Russia [...]. In a word, [...] even if it is a bit old fashioned: it was a fine day in August 1913."[103] The weather as the subject of a visual language, which before Romanticism had already served to highlight the affinity between humans and nature, became in the nineteenth and twentieth centuries a *metaphora continua* which, as a literary trope, could sometimes better describe a person's inner life than a soliloquy or other stylistic devices.[104]

5 Experiment, Measure, and Number

The history of meteorology is often described as a natural consequence of scientific method, with the goal to give detailed predictions. Through the 18th century, this is only true for a small portion of meterological texts. The occupation with weather phenomena was rather a challenge of natural philosophy, which within astronomy covered the mathematically predictable supralunar world; atmospheric phenomena of the sublunar world, however, have always been seen as chaotic and unpredictable. Furthermore, the natural philosophic analysis of alchemical and astrological research was common, even though critised by theology, philosophy, and natural science. A seperation of old and new forms of science is, for most authors, not useful and not possible, but misleading, as the example of Descartes shows.[105]

102 Pushkin, Aleksandr S., *Izbrannye proizvedenija v dvukh tomakh. Tom pervy: stikhotvoreniya, skazki, poëmy* [Selected Works in Two Volumes: Poems, Fairy Tales, Verse Novels] (Moscow: Chudojestvennaja Literatura, 1968), 165.

103 Musil, Robert, *The Man without Qualities* (London: Picador, 1997), 3; Becker, *La pluie et le beau temps dans la littérature Française*, 34-37.

104 Erlebach, *Natur und Spiritualität*, 153-154.

105 Zittel, Claus, *Theatrum philosophicum. Descartes und die Rolle ästhetischer Formen in der Wissenschaft* (Berlin: Akademie, 2009), especially 189-228; Meinel, Christoph, "Natur als moralische Anstalt. Die Meteorologia philosophico-politica des Franz Reinzer, S.J, ein

In the fourth century, Aristotle proposed the first coherent scientific system of meteorology embedded in (geo-)physics as a whole.[106] Up until René Descartes' (1596-1650) *Discours de la méthode* (1637), which introduces an optical, geometrical, and meteorological discourse in its explanation of method, Aristotle's central statements held.[107] But from this time on, a meteorology aimed at the reformulation of the natural sciences based on experiment, using instruments to take measurements and quantification, takes root.

In the first place, Aristotle outlined his cosmological and physical ideas as the basis for explaining immediate meteorological events. Here, it is important to understand the assertion that the natural position that the sublunary spheres of the four elements (fire, air, earth, and water) tend to occupy, if they remain unhindered, is calculated according to the weight of the body of a given element. Air—we would say today: and other gases too—continues to rise upwards so long as it is not prevented from doing so: for example, by combining it with earth or water. Fire is to be understood as dry air, not as a naked flame.[108] Friction caused by contact with other materials can lead to combustion, as shown by the example of shooting stars. The earth and its immediate surroundings do not exist in a pure form: they are thoroughly mixed but compact. That is, contrary to the assertions of Atomists such as Democritus, there is no such thing as a vacuum. A central principle of Aristotelian physics is therefore: *omne quod movetur, ab alio movetur* (everything that is in motion is in motion because of a specific cause, or: *every process has its own specific source*).[109] Motion, for Aristotle, is also every kind of material change.

Movements in the sublunary zone are sustained through the ongoing rotation of the superlunary area around the earth. It is the course of the sun in

---

naturwissenschaftliches Emblembuch aus dem Jahre 1698," *Nuncius: Annali di Storia della Scienza* 2 (1987), 37-94; id., "Les météores de Froidmont et les météores de Descartes" in Bernès, Anne-Catherine ed. *Libert Froidmont et les résistances scientifiques* (Haccourt: Association, 1988), 105-129.

106 Strohm, Hans, "Einleitung," in Aristoteles, *Meteorologie. Über die Welt, Aristoteles Werke in deutscher Übersetzung*, vol. 12, ed. Hans Strohm (Berlin: Akademie, 1984); Körber, *Vom Wetteraberglauben zur Wetterforschung*, 58-59; Fierro, *Histoire de la météorologie*, 22-25; Taub, *Ancient Meteorology*, 77-115.

107 Descartes, René, *Discours de la méthode pour bien conduire sa raison, & chercher la verité dans les sciences* (Leyde: Ian Maire, 1637), accessed March 27, 2017. <http://reader.digitale-sammlungen.de/resolve/display/bsb10860336.html>.

108 Strohm, "Einleitung," 12-16.

109 Wagner, Hans, ed., *Aristoteles, Physikvorlesung, Aristoteles Werke in deutscher Übersetzung*, vol. 11 (Darmstadt: Wissenschaftliche Buchgesellschaft, 1967), 231 (256a3).

particular that is responsible for both the creation and the dispersion of the water and air layer that surrounds the earth.

> The exhalation of water is vapour: air condensing into water is cloud. Mist is what is left over when a cloud condenses into water, and is therefore rather a sign of fine weather than of rain; for mist might be called a barren cloud.[110]

Subsequently, the role of evaporation in the water cycle is explained as the exchange of matter between water and air that takes place under the influence of the sun's rays. Other meteorological events result from this cycle through the presence or absence of warmth or cold, such as snow, ice, hail, dew, and rain, whereby wind direction and regional conditions are also incorporated into Aristotle's reflections.[111]

Further to his thoughts on evaporation and the water cycle, Aristotle also considers the earth's hydrology, thereby assuming that the earth itself is cold. The moisture that comes from the air collects in the ground and reappears on the surface as rivers or provides, as a result of the well builder, a source of water for human consumption. In this case too, it is the sun and the regional and seasonal variations in the strength of the sunlight that is responsible for varying levels of water, which determines in turn the scope for agricultural yields.[112]

The existence of wind is linked to movements in the sky. The rise or fall of moisture in the air is a result of evaporations (*anathymiaseis*) hailing from the earth, which the wind sets in motion. This theory of evaporations is embedded in the interplay of sun rays and wind direction, the main directions being north and south, while winds blowing from east to west and vice versa are anyway dependent on the course the sun takes. Here, Aristotle can provide grounds for numerous weather phenomena, including regional differences in the probability of rain. Wind direction is explained in the same way, according to the status of the sun, whether with regard to the principal directions—those that blow from the rising and from the setting sun—or those in between, which are associated with the solstices.[113]

Here is not the place to discuss further explanations of, for example, the creation of the sea, the various conditions of salt and sweet water, earthquakes,

110 Webster, E.W., trans., "Meteorologica," in *Works of Aristotle*, vol. 3, ed. W.D. Ross (Oxford: Clarendon Press, 1931), 346b 32-36.

111 Ibid.; Strohm, "Einleitung," 25-29.

112 Ibid., 30-39.

113 Strohm, "Einleitung," 53-63.

thunderstorms, rainbows, halos and parhelia. It becomes clear that this Aristotelian didactic material constitutes a case of a scientific system based on systematic observation guided by theory and the experience gained as a result—a system that can even deliver forecasts, as well as conclusive explanations of daily weather. It can therefore be considered no surprise that this system was widely adopted and in some respects elaborated upon later.

René Descartes continued to work within Aristotle's cosmological system, albeit without developing in detail a cosmology in his meteorology—with notable exceptions: The basis of all weather phenomena is the different forms and physical properties of particles forming the elements, i. e. matter; the Cartesian universe is without entelechy, all movements and processes can be explained mechanically by the movements of particles. Descartes concentrated at the beginning[114] on the elements and their composition[115] and thus produced reflections thoroughly in keeping with particulate theory, though these do not provide for the possibility of a vacuum as did Democritus or Lucretius. Descartes also assumes that vapours and evaporations, under certain conditions—sun rays, mechanical motion—rise upward and downward through the air.[116] An air current itself is to be understood as a mechanical process effected by warmth, and not caused by the orbit of the stars.[117] Descartes explained the rainbow as the refraction of sunlight in water, an insight that he was able to obtain through his previously developed theory of optics that was based on numerous experiments.[118]

Descartes remained, like Robert Fludd and others, under the influence of the outlook of his times, when blood, milk or iron rain were still cited as marvels.[119] Equally, he continued to work within the framework of the four elements that would shortly thereafter be put into question and rejected by the sceptical chemist Robert Boyle, albeit without offering any supporting evidence. It was Antoine Laurent de Lavoisier (1743-1794) who first provided such evidence that has left no room for any doubt. Lavoisier showed that water and air cannot be elements since they consist respectively of hydrogen and oxygen, and of oxygen and nitrogen.[120] But this was not the first time that many of

114 Zittel, Claus, ed, *René Descartes, Les météores / Die Meteore* (Frankfurt a.M.: Klostermann, 2006), esp. 7-9, 316-331.

115 Descartes, *Discours de la méthode*, 159-164.

116 Ibid., 165-173.

117 Ibid., 203-215.

118 Descartes, *Discours de la méthode*, 250-270.

119 Ibid., 246.

120 Lavoisier, Antoine Laurent de, *Traité élémentaire de chimie: Présenté dans un ordre nouveau, et d'après les découvertes modernes*, 2 Vols. (Paris: Deterville, IX/1801), T. 1, 33-56,

Aristotle's basic assumptions had proven to be unsupportable in the natural sciences, something that had already been established at the beginning of the seventeenth century.[121]

Historically, in all natural sciences, it is the transition to models of measurement that remains decisive: that is, the transition from qualitative empiricism to quantitative measurements taken using instruments able to provide standardised results, which is to say, results that are as independent of time and space as possible. A prerequisite for all such measurements is a theory about the involved physical phenomena and experimental equipment that demands close cooperation between science and skilful handicraft, such as embodied in the person of Leonardo da Vinci (1452-1519).[122] This transition begins in the seventeenth century and provides the benchmark for every subsequent development in the area of the natural sciences since the mid-eighteenth century.

It should, however, be remembered that, at least until the mid-nineteenth century, the emerging disciplines of engineering and natural science led a marginal existence in the context of people's consciousness generally.[123] Only with the successful application of technology as well as increasing affluence could the resulting commercial developments driven by such breakthroughs replace the old forms of belief through an often-firm belief in the universal validity of the methods and predictions of natural science.

Meteorology took part in this development too, as measuring instruments had evolved and been tested to the extent that comparable results from different locations and times could be accomplished. Between 1640 and 1670, all of the relevant instruments for the measurement of weather events up into the twentieth century had been invented: thermometer, barometer, hygrometer (humidity), pluviometer (udometer, rain gauge), anemometer (wind force and wind speed).[124]

Meteorology made special progress where it took root institutionally through research associations or, as in the nineteenth century, through

---

87-100, 192, accessed March 27, 2017. <http://www.mdz-nbn-resolving.de/urn/resolver.pl?urn=urn:nbn:de:bvb:12-bsb10073155-7>; Greenberg, Arthur, *From Alchemy to Chemistry in Picture and Story* (Hoboken: Wiley, 2007), 172-176.

121 Weichenhan, Michael, "Ergo perit coelum...:" Die Supernova des Jahres 1572 und die Überwindung der aristotelischen Kosmologie(Stuttgart: Steiner, 2004), 50-57.

122 Mittelstraß, Jürgen, Neuzeit und Aufklärung: Studien zur Entstehung der neuzeitlichen Wissenschaft und Philosophie (Berlin: de Gruyter, 1970).

123 Meinel, *Meteorologia*, 59.

124 Fierro, *Histoire de la météorologie*, 61-68; Körber, *Vom Wetteraberglauben zur Wetterforschung*, 103-136; Middleton, William E.K., *Invention of Meteorological Instruments* (Baltimore: Johns Hopkins, 1969).

permanent weather observation institutes, thus guaranteeing that the discipline could rely on precise and comparable data. An early association of this kind was the *Accademia del cimento* in Florence, which collected data from throughout Europe from 1657 to 1667 and contributed not only to the standardisation of the instruments used but also to the standardisation of observational procedures and, besides temperature, measured humidity and cloud cover.[125] Robert Hooke proposed in the 1660s plans for an observation network with a standard template for data. The *Societas Meteorologica Palatina* in Mannheim was—and remains—particularly well known. Founded in 1780 by Elector of Bavaria Charles Theodor, it used standardised instruments and regular hours of observation—that is, the so-called Mannheim hours of 7:00 am, 2:00 pm and 9:00 pm—and data collection points at 39 stations, of which 36 were in Europe, one in Greenland, and two in North America. Measurements were taken from 1780 to 1795 and documented in the *Ephemerides Societatis Meteorologicae Palatinae*.[126] Decisive steps were taken in the form of the publication in 1803 of a first attempt at cloud classification.[127] This classification was further developed in the nineteenth and twentieth centuries and published in cloud atlases.

The atmosphere's water cycle was first understood in the eighteenth century, when Charles le Roy (1726-1779) described in detail the process of evaporation and Jean André Deluc (1727-1817) recognised water vapour as a gas. The concept of relative humidity also stems from Deluc.[128] The realisation that a constant wind (so-called trade winds) can be caused by the rotation of the earth would first lead to the insight in the context of a heliocentric world outlook that the earth did indeed revolve. In *Dialogo sopra i due massimi sistemi del mondo Tolemaico e Copernicano* (1632), Galileo Galilei (1564-1642) had expressed just such an assumption in support of his heliocentric argumentation. Gustave Gaspard Coriolis (1792-1843) and Siméon Denis Poisson

125 Hellmann, Gustav, ed., *Evangelista Torricelli, Esperienza dell' argento vivo. Accademia del Cimento, Istrumenti per conoscer l'alterazioni dell' ari. Neudrucke von Schriften und Karten über Meteorologie und Erdmagnetismus*, vol. 7 (Berlin: Asher, 1897); Körber, Vom Wetteraberglauben zur Wetterforschung, 116-118.

126 Körber, *Vom Wetteraberglauben zur Wetterforschung*, 133-136.

127 Howard, *On the Modification of Clouds*.

128 Luc, Jean André de, *Neue Ideen über die Meteorologie*, 2 Vols. (Berlin, Stettin: Nicolai, 1797); Körber, *Vom Wetteraberglauben zur Wetterforschung*, 138-139, accessed March 27, 2017. <http://www.mdz-nbn-resolving.de/urn/resolver.pl?urn=urn:nbn:de:bvb:12-bsb10134206-1>; <http://www.mdz-nbn-resolving.de/urn/resolver.pl?urn=urn:nbn:de:bvb:12-bsb10134207-6>.

(1781-1840) were the first to describe the repellent force of the earth's revolutions (today known as the Coriolis effect) as caused by inertia.[129]

Observations of air pressure and of winds are important as the basis of weather forecasts, whereby global conditions (the rotation of the earth, air currents from the poles to the equator) must also be incorporated. As concerns the modelling of this dynamic approach, Heinrich Wilhelm Dove (1803-1879) was the authoritative figure of the nineteenth century with his *Gesetz der Stürme* (1840, *On the Law of Storms*). Carl Anton Bjerknes (1825-1903) marked the transition to the modern meteorology of the twentieth century in a work that he penned with his colleague entitled *Dynamische Meteorologie und Hydrographie* (1910/11, *Dynamic Meteorology and Hydrography*). His son Vilhelm Bjerknes (1862-1951) called the precalculation of future conditions a crucial criterion of precise physics of atmosphere and, therefore, of precise natural science. This claim was held up by Lewis Fry Richardson (1881-1953) and his mathematisation of weather prediction.[130] Proposals for the mapping of weather data made at the beginning of the nineteenth century were also important for gaining insights into how the data were interrelated. However, precedents for this date back to the seventeenth century.[131]

A prerequisite for today's observations in the context of an international network was the introduction of the telegraph in the nineteenth century and the organisation of national and then international cooperation. Following the catastrophic storm during the Crimean War in 1854, a state weather service was introduced in France in 1855. Austria (1865), Norway and Italy (1866), the USA (1869), Canada (1871), Denmark and Sweden (1873), and Russia (1874) followed suit.[132] In 1873, the International Meteorological Organisation was founded in Vienna, and complemented in 1879 by a permanent International Meteoro-

129 Galilei, Galileo, *Dialogo sopra i due massimi sistemi del mondo Tolemaico e Copernicano.* [...] (Fiorenza: Giovanni Batista Landini, 1632), 433-434; Körber, Vom Wetteraberglauben zur Wetterforschung, 166-168.

130 Lynch, Peter, *The Emergence of Numerical Weather Prediction: Richardson's Dream* (Cambridge: University Press, 2006), 4-9, 248-250.

131 Hellmann, Gustav, ed., E. Halley, A. von Humboldt, E. Loomis, U.J. Le Verrier, E. *Renou: Meteorologische Karten 1688, 1817, 1846, 1863, 1864, Neudrucke von Schriften und Karten über Meteorologie und Erdmagnetismus*, vol. 8 (Berlin: Asher, 1897); Hellmann, Gustav, ed., E. Halley, W. Whiston, J.C. Wilcke, A. von Humboldt, C. *Hansteen: Die ältesten Karten der Isogonen, Isoklinen, Isodynamen 1701, 1721, 1768, 1804, 1825, 1826, Neudrucke von Schriften und Karten über Meteorologie und Erdmagnetismus* vol. 4 (Berlin: Asher, 1894).

132 Körber, *Vom Wetteraberglauben zur Wetterforschung*, 195-198.

logical Committee (IMC). In 1950, the World Meteorological Organisation (WMO) became the successor organisation.[133]

The development of a meteorology oriented toward the natural sciences had, from an organisational perspective, found a permanent platform. This can be considered from a scientific perspective as an effort that continues today to optimise the models for complex weather patterns, so that forecasts, particularly for such difficult and of course dangerous weather situations, become increasingly precise—for which today's computer technology is an indispensable precondition. Scarcely any other discipline demands more in terms of the processing of data and the development and assessment of mathematical models.

Keeping records of weather data over extended periods of time first facilitated—it appears when looking back—the kind of empiricism necessary for the systematisation of knowledge. The dissemination of antique knowledge of the weather in printed form alone brought no new insight. Only the instrument-supported perception of geophysical connections led to something more, along with the long-term documentation of weather data and the incorporation of experiment, empiricism and theory—entirely in accordance with a Baconian approach. Systematisation of this kind was a prerequisite for written records organised through scientific networks—today, knowledge stored using media other than paper.

This applies too to the repeatedly and long neglected area of so-called biometeorology, which concerns the effects of weather events on organisms—plants and animals; that is, it pursues an old Hippocratic question with today's much more developed methods.[134] Just as phenological series have been upheld in the modern era and the responses of plants as well as animals to changes in the weather remain part of traditional knowledge, so also will systematic research along these lines certainly provide insights as to how to facilitate a stronger link between meteorological research and the behaviour of plants and animals as influenced by the weather. Clearly, such research is of great use to agricultural as well as in terms of medicinal insights. Here, the

133 Ibid., 200-202.

134 Matzarakis, Andreas, and Helmut Mayer, ed., *Proceedings zur sechsten Fachtagung BIOMET des Fachausschusses Biometeorologie der Deutschen Meteorologischen Gesellschaft* (Freiburg: Meteorologisches Institut der Universität Freiburg, 2007); Köpke, Peter et al., "Solare UV-Strahlung und ihre Wirkung auf den Menschen," *Promet* 33 (2007): 95-108; Bucher, Klaus, and Eva R. Wanka, "Das Problem der Wetterfühligkeit," *Promet* 33 (2007): 133-139.

cooperation between geographers and biologists will bring further practical insights.[135]

Particularly with regard to environmental history, which has examined during the last thirty years, among other things, the effect of climate on history, systematic models have been developed for a long time in order to bring to light feedback processes between climate, biota and societal developments. In response to global warming, heavily promoted research in paleoclimatology, paleozoology and paleobotany has made considerable progress in recent decades as concerns the long-term development of the weather in historical—primarily with reference to written documents—and prehistoric times.

What we are concerned with here is the systemic relation of climate to numerous other (dependent and independent) variables. The first reflections in this respect were already made long ago, particularly in relation to the problem of famine in history.[136] More recently, models intended to draw natural and socioeconomic variables into relation with one another have been discussed with reference to the concept of vulnerability or of risk.[137] By 1985, Christian Pfister and Daniel Krämer, along with Jan de Vries, are already conscious of how difficult it is to link socio-cultural factors to such natural, and particularly climatic, factors. It is scepticism that caused John von Neumann to attribute a high level of difficulty to the modelling of regular relations among phenomena that cannot be grasped in terms of algorithms. Moreover, the remaining methodological differences between natural science,

135 For example: Fels, Edwin, *Der wirtschaftende Mensch als Gestalter der Erde – Erde und Weltwirtschaft: Ein Handbuch der Allgemeinen Wirtschaftsgeographie*, vol. 5 (Stuttgart: Franckh, 1967); Frankenberg, Peter, *Ähnlichkeitsstrukturen von Ernteertrag und Witterung in der Bundesrepublik Deutschland, Erdwissenschaftliche Forschung*, vol. 17 (Wiesbaden: Steiner, 1984).

136 Grigg, David B., *Population Growth and Agrarian Change: A Historical Perspective* (Cambridge: University Press, 1980), 45-46; Kates, Robert W., Jesse H. Ausubel, and Mimi Berberian, ed., *Climate Impact Assessment: Studies of the Interaction of Climate and Society, Scientific Committee on Problems of the Environment* (International Council of Scientific Unions), vol. 27 (Chichester: Wiley, 1985); Komlos, John, *Nutrition and Economic Development in the Eighteenth-century Habsburg Monarchy: An Anthropometric History* (Princeton: Princeton University Press, 1989), 33, 162.

137 Pfister, Christian, and Daniel Krämer: "The Relaunch of Historical Climate Impact Research—A Timely Challenge for Historical Climatology." Paper presented at the conference "Historical Climatology—Past and Future," German Historical Institute, Paris. Paris, September 5-6, 2011; Krämer, Daniel, "Vulnerabilität und die konzeptionellen Strukturen des Hungers: Eine methodische Annäherung," in *Handeln in Hungerkrisen: Neue Perspektiven auf soziale und klimatische Vulnerabilität*, ed. Dominik Collet, Thore Lassen and Ansgar Schanbacher ed (Göttingen: Universitätsverlag, 2012), 45-65.

which employs quantitative procedures, and the *Geisteswissenschaften* and social sciences, which present arguments—not exclusively but to a far greater extent discursively, as did Aristotle—cause insoluble problems.[138]

Be this as it may, there is a related, albeit somewhat minor observation to be made with reference to today's situation: though weather catastrophes cannot be avoided, once they do happen, measures aimed at mitigating the consequences, curbing price fluctuations and improving transport infrastructure must already have been taken long in advance.[139] However, it is precisely these measures that have been more and more extensively realised in Europe since at least the seventeenth century in order to avoid climatic events directly endangering social stability. Political actors were always conscious that retaining their power depended on sufficiently catering for the wellbeing of their subjects, and behaved accordingly.[140]

6 The Immeasurable

Alexander von Humboldt (1769-1859) was probably the last geographer and meteorologist to know the world of the old natural history inside out, the world that met its demise with the passing of the 19th century. In 1845, Humboldt had expressed the following concern in his monumental work *Cosmos*, in which he linked both knowledge worlds, old and new, together:

> It is almost with reluctance that I am about to speak of a sentiment, which appears to arise from narrow-minded views, or from a certain weak and morbid sentimentality—I allude to the 'fear' entertained by some persons, that nature may by degrees lose a portion of the charm and magic of her power, as we learn more and more how to unveil her secrets, comprehend the mechanism of the movements of the heavenly bodies and estimate numerically the intensity of natural forces. [...] The mere accumulation of unconnected observations of details, devoid of

138 Mathieu, Jon, "Klimawandel und Wirtschaftsgeschichte der Vormoderne: Zur Methodendiskussion," *Schweizerisches Jahrbuch für Wirtschafts- und Sozialgeschichte* 30 (2015): 21-33.

139 Pfister, Krämer, "The Relaunch of Historical Climate Impact Research," 11.

140 Göttmann, Frank, "Die Versorgungslage in Überlingen zur Zeit der Hungerkrise 1770/1771," in *Vermischtes zur neueren Sozial-, Bevölkerungs- und Wirtschaftsgeschichte des Bodenseeraumes*, ed. Frank Göttmann (Konstanz: Gorre, 1990), 75-134; Pfister, Christian, "Food Supply in the Swiss Canton of Bern, 1850," in *Hunger in History: Food Shortage, Poverty, and Deprivation*, ed. Lucile F. Newman (Cambridge: Blackwell, 1990), 281-303.

> generalisation of ideas, may doubtlessly have tended to create and foster the deeply rooted prejudice that the study of the exact sciences must necessarily chill the feelings, and diminish the nobler enjoyments attendant upon a contemplation of nature.[141]

Humboldt underestimates the fact that the history of the human perception of both immediate weather events and of weather conditions over an extended period of time is at the same time a history of the separation of the human from nature.[142] Insofar as this separation is received as painful, this history could reveal a path that overcomes this separation, that is, brings a new form of reconciliation. The question is whether this path seems navigable today or must instead be seen as an intellectual aberration. Humboldt is, in any event, certain:

> Nor let it be feared that the marked predilection for the study of nature, and for industrial progress, which is so characteristic of the present age, should necessarily have a tendency to retard the noble exertions of the intellect in the domains of philosophy, classical history and antiquity, or to deprive the arts by which life is embellished of the vivifying breath of imagination. Where all the germs of civilisation are developed beneath the aegis of free institutions and wise legislation, there is no cause for apprehending that any one branch of knowledge should be cultivated to the prejudice of others.[143]

The history of the weather outlook shows that at all times, weather events were perceived as chaotic. Michel Serres speaks in *Atlas* (1994) of events that he deems to have been robbed of all simple regularities, and of a swarm of chance occurrences. And Arden Reed asserts in *Romantic Weather* (1983) that clouds are cosmic monsters, unpredictable, random, chaotic.[144] None other than Hermann Flohn, a founder of German climate history, draws attention to an interview with John von Neumann, in which Neumann is asked to comment on the use of computers for long-term weather forecasts. Neumann responds

141 Humboldt, Alexander von, *Cosmos: A Sketch of the Physical Description of the Universe*, vol. 1, trans. E.C. Otte, (Baltimore: Johns Hopkins University Press 1997), 39-40. First published as *Kosmos: Entwurf einer physischen Weltbeschreibung*, 5 Vols. (Tübingen: Cotta, 1845-1862), 19, 21.

142 Vasak *Météorologies*, 11-12.

143 Humboldt, *Cosmos*, 54; German original, 24.

144 Becker, *La pluie et le beau temps dans la littérature Française*, 24-25.

that he considers the calculation of weather events the second most difficult task for a computer. As for the most difficult of all, he is supposed to have answered: human behaviour.[145] The *Geisteswissenschaften* are well accustomed to natural science necessarily insisting, with an imperious gesture, that predicted consequences are certain to occur. And the *Geisteswissenschaften* can relax as it observes how natural science, in this way, reaches its limit, that is, fails by its own standards.

But it also fails—as was already known, even before Ludwig Wittgenstein (1889-1951) formulated it thus not only in *Tractatus* but above all in his *Philosophical Investigations* (*Logische Untersuchungen*)—in that natural science cannot solve societal problems. Weather conditions were for a long time an existential issue for whole populations, and still are in many regions of the earth, which were also on average much poorer even than today, that is, they had less bodily reserves and food stocks. Today we live in a time in which a majority of the world population has access to at least twice as much food as is actually consumed. The rest is thrown away and is no longer available as edible food.[146] The infrastructural deficit that, during pre-modern times (before about 1850) could be made jointly responsible for some supply crises, including on a smaller regional scale, exists today on a global level, to say nothing of the mental deficit of the rich that seems to stop them giving up something of their wealth for the poorest.

This does not render work in the natural sciences obsolete but does reveal its limits, which can only be overcome through societal negotiation. Aristotle separates natural science and the politico-social world. He founds this separation upon the difference between *epistëmë* (Gr., "understanding, science") and *doxa* (Gr., "opinion"), between natural science and socio-political communication. While natural science alone is in a position to provide rational and true insight, since it concerns things that are necessary in and of themselves, socio-political communication concerns things about which there are different opinions. Aristotle considers both kinds of knowledge and communication as being of equal value. This must be insisted upon in today's debates too.[147]

---

145 Oeschger, Hans, Bruno Messerli, and Maja Svilar, ed., *Das Klima. Analysen und Modelle: Geschichte und Zukunft* (Berlin: Springer, 1980), 3-4.

146 Stuart, Tristram, *Waste: Uncovering the Global Food Scandal* (London: Penguin, 2009).

147 Kopperschmidt, Josef, "Das Ende der Verleumdung: Einleitende Anmerkungen zur Wirkungsgeschichte der Rhetorik," in *Rhetorik Band II: Wirkungsgeschichte der Rhetorik*, ed. Josef Kopperschmidt (Darmstadt: Wissenschaftliche Buchgesellschaft 1991), 1-33; Campe, Rüdiger, *Spiel der Wahrscheinlichkeit: Literatur und Berechnung zwischen Pascal und Kleist* (Göttingen: Wallstein 2002), 135-139, 211-212, 253-283, 444-445.

CHAPTER 2

Theological Perspectives in the Ethical Debate about Climate Change

Markus Vogt

Abstract

Christian ethics can draw on a very rich tradition that aims at translating ethical arguments into ethical attitudes by equally addressing hearts and minds, deep-seated hopes, and daily life. That is why religious and ecological perceptions can enhance and complement each other in the ethical debate about climate change. The following essay shows a perspective that does not look at the relevant aspects in isolation, but rather in their entirety and unity. Christian ethical thinking adds a special value to the discussion of responsibility for nature in times of climate change, overpopulation, and scarcity of resources. For example, this chapter shows a way to see climate change as a "sign of the times"[1] and gives an answer to the question of which competencies religions have in the diagnosis of the greatest challenges of the present time.

1 Introduction

1.1 *Spirituality and Ethics in the Face of Rationalism*

The distinct quality of Christian ethics in a pluralistic society is not primarily derived from an additional contribution of arguments but rather lies in the incorporation of a spiritual dimension, inspiring considerable potential for the motivation of ethical behaviour. The rich tradition which Christian ethics can

1 "Sign of the times" is a key term of Christian Social Ethics since the Second Vatican Council (1962-1965). It means that the Church has to 'read' every time anew the challenges of present time in the light of the gospel, because that is a precondition of a really living and rescuing faith: "The joys and the hopes, the griefs and the anxieties of the men of this age, especially those who are poor or in any way afflicted, these are the joys and hopes, the griefs and anxieties of the followers of Christ" (as the beginning of the document "Gaudium et spes" says, which was promulgated in 1965 as a constitution of the Church in the modern world). See: Hünermann, Peter, ed., *Das Zweite Vatikanische Konzil und die Zeichen der Zeit heute: Anstöße zur weiteren Rezeption* (Freiburg: Herder 2006).

 | DOI 10.1163/9789004356825_004

draw on aims at translating ethics into an ethos by equally addressing hearts and minds, deep-seated hopes, and daily life.

Leonardo Boff criticises the anthropological and ethical traditions of modernity for not moving beyond rationality:

> Without mysticism and its institutionalization in the different religions, ethics would degenerate to a cold catalogue of regulations and the codes of ethics would become processes of social control and cultural paternalism.[2]

1.2 *Ethical Analyses of the Spiritual Type of Knowledge*

Spirituality is a type of knowledge that draws attention to the connection between ideas and emotions. It enables us to understand the manifold qualities of nature beyond their physical, quantifiable features. There are elementary approaches to knowledge based on everyday experiences, which are in some aspects more creative and richer than knowledge based on scientific experiments that are in line with the modern ideal of objectivity.[3] Sensual experiences constitute understanding on a different level than quantifiable measurements. Bearers of this kind of multi-perspective knowledge can be predominantly found in the disciplines of theology and the arts.

Reflection on the manifold kinds of perception, knowledge and understanding, including the emotional aspects of our relationship to nature, leads to deeper spheres of human self-perception and, therefore, provides new energies of ethical and religious thinking and acting.

Many environmentalists insist that nature has intrinsic value. This requires a perspective that not only endorses a factual and scientifically quantifiable nature, but also its beauty, sense, and symbolism. It requires an aesthetical and spiritual sensibility that does not look at things in isolation, but rather in their entirety and unity. This is how ecological and religious perceptions can enhance and complement each other.

---

2 Boff, Leonardo, *Ethik für eine neue Welt* (Düsseldorf: Patmos, 2000), 11. Quotes in the following essay are given in English language; if the original is in German, the quote is given in the author's own translation.

3 Hofmeister, Georg, *Ethikrelevantes Natur- und Schöpfungsverständnis: Umweltpolitische Herausforderungen, naturwissenschaftlich-philosophische Grundlagen, schöpfungstheologische Perspektiven, Fallbeispiel: grüne Gentechnik* (Frankfurt a.M.: Peter Lang, 2000), 77.

1.3 *The Gap between Knowledge and Belief*

Responsibility for nature in times of climate change, overpopulation and scarcity of resources is not primarily a problem of knowledge, but a problem of conviction and belief: we do not believe what we know about climate change and environmental problems, because we cannot sufficiently imagine what it means for us, for the fate of people all over the world and for life on earth in general. We are not able to react adequately because we have never experienced such a deep and complex change in living conditions. For most decision makers, the crucial consequences are too far in the future.

The lack of readiness to follow the decisions of the Climate Conference in Paris (December 2015) shows that we are "atheists of the future".[4] In order for us to realise what climate change means and to react adequately, it is necessary to translate our scientific data into descriptions of what they mean for society in terms of suffering, values and lifestyles. Thus, the cultural sciences (including theology) would be of crucial help towards overcoming that deep gap between knowledge and belief in this very specific context.

I propose the establishment of a think tank for culture and religions at the International Panel on Climate Chance (IPCC), enabling us to gain a better understanding of the spiritual and cultural knowledge that is necessary to motivate people in different cultural contexts to contribute to the protection of our climate and to the adaptation towards environmental change.

2 The Biblical Creation Mandate

2.1 *The Belief in Creation as Crisis Management*

The biblical faith in creation developed at a time of political and social crisis: the times of Exodus and the Babylonian Exile (approximately in the 6th century BCE). The act of creation as the beginning of all things inspires a hope for the care of God and the survival of his creatures that has the power to transcend all phenomena of crisis.[5] This beginning has its primary place in the hope for salvation.

---

4 Sloterdijk, Peter, "Das 21. Jahrhundert beginnt mit dem Debakel vom 19. Dezember 2009," *Süddeutsche Zeitung*, December 21, 2009, 10; Sloterdijk related this to the Climate Conference in Copenhagen; the conference in Paris seemed to be successful, but the implementation failed because the "mental infrastructures" did not change.

5 Ganoczy, Alexandre, *Theologie der Natur* (Zürich: Benziger, 1982), 11.

2.1.1 Unity of the Mission to Dominate and the Mission to Guard Nature

The so-called dominion mandate in Genesis 1:28—"fill the earth and subdue it"—has frequently been interpreted as the cultural-historical program for environmental destruction.[6] In fact, the Hebrew verbs *rdh and kbs* can be interpreted as trample, stomp, subdue, or even rape. In Genesis 1:28, however, they have to be interpreted in the context of royal terminology. Consequently, rather than despotic domination, the term should be interpreted as royal rule in the sense of responsible care—just as a good king would care for his subjects.[7]

This interpretation is backed by the biblical composition of the first and the second creation account: "God put them into the Garden of Eden to tend and to watch over it" (Genesis 2:15). The "domination mandate" has to be seen in the light of the "gardener" or "stewardship mandate": man is authorised to rule over nature by responsibly cultivating, conserving, shaping and protecting God's creation.

2.1.2 Godlikeness and Earthly Nature of Man

The mandate to dominate (Genesis 1:28) is theologically based on the notion of man as "God's image", which entails both freedom and responsibility. Humans are not directly or primarily ascribed a sacred status, but rather a social one which includes a sacred dimension: they shall serve each other, which is the best way to serve God. They should develop their skills and talents freely and responsibly as a way to honour God.

Consequently, the verb *rdh* has to be interpreted as "to dominate" according to the ideal of God: to guide and to tame. The mandate to dominate implies a hierarchical partnership between men and animals according to the notion of domination by providence.[8]

Man is conceptualised as "earthling" (Adam), as an earthly creature, which amounts to an essentially ecological designation. According to the biblical view, he is "dust" and a creature that belongs to the earth. This applies to all human beings and implies that man is embedded in nature—there is no privileged position for him. Only the reference to God, the capacity to experience God as a gift, distinguishes him from nature. This capacity to experience God

---

6 Amery, Carl, *Das Ende der Vorsehung: Die gnadenlosen Folgen des Christentums* (Hamburg: Rowohlt, 1972). For a critical answer to Amery, see: Rappel, Simone, '*Macht euch die Erde untertan': Die ökologische Krise als Folge des Christentums?* (Paderborn: Schöningh, 1996).

7 Löning, Karl, and Erich Zenger, *Als Anfang schuf Gott: Biblische Schöpfungstheologien* (Düsseldorf: Patmos, 1997).

8 Welker, Michael, *Schöpfung und Wirklichkeit* (Neukirchen-Vluyn: Neukirchener, 1995), 101-106.

enables humans to love without reserve and to assume responsibility for others—hence, it distinguishes man from nature and at the same time connects him more deeply with it.

2.1.3 Nature as God's New Contractual Partner

After the expulsion from paradise, we no longer live in a perfect world, but rather in a world often shaped by conflicts and suffering. Nevertheless, the world is still under God's protection, which is symbolised by the rainbow. In the narrative of Noah, not only man but also nature is called God's contractual partner: "I have set my rainbow in the clouds, and it will be the sign of the covenant between me and the earth" (Genesis 9:13). All aspects of the composition of the story are based on the understanding of the intrinsic value of nature. Nature is seen as a subject.

This reading is not compatible with modern European law, which is mainly based on ancient Roman legal traditions that clearly distinguish between man and nature. In the world of modern (juridical and scientific) thinking, animals and plants are just objects; in the world of the poetic language of the Bible, they are considered subjects.[9]

2.2 *Consequences for an Ecological Doctrine of Creation*

2.2.1 God and the World: A Difference that Inspires Creativity

The Christian view is not that of a deification of nature; it is not pantheistic. God and nature are distinct. The difference between creator and creation alleviates human beings of the burden of a supposed divineness.[10] We are not God; we do not need to be perfect. Our way of life is a continuous search with a lot of uncertainties, detours and impasses. It is this distinction that enables a free relationship between the creator and his creatures. We are compelled to be critical if a man claims to be perfect, a king claims to be like God, or a society or a specific community is claimed to be "heaven on earth". This critical distance allows for and inspires creativity.

This difference between creator and creation is absolutely crucial in the ecological context: it is a very common perception (especially in the global movement of "deep ecology", which started in the 1960s in Norway) that nature in itself has divine and holy characteristics. The consequence of this deifica-

9 For such an approach in modern philosophy and politics, see: Latour, Bruno, *Das Parlament der Dinge* (Frankfurt a.M.: Suhrkamp, 2010).

10 Die deutschen Bischöfe, "Handeln für die Zukunft der Schöpfung," in *Erklärungen der Kommissionen*, vol. 19, edited by Sekretariat der Deutschen Bischofskonferenz (Bonn: Sekretariat der Deutschen Bischofskonferenz, 1998).

tion of nature is that human beings are not allowed to modify it. Technical intervention, such as in agriculture, genetic engineering or scientific experiments with plants and animals would be an absolute taboo. Nature would be seen as an ideal order, at every time in a perfect balance.

Christian ethics, on the other hand, considers both the marvellous and the imperfect sides of nature. Nature is an order with positive and negative sides, life and death, harmony and conflict, happiness and grief. Our mission is to identify the good aspects of nature in us and nature around us, to cultivate them and to accept with humility and in solidarity even those aspects of life that seem to contradict our wishes.

This understanding of nature, which includes harmony as well as disharmony, is compatible with modern evolutionary approaches that understand nature as an order with astonishing adaptation capabilities as well as an order of never-ending conflict (struggle of life). In the long run, evolution is the sequence of unbalanced systems, and exactly these situations of imbalance incite progress.[11]

If nature is seen as a perfect order, the emergence of human civilisation would have to be considered a disturbing element. The tremendous dynamics of modern society disturb the order of nature on earth. We would have to consider ourselves as the "biggest catastrophe of nature" and acknowledge that the environment would benefit from the extinction of the human race. This would be the ultimate consequence of any biocentric or ecocentric view.

The Christian worldview fundamentally differs from these perceptions. From a Christian point of view, values are not predetermined by nature, but are rather a product of communication, culture and life. Values are not naturally given, but created by human discourse and by our relation with each other, nature and God. Nevertheless, there are quite fundamental differences within the Christian traditions: in Catholic theology, the trust in the good sides of nature is stronger (paradigmatic for this is the approach of Thomas of Aquino); the protestant tradition is more aware of sin as a dominant part of (human) nature (representative for this are the theology and ethics of Augustine and Luther).

2.2.2 The Belief in Creation and Perception of Time

The biblical understanding of the world is not cyclical; it cannot be confined to the circular flows of nature, and it is not limited to the interminable repetition of the same structures. Rather, it endorses a perspective that looks to the future

11 Reichholf, Josef, *Stabile Ungleichgewichte: Die Ökologie der Zukunft* (Frankfurt a.M.: Suhrkamp, 2008).

as something truly new, often surprising and unforeseeable, while at the same time considering history as something from which we can learn and which we do not necessarily have to repeat. This is the basic precondition for the Judeo-Christian hope to find salvation: not by being redeemed from material life, body and natural constraints, but to find salvation in the redemption and transformation of reality.[12]

Many might think that this would distinguish Christian and Buddhist approaches to the relationship between God, man, and creation. Buddhist thinking is focused on overcoming the flesh and materialism, and it pictures Nirvana as the antithesis of the empirical world of creation. Upon closer inspection, however, the differences become much less categorical: Christians also distance themselves from materialistic approaches, and Buddhists do not conceptualise the empirical reality of creation as purely negative.

Nevertheless, the conceptualisation of time is crucial for the understanding of the theology of creation and the relationship between God, nature, and man: if the world is an infinite cycle of repetition, salvation as the encounter between God and man has to be thought of as something taking place outside of nature and outside of human life in space and time. In contrast, Christians think of time as history, eternity as the source and unity of time (not just the negation of time) and the kingdom of God not as a second world behind our world, but as the hidden and mysterious presence of God in the midst of creation and human existence.

2.2.3 Creatio Continua: Creation as a Process

Despite the clear demarcation between the creator and the creature, eternity is a hidden aspect in time, which is understood as *creatio continua* in theology. In opposition to the common reading, the narrative of Genesis 1:11 is not just a speculation about the very beginning of the world in terms of a *creatio prima*, but rather a narrative description of the general features of the process of creation as a *creatio continua.* Creation is a perpetual creative process, a constant evolution.[13] God is present in the process of creativity and in the emergence of new life.

2.2.4 The Tension between the Redeemed and the Unsaved World

The Christian understanding of creation is shaped by the tension between the unsaved world and the anticipation of a reconciled reality. The same tension

12 Ganoczy, *Theologie der Natur*, 53.

13 Vogt, Markus, *Prinzip Nachhaltigkeit: Ein Entwurf aus theologisch-ethischer Perspektive* (München: oekom 2013), 305-346.

characterises the notion of nature: harmony and conflict, order and chaos, life and death, growth and decay. Given this inherent tension, ethics cannot be derived directly from nature, but rather needs to draw on historic, cultural, and religious categories. The Christian understanding of creation precludes a naturalistic ethics of nature. At the same time, the notion of creation needs to be based on a reflection of nature as an open, process-oriented and ethically ambivalent order, so as not to remain an abstract concept.

In the tension between the redeemed and the unsaved world, the Christian ethics of nature has a complex character: values are not perfectly given in the order of nature, but rather we have to search for them in the process of cultural interpretation and formation of nature. The 'grammar of nature' is relevant for ethics: it defines the limits, conditions and duties of ethics. But natural law and ethics are not the same. Ethics is related to freedom as the enabling precondition of morality; it therefore concerns the decisions pertaining to a subject and does not necessarily arise as a consequence of natural law. Christian ethics is not a naturalistic concept.[14] Nevertheless, it has to learn to be more conscious of and responsible for nature. The ecumenical and interreligious dialogues about sustainable development are chances for such a process of learning and rethinking the relationship between faith and concepts of nature.[15]

2.2.5 Core Ideas of Christian Ethics Concerning Creation

Against this background, the ethical and practical relevance of the belief in creation can be summarised in three core concepts:[16]

- *Man as the image of God*: In the Christian tradition, human beings as free and responsible subjects are the temple of God. Man can conform to God's value system if he acts freely, independently and dutifully. This is the basis for man's unique dignity and his responsibility for all living creatures. The emphasis on the inviolable dignity of man in God's own likeness does not preclude recognition of the intrinsic value of nature, but rather constitutes its logical precondition. This is the real epistemological meaning of "anthropocentrism" in a biblical sense—it does not imply that only human beings

14 Concerning this wide field of theological, philosophical and ethical debates, see: Griffin, David R., *Religion and Scientific Naturalism* (Albany: State University of New York Press, 2000); Habermas, Jürgen, *Zwischen Naturalismus und Religion* (Frankfurt a.M.: Suhrkamp, 2005).

15 Bundesministerium für wirtschaftliche Zusammenarbeit und Entwicklung, ed. *Voices from Religions on Sustainable Development* (Berlin: BMZ, 2016).

16 Cf. Die deutschen Bischöfe, *Handeln für die Zukunft der Schöpfung*, No. 56-84.

matter.[17] It is not more and not less than an epistemological precondition of knowledge and ethical evaluation.
- *Community of God's creatures* (Ger. "Mitgeschöpflichkeit"): Man and all living creatures are united as creations of God. This prohibits us from considering our common creatures as a mere means to an end. Respect for all other creatures is the consequence of God's love. Man shares a common destiny with all other living creatures, each of which holds its own place in God's creation.[18] The term "Mitgeschöpfe" (fellow creatures) was included in the German law for animal protection during the last modification in 1990. However, we are far from respecting and implementing it in reality, e.g. in practices of modern farming.
- *Reverence:* The Christian belief in creation emphasises an attitude of reverence which is able to constantly rediscover and protect the beauty of creation in the face of sorrow and conflict. Reverence for creation implies reverence for the creator as well as respect for the given limits, measures and dimensions of creation. Albert Schweitzer (1875-1965), the Protestant theologian, jungle doctor, musician, and scientist, appreciated the attitude of reverence as the fundament of a new, nature-based ethics.

If we want to reconceptualise environmental ethics, we have to recognise the unity of these three aspects instead of playing them off against each other.

2.3 *Climate Change as a "Sign of the Times"*

A theology of the "signs of the times"[19] interprets the present in prophetic terms as God's calling: it addresses the challenges and upheavals of each historical situation in order to seek within these situations the hidden presence of

---

17 Vogt, *Prinzip Nachhaltigkeit*, 252-259; Franziskus, *Encyclical Laudato Si': On Care for Our Common Home* (Vatican: Vatican Press, 2015), No. 115-136.

18 Löning, and Zenger, *Als Anfang schuf Gott*, 142-146.

19 The theology of the signs of the times initially emerged in the 1920s within a Protestant context. After formative development in the field of Catholic theology, amongst others, for instance through the French "Nouvelle Théologie", it had a major influence on the program of the Second Vatican Council ("Opening up to the modern world") as well as the socio-ethical Pastoral Constitution "Gaudium et spes"; cf. Sander, Hans-Joachim, "Theologischer Kommentar zur Pastoralkonstitution über die Kirche in der Welt von heute Gaudium et spes," in *Herders Theologischer Kommentar zum Zweiten Vatikanischen Konzil*, vol. 4, ed. Peter Hünermann, and Bernd Jochen Hilberath (Freiburg, Basel, Wien: Herder, 2005), 581-886; Ruggieri, Giuseppe, "Die Zeichen der Zeit: Herkunft und Bedeutung einer hermeneutischen Chiffre der kirchlichen Praxis," in *Das Zweite Vatikanische Konzil und die Zeichen der Zeit heute*, ed. Peter Hünermann (Freiburg: Herder, 2006), 61-70.

God who has revealed himself and newly reveals himself as our companion. The texts by the Second Vatican Council postulate God's acting throughout history.[20] Accordingly, history is considered to be a dialogue between God and the Church on the latter's peregrination through time. Faith is *interpretatio temporis*, i.e. interpretation of existence through time, and not simply adherence to archaic truths. The signs of the times are traced back to the ministry of the Holy Spirit in history.

> In deciphering and interpreting the signs of the times, the Church then perceives direction and guidance by the Holy Spirit who challenges the Church to embark upon new paths and new ways of behaving within their faith.[21]

The signs of the times raise questions about life. The answers lead to Revelation becoming accessible in a liberating and meaningful manner. By listening to man's experiences and contributing to the interpretation of the signs of the times and the quest for answers, faith becomes alive and acquires a contemporary testimonial character.[22]

According to the Bible (Matthew 16:3: *semeia ton kairon*; Luke 12:56: *ton kairon touton*), the "signs" are not factual historical reality as such but rather God-given *kairos* manifesting itself in reality. In ecological terms one can conclude the following: the challenges of the ecological crisis can only be considered "signs of the times" in the context of the upheavals and departures towards a new awareness; the signs manifest themselves in this new awareness and subsequently need to be interpreted from a theological perspective before being integrated by the Church. In the course of these developments, a new awareness of global neighbourhood and a sense of man and his fellow creatures being companions in fate has emerged. Many people are searching in new ways for the normative meaning of nature and for "God's imprints"[23] therein. Many are also anxiously apprehending the ecological destruction of the chances of survival of the poorest members of society and of opportunities in life for future generations. An interpretation of the ecological crisis and the

---

20 Cf. Hünermann, Peter, "Gottes Handeln in der Geschichte: Theologie als Interpretatio temporis," in *Freiheit Gottes und der Menschen: Festschrift für Thomas Pröpper*, ed. Michael Böhnke (Regensburg: Pustet, 2006), 111; see also Ruggieri, "Die Zeichen der Zeit," 68-69.

21 Hünermann, "Gottes Handeln in der Geschichte," 110.

22 Ibid., 133.

23 Vogt, Markus, "Naturverständnis und christliche Ethik," in *Berichte der* ANL*: Wir und die Natur – Naturverständnis im Strom der Zeit*, vol. 25, ed. Bayerische Akademie für Naturschutz und Landschaftspflege (Laufen: ANL, 2001), 103-111.

global environmental movement as signs of the times is premised on certain religious experiences that are made within this context,[24] for example, the rediscovery that man is an integral part of creation, a rediscovery that has been inspired by ecological challenges. The ecological crisis is only considered to be a theological sign because it has given and is still giving rise to the discovery of forgotten and new dimensions of the belief in creation. It is a challenge for faith and the Church to fathom God's hidden 'message' in the ecosocial crisis and to interpret it in the light of the gospel.

A set of criteria is necessary in order to distinguish the historical exegesis of existence within the context of faith from an adaptation to the *zeitgeist* and in order to purify it from historical contingencies and ideological influences:[25]

- Signs of the times are those phenomena that shape an era due to their universality and frequency. Signs of the times are characteristic for the distinctively new conflict situations within each historical context. Moreover, they refer to an epochal process of change that is historically significant. They do not only concern individual groups and their interests but mankind as a whole. They are universally meaningful for the development and future of mankind. Theologically they aim for a pastoral *aggiornamento* in order to see the traces of the future in the present.
- Signs of the times relate to essential questions of human existence that express the hardships and longings of a certain time. They are not man's projections of desire but rather emanate from experiences of suffering, failure and brokenness in which the longing for God's salvific intervention becomes apparent *sub contrario*. The signs of the times are an outcry in which the ministry of the Holy Spirit, the entire profoundness of cross and resurrection, and the experience of suffering and hope manifest themselves. The signs of the times relate to questions concerning fate and salvation that encompass man's entire life and self-conception.
- Signs of the times are not historical facts and natural phenomena as such but rather the ensuing changes in man's awareness. Triggered by emergency situations and challenges, man's awakening to new ways of understanding, new guidelines and standards shapes the referential character of signs. They

---

24 On the basic meaning of the category "religious experience" for the theological understanding of the signs of the times cf. Ruggieri, "Die Zeichen der Zeit", 70.

25 The following attempt to create a set of criteria of the signs of the times has been significantly inspired by Hünermann, "Gottes Handeln in der Geschichte", Sander, "Theologischer Kommentar zur Pastoralkonstitution," and Ruggieri, "Die Zeichen der Zeit." No claim is made to completeness.

are experiences through which God's spirit expands into new forms of expressing faith and humanity. To 'experience the experience', which can transform an emergency situation into a salvific experience, is a constitutive element. The complex relation between faith, religious experience and the exegesis of existence is the hermeneutical key for a theology of the signs of the times.[26]

- The signs of the times refer to crises that necessitate a decision by man. They enable a new level of differentiation between justice and injustice. Moreover, they offer freedom in that humans can get rid of old boundaries and dependencies and choose between salvation and misery. Signs of the times cannot be properly identified from the distance of neutral observation; it takes someone who demonstrates faith, hope and active sympathy for the fate of those people that are suffering.[27] From a theological perspective, catastrophes, trends and awareness-raising processes can only be perceived as signs through the processes of repentance and distinguishing between the spirits.

According to these criteria, the ecological crisis as such is not a sign of the times unless Church and society experience God within this crisis. Church and society must learn anew that a "culture of life" (John Paul II)[28] can only grow when man respects nature in and around him as a gift from God with an intrinsic value. From a theological perspective, these crisis phenomena are ambivalent in so far as taken individually they tend to lead to anxieties that might manifest themselves in violent reactions. A crisis phenomenon becomes a "sign" (Gr. *semeion*) when it reveals the possibility for man to make a decision for God and life. In its original literal sense "to decide" (*krinein*), a crisis is not primarily something negative, rather it is a decision situation that can lead to essential elucidations. Under these premises a theological diagnosis of the times with regard to the phenomena that are emerging in the course of the ecosocial crisis generally goes beyond scaremongering worst-case scenarios. In fact, it also perceives signs indicating new departures that enable it to transform radical change and upheaval into the right moment (*kairos*) of liberation

26 Cf. Hünermann, "Gottes Handeln in der Geschichte," 117-120; Ruggieri, "Die Zeichen der Zeit," 70.

27 According to Ruggieri, the option for the poor is essentially part of the theology of the signs of the times and it is "one of the most frequently and even deliberately forgotten teachings of the Council" (Ruggieri, "Die Zeichen der Zeit," 68).

28 Johannes Paul II, *Evangelium vitae: Encylica about the Gospel of Life* (Vatican: Vatican Press, 1995), No. 21, 28, 50, 77, 82, 87, 92, 95, 98, 100.

from false notions. [29] Within the context of ecology, the modern age is often construed as a history of decay, but this view is not compatible with the theology of the signs of the times. The theological reason for the repudiation of the theory of decay is not an optimistic downplaying of problems but rather the trust that God is present even in the upheavals and errors of human history.

Our relationship to God constitutes the theological point of reference in this context of challenges; this relationship cannot be separated from our relationships to our fellow humans and fellow creatures, to nature and history; they are all intertwined. The focus is specifically limited to the question of justice that underlies the given social, ecological, and economical crises. This question naturally implicates the question regarding man's role in nature as well as his relationship to fellow humans and fellow creatures since in the Judeo-Christian faith and in Islam the relationship between creatures also involves the relationship to God. God does not encounter man beyond the creatural reality but rather within it and through it.

2.3.1 Hearing Creation's Outcry—Religious Experiences with and within the Ecological Crisis

An approach to ecological questions that is shaped by the theology of the signs of the times understands the ecological crisis as an existential questioning of the faith that creation is a blessed habitat for all creatures. In a pressing manner, the question regarding God's liberating and promising presence within his creation is greatly relevant today. The Christian promise of salvation and liberation for all people and the whole of creation (Romans 8) is not reconcilable with indifference towards the fate of fellow creatures. The Church truly fulfils

29 This argumentative structure can be found in almost all of the mostly short but overall relatively numerous statements by Pope John Paul II on ecological questions. He unfailingly connects the description of the crisis to the growing awareness for ecological questions, a development that he praises as being a departure into a new era. It is only in this framework that the crisis phenomenon becomes a "sign of the times"; cf. for example Johannes Paul II. (John Paul II.), "Friede mit Gott dem Schöpfer, Friede mit der ganzen Schöpfung: Botschaft von Papst Johannes Paul II. zur Feier des Weltfriedenstages am 1. Januar 1990," *L'Osservatore Romano* 50 (1989): No. 1. This structure is by no means specifically Catholic; it also shapes the approach of the ecumenical movement worldwide; cf. the following introduction, which is significant for this approach: Reuver, Marc, Friedhelm Solms, and Gerrit Huizer, ed., *The Ecumenical Movement Tomorrow: Suggestions for Approaches and Alternatives* (Kampen, Geneva: Kok Publishing House and WCC Publications, 1993). It starts off by stating: "Crisis upon crisis affects today's world. This, however, is nothing new. Humankind throughout history has seen crisis periods of various kinds. They often were the forerunners of a new era." (Ibid., 9).

its calling when it partakes in the quest for solutions "for the most urgent questions of our times" (Gaudium et Spes 10) and identifies these questions as signs of experiencing God.

Against this backdrop, an ecology-oriented theology of the signs of the times perceives the ecological crisis in a very specific way as "revelation",[30] i.e. as God's calling to his Church. Whoever hears the outcry of this maltreated creation that crucifies the cosmic Christ every day anew[31] does not seek redemption beyond this creatural reality but with and within it. Christian hope is not directed towards redemption from creatural reality but toward its salutary transformation.[32] In times of crisis, believers in God's promise of a future for mankind and the entire creation have to bring to mind God's auspicious power through the testimony of love. There can be no future for mankind beyond the responsibility for his fellow human beings and fellow creatures.

In the context of ecology, a new kind of interest in religious questions is emerging.[33] This constitutes a sign of the times in that it is driven by a quest for sustaining life and thereby promising a future. At the same time, elements from religious thinking in the context of ecology have to be critically exposed in the modern environmental movement, particularly when equilibrium or harmony models are normatively adapted and misinterpreted as "doctrines of salvation".[34] Nature is increasingly occupying a new and very significant spot

---

30 The Indian theologian Raimundo Pannikar characterises the ecological crisis as a "revelation" in the sense of a liberating liminal experience; cited according to: Rosenberger, Michael, *Was dem Leben dient: Schöpfungsethische Weichenstellungen im konziliaren Prozeß der Jahre 1987-89* (Stuttgart: Kohlhammer, 2001), 275.

31 Cf. Fox, Matthew, *Vision vom Kosmischen Christus: Aufbruch ins Dritte Jahrtausend* (Stuttgart: Kreuz, 1991).

32 Cf. Romans 8:21: "Because the creature itself also shall be delivered from the bondage of corruption into the glorious liberty of the children of God" (King James Version). On the theological interpretation of the relationship between belief in creation and eschatology cf. Pannenberg, Wolfhart, *Systematische Theologie*, vol. 2 (Göttingen: Vandenhoeck & Ruprecht, 1991), 163-201; Auer, Johann, and Joseph Ratzinger, *Kleine Katholische Dogmatik Bd. III: Die Welt – Gottes Schöpfung* (Regensburg: Pustet, 1975), 110-164; Link, Christian, *Schöpfung: Handbuch Systematische Theologie*, vol. 7.1, and 7.2 (Gütersloh: Gütersloher Verlagshaus, 1991).

33 Gabriel, Karl, *Christentum zwischen Tradition und Postmoderne, Quaestiones Disputatae*, vol. 141 (Freiburg: Herder, 1994), 157-163. Cf. Taylor, Charles, *A Secular Age* (Cambridge: Harvard University Press, 2007), 299-321.

34 Trepl, Ludwig, "Ökologie als Heilslehre: Zum Naturbild der Umweltbewegung," *Politische Ökologie* 25 (1991): 30-45; Sachs, Wolfgang, *Nach uns die Zukunft: Der globale Konflikt um Gerechtigkeit und Ökologie* (Frankfurt a.M.: Brandes & Apsel, 2003), 112-134.

in our construction of meaning.[35] Society's desire for devotion has partially shifted from religion to nature.[36] All this has to be critically exposed and constructively expanded towards a truly transcendental horizon. Such a horizon cannot be found in ecology. However, being a truth-seeking movement that emanates from the breaches and turmoil of our times, the environmental movement can offer multifarious, fascinating, and inspiring bridges towards this transcendental horizon. The ecology of time, i.e. the rediscovery of the dimensions and rhythms of creation, could be a good starting point.[37] In the context of ecology, there is a quest for the origins, conditions, and sustainment of life. This quest can be interpreted as a modern variant of the question of transcendence and should be critically expanded.

When dealing with these issues theology has to remain true to its fundamental values, i.e. defend man's dignity within the frame of his mission concerning integral development.[38] Theology's ecological dimension, which also includes the recognition of the intrinsic value of fellow creatures, has to be established anew today. The theological perspective on ethical, anthropological, and human-ecological contexts can be an important corrective, especially in today's society, which is segmented into parties of different interests and worldviews. On the other hand, the awareness of the complex interconnectedness between different areas of society and science, an awareness that has grown within the context of ecology, constitutes a challenge for further developments in the field of theological ethics.

---

35 Cf. Rink, Dieter, Monika Wächter, and Thomas Potthast, "Naturverständnisse in der Nachhaltigkeitsdebatte: Grundlagen, Ambivalenzen und normative Implikationen," in *Naturverständnisse in der Nachhaltigkeitsforschung*, ed. Dieter Rink, and Monika Wächter (Frankfurt: Campus, 2004), 11.

36 "Nature Is Today's False God, and Ecology Increasingly Replaces Theology," Bolz, Norbert, and David Bosshart, *Kult-Marketing: Die neuen Götter des Marktes* (Düsseldorf: ECON, 1995), 81.

37 Henrici interprets the order of time as the central motif of the Jewish-Christian tradition; cf. Henrici, Peter, "Il senso cristiano del tempo: Prospettive per un nuovo ordinamento del lavoro e del riposo in una società," ed. Ufficio Nazionale per i Problemi Sociali e il Lavoro, *Notiziario* 4 (2002), 130-150.

38 On the topos of holistic and integral development cf. Pope Paul VI., *Encyclical "Populorum progressio"* (Vatican: Vatican Press, 1967), 6-11; on the recognition of the dignity of fellow creatures as expression and consequence of a definition of man's dignity that is based on theocentrism and interpreted from the perspective of the ethics of responsibility, cf. Münk, Hans, "Die Würde des Menschen und die Würde der Natur: Theologisch-ethische Überlegungen zur Grundkonzeption einer ökologischen Ethik," *Stimmen der Zeit* 1 (1997): 17-29.

Methodologically, the theology of the "signs of the times" most notably manifests itself in the triad "see–judge–act" (Gaudium et Spes 4). For ethics, this means a reversal of the traditionally deductive methodology: it does not start with the deduction of norms and postulates from theological axioms but rather sets in with a description of the situation. Ultimately, there is an unbreakable hermeneutic circle between the epistemological preconditions, the theological and ethical axioms as well as the description of reality, but the shift of emphasis in the ethical methodology through the triad "see–judge–act" should not be ignored.[39]

2.4 *Ecological World-Ethos: Which Competencies Do Religions Have?*

2.4.1 The Worldwatch Institute's Study on Religion and Sustainability

There is a growing conviction that religions will play a key role in establishing a broad-based consensus and a deeper understanding of the ethical basic principles of a sustainable society: religions are challenged to engage in the socio-political dialogue by contributing ideas on the basic option for a comprehensive responsibility for creation, a reconsidering of the relation between man and nature, the power of faith for personal change and a return to the essentials of life beyond consumerism and egoistic mentality.[40]

The renowned Worldwatch Institute assumes that this 'change of course', i.e. a change of policy of the global community towards a sustainable development, can only succeed if religions consider and live up to their role and responsibility in this endeavour. Gardner, author of the main report, names five outstanding assets which religious institutions and leaders can contribute to a sustainable world: the capacity to shape cosmologies (worldviews), moral authority, a large base of adherents, significant material resources, and community-building capability.

I would like to amend the list of assets for sustainability drafted by Gardner based on my point of view that is shaped by Catholicism and social ethics:

- All religions identify themselves through long-term thinking. On this basis, they have a very fundamental approach to the concept of sustainability.
- The Christian Church is the oldest global player on earth and the Catholic Church administrates an influential global institution and network;

39 Cf. Vogt, Markus, "Empirie in der Ethik," in *Interdisziplinarität der Christlichen Sozialethik*, ed. Peter Schallenberg, and Arnd Küppers (Paderborn: Schöningh, 2013), 405-424; Lonergan, Bernard J.F., *Method in Theology* (London: Darton, Longman and Todd, 1972).

40 Gardner, Gary, "Engaging Religion in the Quest for a Sustainable World," in *State of the World 2003*, ed. The Worldwatch Institute (New York: Norton & Company, 2003), 152-176.

therefore, it has a specific duty and possibility to fight for a globalisation of solidarity.

- Christian anthropology does not measure the value of a human being on the basis of goods consumed or produced. Consequently, it can empower people to handle these goods in a modest, just and responsible way. In Asian religious traditions, asceticism is even stronger than in the Christian tradition.
- Belief in creation does not only aim at issuing moral appeals; it also enables communication that strives to create and communicate values as well as to understand ecological responsibility as an integral part of man's self-respect. What we need are not just moral appeals but also a new cosmology—or the remembrance of old religious cosmological traditions.
- The distinct quality of the Christian point of view regarding environmental matters lies in its root in cultural and social contexts. Environmental protection and human protection form a unit in Christian ethics.

An awareness of these interlinked dimensions of Christian faith and ecological challenges could enable religious people and environmentalists towards a new and deep encounter. It would touch not only moral and ecological questions but also the very centre of the relation between God, man and nature itself: in the light of the existential experience of the ecological crisis, the question of God imposes itself in a new way. The ecological crisis is a revelation of God in our time. The crucified Christ can be experienced on a cosmological level in the "crucified creation" in our time.[41] There is no relation to God beyond our relation to creation. Therefore, responsibility for creation is a practical test for Christian belief in our times.

2.4.2 Seminal Figures in the Christian Spirituality of Creation

Christian ethics can draw upon a very rich tradition of spirituality, which is essentially a spirituality of creation. Francis of Assisi (1181-1226), who exemplified an unprecedented expansion of Christian care and love for all non-human creatures by his brother/sister relationship with the sun, the moon, water, fire, bees, lambs, bugs, flowers, birds and fish, is probably the most well-known

41 Fox, Matthew, *The Coming of the Cosmic Christ* (San Francisco: Herper, 1988). The former Dominican and Catholic priest is very radical in his thinking and actions. He postulates a new, ecologically motivated reformation: Fox, Matthew, *A New Reformation: Creation Spirituality and the Transformation of Christianity* (Rochester: Inner Tradition, 2006).

representative. Francis' "defiant sympathy for creation and its creatures",[42] which was also aware of misery and death as part of life, constitutes a radical contrast to the naïve projection of the human longing for an ideal world regarding the concept of nature.

Another paradigm of Christian devoutness to the cosmos is the rather intellectual-contemplative side represented by Thomas Aquinas (1225-1274), who, based on his reflection on the theology of creation, illustrated the divinely ordained independence of empirical reality and thus succeeded in integrating Aristotle's philosophy into the Christian faith. For Aquinas, the order of creation and the order of salvation are complementary processes of God's affirmation of the world. This is why the affirmation and appreciation of the reality of creation in all its aspects is Aquinas's basic moral attitude.[43] The tradition of natural law, as coined by Aquinas, is the main starting point for Christian environmental ethics. The main intention of its curious and thorough attention to the empirical reality has strongly influenced modernity.

Christian spirituality not only plays out in the retreat into one's self and the striving for self-perfection, but also in the capability for curiosity, sympathy, responsibility, and love. Especially in its vast diversity religious spirituality constitutes a source of socio-ecological transformation.[44]

3 A New Understanding of Welfare and Progress

3.1 *The Conflict between Climate Protection and Social Justice*

Justice and peace cannot be realised in the 21st century without climate protection. But there is a profound conflict between climate protection and the fight against poverty, as the well-known and financially viable methods of economic development are to a large extent dependent on access to fossil fuels. However, there is no additional capacity in the atmosphere for the CO_2 that would be emitted by developing countries if they were to develop along the same lines as the industrialised nations.

---

42 Ganoczy, *Theologie der Natur*, 89-94; Werner, Hans-Joachim, *Eins mit der Natur: Mensch und Natur bei Franz von Assisi, Jakob Boehme, Albert Schweitzer, Teilhard de Chardin* (München: C.H. Beck, 1986), 13-37.

43 Mertens, Gerhard, *Umwelten: Eine humanökologische Pädagogik* (Paderborn: Schöningh, 1998).

44 oekom e.V. – Verein für ökologische Kommunikation, ed. *Religion & Spiritualität. Ressourcen für die Große Transformation* (München: oekom, 2016).

The technical possibilities for fighting poverty and protecting the climate as well as the integration of these two aims are in theory relatively good. Realising these aims is primarily a question of overcoming political and institutional obstacles, as the necessary investments can only be made under conditions which facilitate a fair, cooperative and long-term sharing of the burden. Currently, from the point of view of the developing countries, there are hardly any consensual and attractive suggestions on the table for a fair 'burden sharing' in terms of climate protection.

The particular nature of ethical problems that arise as a result of climate change lies in the long distance between initiators and victims. This distance can be defined in three ways: climate change is having a profound and negative impact on (1) future generations, (2) poorer countries in the southern hemisphere, (3) and the habitats of fauna and flora, and thereby the relationship between humans and nature. It can be regarded as a threefold externalisation of the costs of our way of life: the price for our way of life will have to be paid by the future generations, the poor and the environment. On the basis of this analysis, the German Conference of Catholic Bishops has referred to climate change as the "crossroads of global, intergenerational and ecological justice".[45]

Climate protection needs a code of ethics which sheds light on potential causes for injustice, analyses dilemmas and provides firm criteria on which to base political decisions. We have to speak about philosophy of nature, anthropology and the complexity of human wishes, hopes and conflicts and the cultural reasons and obstacles pertaining to the changing of our behaviour.[46] Without an understanding of the cultural and religious dimension of man's responsibility regarding climate change, the political negotiations will not stand the slightest chance of changing society.

Climate protection is a question of solidarity on a long-term and global level. It is a crucial test, especially for the religions regarding their readiness to assume responsibility for creation. Climate protection is the 'moral stress test' for contemporary society.

3.2 *Newton's Concept of Nature and the Understanding of Progress*

Our current model of progress is based on Newtonian mechanics and philosophy of nature, which sees time and space as empty vessels and as something

45 Deutsche Bischofskonferenz, "Climate Change: A Focal Point of Global, Intergenerational and Ecological Justice," *Commission for Society and Social Affairs and Commission for International Church Affairs* 29 (Bonn: DBK, 2007).

46 Vogt, Markus, *Climate Justice* (München: Rachel Carson Center for Environment and Society, 2010).

lacking both, direction and structure as well as a beginning and an end. Time and space are merely obstacles to be overcome. Our accelerated society, which is managing to use up a myriad of resources at a breakneck speed and defines the pace of our lives by the maxim 'anytime always everything', is a consequence of our interpretation of nature.[47] Christian belief in creation leads us to search for alternatives to this view of nature, and it can base its nature philosophy on process theology. Sustainability needs new concepts of time and space, and thus a cosmology, which draws on the knowledge derived from Albert Einstein's theory and from new theories on the development of complex adaptive systems.

The environmental crisis is not just a challenge for political negotiations and technical innovation; it is also a question of changing society's values. "Faster, higher, further" has proved to be an inadequate ideal of progress. The current situation demands individual and collective answers to genuinely ethical questions about the goals, limits and conditions of our lifestyle. How much is enough? What are the priorities in striving for progress? How can we ensure fair chances for people all over the globe? How can we ensure that long-term interests are properly represented in the democratic system?

In the search for answers to these questions, which are profoundly significant for the threefold goals of fighting poverty, saving energy and protecting the climate, religions can make a substantial contribution. Their competence is based especially on the fact that they embed moral claims into a cosmology, a deep anthropological understanding of human behaviour and into a symbolic and ritual communication that has more chances to change the behaviour of people. The contemporary paradigm of progress as unlimited growth needs to be replaced by a concept of development governed by the cycles of resources and the rhythm of nature. Long-term economic success needs to be measured by how well it is integrated into the whole, i.e. the economy of creation.

3.3 *The Hope in God and the Belief in Political Utopia*

The modern model of progress has come to its end. This fact has yet to sink into the consciousness of many. Some interpretations of sustainability do not bring clear awareness of that situation, but rather help to artificially keep up our self-delusion. Sustainability with its far-reaching promise of a global, eco-social and economic approach has become one of the central utopias of the 21st century. In the 20th century, we had to experience the negative side of the deep ambivalence in the blind belief in utopias. Seen from a theological

47 Vogt, Markus, *Prinzip Nachhaltigkeit: Ein Entwurf aus theologisch-ethischer Perspektive* (München: Oekom, 2009), 305-346.

perspective, sustainability demands a rejection of the utopia that politics, science and economic progress will solve all problems.[48]

Even the agreements reached at the UN conference in Rio (1992) do not clearly address the limits of growth and simply paper over the cracks of these existential boundaries. We are promised a utopian global management of ecological and social problems, while behind the scenes the same old models and power networks are pursued.[49]

Sustainability has degenerated into a disguise of the traditional prosperity model, which according to the trickle-down principle makes the supply and accommodation of the poorest in society dependent on growth while actually generating a surplus for the society's rich people. The experience of the last two decades clearly shows that this is a misleading promise. 'Green-washing' is not enough; we need a deep shift in values and models of economy.

Currently, the 'green-washing' model of sustainability is excessively communicated and often willingly believed while it is indeed very questionable. We must realise that CO_2 emissions are still increasing, especially in China and India, and the chances to reach the target of two degrees Celsius global warming in climate politics are declining. The methane emissions from the melting permafrost have exceeded various worst-case scenarios, and we are well on the way to accelerating this process even further.

Churches and religious communities must tell the truth—even if nobody likes to hear it. Postmodern society is bound by the promise of everlasting growth and progress without believing it. Christian faith, on the other hand, has nothing in common with a belief in progress. It is a hope which is quite different from the expectations of prosperity that we have become used to in Western society and which now seem to spread very quickly also throughout Asiatic societies. Christian faith is a way of managing contingency in the face of the ambivalence of progress and setbacks, security and risk, joy and suffering, life and death. We have to accept that our life is finite, that it seems quite opaque and not fully understandable at times. The Christian religion and also all world religions accept the limits of human ability and life without downsizing the mind and the hope of a small-scale world.

---

48 The critique of the techno-centric concepts of sustainable development is quite common in the religious environmentalism; a critical view upon the utopian promises of sustainable development is still not very representative; cf. Bundesministerium für wirtschaftliche Zusammenarbeit und Entwicklung, ed. *Voices from Religions on Sustainable Development* (Berlin: BMZ, 2016).

49 Reis, Oliver, *Nachhaltigkeit – Ethik – Theologie: Eine theologische Beobachtung der Nachhaltigkeitsdebatte*, Forum Religion & Sozialkultur, vol. 18 (Münster: LIT, 2003).

If we assume, in line with leading sociologists of our time,[50] that managing contingency as described above is a primary function of religion, then it is also on sustainability. Coping with contingency is necessary in order to not react with ecological gloom and doom scenarios or with a reintroduction of the utopia of perpetual growth to the shattering of faith in man's progress which is the starting point of debates on climate change and sustainability. The religious dimension of hope liberates society from a blind belief in the political promise of actually being able to cope with ecological and social problems. God's place is not outside of nature. God is nature's mystery. Therefore, it is man's destiny to seek him and to make room for God in nature, society, and our minds.

---

50 Lübbe, Hermann, "Kontingenzerfahrung und Kontingenzbewältigung," in *Kontingenz: Poetik und Hermeneutik*, vol. 17, ed. Gerhart von Graevenitz, and Odo Marquard (München: Wilhelm Fink, 1998), 35-47; Luhmann, Niklas, *Die Religion der Gesellschaft* (Frankfurt a.M.: Suhrkamp, 2000).

PART 2

Past

CHAPTER 3

Long- and Short-Term Central European Climate Development in the Context of Vulnerability, Food Security, and Emigration

Rüdiger Glaser, Dirk Riemann, Steffen Vogt, and Iso Himmelsbach

Abstract

Climate reconstructions for Central Europe based on data from societal and natural archives from the last millennium are presented and discussed. The long-term relationship between climate change and food security is addressed by comparing the long-term climate development with harvest yield reports and information on famines and food crises. In three case studies covering the period 1815-1847, a more detailed analysis highlights the short-term interaction between climate, vulnerability, and emigration.

1 Introduction

Knowledge of the historical climate improves the understanding of the processes underlying the climate system by extending the perspective beyond the period for which instrumental weather data is available. The associated insights are particularly valuable since they permit an evaluation of climate variability outside of the predominant influence of the greenhouse effect during recent decades. Being the documentary-records-based branch of paleoclimatology, historical climatology offers detailed information covering specific times and places within certain regions of the earth in some cases extending back the last two millennia. In addition to basic weather information, it also offers authentic insights and comparisons as well as analogies with regard to climate impact, perceptions and societal consequences.

A retrospective examination of the climate and climate change in Europe can be initially founded on standardised, official times series data collected by meteorological instruments, but these data generally extend back only as far as the mid-19th century. Further extrapolations are based on early, non-standardised instrument records, such as are sporadically available in Italy,

 | DOI 10.1163/9789004356825_005

although with many interruptions, back into the 16th century.[1] Even older information can be gleaned from town chronicles, weather journals, annals, harvest records and other written documents that have been passed down. Historical climatology has established itself as an independent research discipline since the 1960s,[2] which saw the beginning of a protracted effort to draw a detailed picture of climate history based on thousands of archived documents.

These data are supplemented particularly in their time dimension by information from natural climate indicators, or proxies, such as tree rings, ice cores, varves, etc. These sources expand our knowledge into times and places for which human-made records are not available and also offer important indications of the interactions between climate and environment.

In multi-proxy approaches, a variety of proxies are combined by means of statistical methods to achieve as accurate a reconstruction as possible of short-, medium- and long-term climate variability. The focus here generally lies on the parameters of temperature and precipitation since these are of central importance for human activities. Moreover, the energy content of the atmosphere can be effectively characterised based on the temperature, making it a useful parameter for describing natural and anthropogenic changes to the atmosphere.

Another important tool for climate analysis can be found in climate models. Since the beginning of numerical modelling the models have been continuously improved, improving the reproduction of the observed and reconstructed variability of the climate. Such models are controlled by reconstructed alterations in natural forcings (CO_2, methane, solar irradiance, sulphites, etc.) in order to match the real processes on the earth as realistically as possible. By comparing the model data with the reconstructed climate parameters, it is possible to obtain a holistic understanding of the processes at work as well as

1 Camuffo, Dario, and Chiara Bertolin, "The Earliest Temperature Observations in the World: The Medici Network (1654-1670)," *Climatic Change* 111.2 (2012): 335-363. doi:10.1007/s10584-011-0142-5.

2 Le Roy Ladurie, Emmanuel, Histoire du climat depuis l'an mil (Paris: Flammarion, 1983); Lamb, Hubert H., *Climate: Present, Past and Future: Climatic History and the Future, Climate: Present, Past and Future*, vol. 2 (London: Methuen, 1977); Pfister, Christian, *Wetternachhersage* (Bern: Haupt, 1999); Alexandre, Pierre, *Le climat en Europe au Moyen Âge* (Paris: Éditions de l'Ecole des Hautes Études en Sciences Sociales, 1987); Glaser, Rüdiger, and Dirk Riemann, "A Thousand-Year Record of Climate Variations for Germany and Central Europe Based on Documentary Data," *Journal Of Quaternary Science* 24.5 (2009): 437-49. doi:10.1002/jqs.1302; Brázdil, Rudolf et al., "European Climate of the Past 500 Years: New Challenges for Historical Climatology." *Climatic Change* 101.1-2 (2010): 7-40. doi:10.1007/s10584-009-9783-z.

to make predictions of possible future climate developments. In order to validate the simulated climate developments, a comparison with time series for various climate parameters extending back as far as possible is therefore an essential prerequisite for optimally reducing the uncertainties in the simulations. Regionally differentiated climate development patterns in Europe can be depicted based on the various available reconstructions and models, and the changes in these patterns can be analysed.

In addition to climate reconstruction, this article addresses the long-term relationship between climate change and food security by comparing the long-term insights into climate development with harvest yield reports and information on famines and food crises. In this way, an empirical research approach at the mesoscale is presented, and the extent to which harvest yield records and food crises follow the long-term climate development trends is investigated in order to establish any causal relationship between these factors.

2 Climate Reconstructions as a Research Subject

A variety of scientific disciplines now concern themselves with climate reconstructions, such as climatology, meteorology, biology, archaeology, geography, and the historical sciences. The climate and weather conditions of the past are elucidated by a broad methodological spectrum of hermeneutic and scientific procedures. In principle, it is possible to differentiate here between societal and natural climate archives.

2.1 *Societal Archives*

A wealth of sources is available for study from the societal climate archive: In addition to annals, chronicles and diaries, there are also illustrative depictions such as leaflets, printed documents and maps that address the topics of weather events and climate conditions.

The structure of the data covering Europe over the last 1,200 years, which itself illuminates key facets of the availability and interpretation of climatological knowledge, can be divided into the following five phases:

1. The oldest, dating back as far as the 8th century and found, for example, in the editions of the Monumenta Germaniae Historica (MGH),[3] tend to

3 *Monumenta Germaniae Historica: CD-ROM 3* (München, Turnhout: Monumenta Germaniae Historica, Brepols Publishers, 2002); Wattenbach, Wilhelm, Ernst Dümmler, and Franz Huf,

be sporadic. They represent descriptions of special isolated events and natural catastrophes such as harsh winters, summer droughts, floods and other natural events such as aurora borealis, earthquakes or volcanic activity. The descriptions are generally very tersely worded.

2. Since the late Middle Ages, nearly uninterrupted descriptions of summer and winter climatic events are available as well as an increasing body of information on the transitional seasons. The descriptions become more differentiated and permit insights into the imagination of the writers, the climate impact on societies and the measures taken in response.
3. As of about 1500, the situation changes fundamentally. The printing process developed by Gutenberg spread quickly, paper production expanded rapidly and, above all, more people were able to read and write. The body of source material is therefore considerably denser, as is the diversity of the forms of expression. In addition to sporadic references in chronicles, we now find systematically maintained diaries and calendars, often in the form of prognostications aspiring to the prediction of weather.
4. As of around 1680, this information is supplemented by instrument measurements, although these are initially still highly individual and inhomogeneous and tend to have been recorded sporadically in experimental fashion.[4]
5. Official measurement networks based on instrument measurements were established starting in the 19th century.

A portion of the historical source material was made accessible in the form of compilations and was used already in the early efforts at climate reconstruction.[5] Due to the lack of critical analysis of the sources, these early attempts have been subject to regular criticism. Historical climatology has now developed extensive standards of critical source analysis and index creation. The derivation of indices is a useful and by now widespread method to derive semi-quantitative time series. The defined number of indices should reflect and depend on the semantic or linguistic differentiation and context, which is

---

Deutschlands Geschichtsquellen im Mittelalter: Frühzeit und Karolinger, 2 Vols (Kettwig: Phaidon, 1991).

4 Hellmann, Gustav, ed. *Meteorologische Beobachtungen vom XIV. bis XVII. Jahrhundert* (Berlin: A. Asher & Co, 1901), 13; Hellmann, Gustav, *Repertorium der deutschen Meteorologie: Leistungen der Deutschen in Schriften, Erfindungen und Beobachtungen auf dem Gebiete der Meteorologie und des Erdmagnetismus von den ältesten Zeiten bis zum Schlusse des Jahres 1881* (Leipzig: Engelmann, 1883).

5 Henning, Richard, "Katalog bemerkenswerter Witterungsereignisse von den ältesten Zeiten bis zum Jahre 1800," *Abhandlungen des Preußischen Meteorologischen Instituts* 2.4 (1904): 93.

given in the written documents by the authors. There are three "unweighted indices" and five, seven, nine, or eleven "weighted indices" in practise. Riemann clearly points out that even a simple three-class index is enough to generate a long-term signal.[6] To get quantitative temperature, precipitation, or other climatological parameters for these indices, statistical methods—likewise variance scaling, linear regression, or canonical correlations—are used to derive transfer functions. The resulting temperature and precipitation "values" can be regarded as intervals or estimates, as in each statistical method there still is a degree of uncertainty, which should be indicated in the results.[7] Impressively large collections of sources have been documented and made publicly accessible in web portals such as tambora.org, RECLIDO and CLIWOC.[8]

The climatic references in the sources are frequently linked with informative notes concerning environmental impacts and societal consequences: for example, references to phenology are made very frequently, in particular with regard to the condition of cultivated crops and their development over time.[9] The nature of the harvest and the associated price developments are also recorded. The observers often take special care to ensure objectivity and comparability, for instance by noting the regular freezing over of a specific pond or combining their assessments with other comparisons. The resulting holistic pictures of earlier climatic, environmental, and social conditions are amenable to productive interpretation.

The particular advantage of this hermeneutic research discipline is that the statements are based on direct observations of weather and weather events with a high resolution, at least over certain periods, and can generally be clearly

---

6 Riemann, Dirk, "Methoden zur Klimarekonstruktion aus historischen Quellen am Beispiel Mitteleuropas," *Freiburger Geographische Hefte* 69 (2012).

7 Glaser, Rüdiger, *Klimageschichte Mitteleuropas: 1200 Jahre Wetter, Klima, Katastrophen* (Darmstadt: Primus, 2013).

8 Riemann, Dirk, et al. "The CRE tambora.org—New Data and Tools for Collaborative Research in Climate and Environmental History," *Geoscience Data Journal* (2016). doi:10.1002/gdj3.30; RECLIDO, "REd para la reconstrucción del CLIma a partir de fuentes DOcumentales," accessed September 13, 2014. <http://pendientedemigracion.ucm.es/info/reclido/>; Climatological Database for the World's Oceans (CLIWOC), "Climatological Database for the World's Oceans 1750-1850," accessed September 13, 2014. <http://pendientedemigracion.ucm.es/info/cliwoc/>.

9 Le Roy Ladurie, Emmanuel. *Histoire du climat depuis l'an mil* (Paris: Flammarion, 1967); Kiss, Andrea, Rob Wilson, and Istvan Bariska, "An Experimental 392-Year Documentary-Based Multi-Proxy (Vine and Grain) Reconstruction of May–July Temperatures for Koszeg, West-Hungary," *International Journal of Biometeorology* 55.4 (2011): 595-611. doi:10.1007/s00484-010-0367-4; Chuine, Isabelle et al., "Historical Phenology: Grape Ripening as A Past Climate Indicator," *Nature* 432.7015 (2004): 289-290. doi:10.1038/432289a.

dated. In addition, the records provide information about the entire course of the year.

Alongside general statements on the temperature, precipitation and the condition of the atmosphere, frequent references are made to weather and climate extremes such as floods, storms and gales and their climatic consequences. Due to the special importance of food security, the entries are very often associated with references to agricultural production and phenology.

2.2 *Natural Archives*

Climate indicators derived from "natural archives", known as (natural) proxies, are also used for climate reconstruction. These are based on the analysis of interactions between the climate and the natural environment and often aggregate climate information over periods of days to years. Information about weather and climate can be found, for example, in the rhythmic sediment layers of lakes, the so called varve clay formations, in the growth rings of stalactites and stalagmites as well as bivalve shells and coral colonies, in the ice layers of glaciers or ice sheets, which often also trap tiny air bubbles that can reveal the atmospheric composition of the past, and in the well-known growth rings of trees.[10] The data on atmospheric chemistry obtained from these sources, especially from ice cores, is also used to adapt climate simulations to real conditions.

Thanks to their precise annual or seasonal resolution, tree rings represent a particularly suitable natural archive that has been studied for some time. With the description of a characteristic growth ring in a number of trees that originated from frost damage in the year 1709, the French researchers Duhamel and Buffon were able to provide the first evidence of the environmental influence on tree growth in the year 1737.[11] However, it was not until the first decades of the last century before knowledge of a climate influence could be leveraged into a reliable methodology for reconstruction of the climate. This was accomplished by the American astronomer A.E. Douglas in the arid regions of the

10 Sirocko, Frank, *Wetter, Klima, Menschheitsentwicklung* (Darmstadt: Wissenschaftliche Buchgesellschaft, 2009); Büntgen, Ulf et al., "2500 Years of European Climate Variability and Human Susceptibility," *Science* 331.6017 (2011): 578-582; Büntgen, Ulf, "A 1052-Year Tree-Ring Proxy for Alpine Summer Temperatures," *Climate Dynamics* 25 (2005): 141-153; Schweingruber, Fritz H., *Jahrringe und Umwelt – Dendroökologie* (Birmensdorf: Eidgenössische Forschungsanstalt für Wald, Schnee und Landschaft, 1993).

11 Fritts, Harold C., *Tree Rings and Climate* (London, New York: Academic Press, 1976).

American Southwest in his study of the relationships between sunspot cycles, climate, and growth rings.[12]

Inspired by Douglas's success, botanist B. Huber began his growth ring research at the University of Tharandt in the 1940s. He and his colleagues developed a well-known growth ring chronology from "Spessart oaks" and wood samples from historical buildings that extended back to 832 CE.[13] In the 1970s, Bernd Becker began creation of a growth ring chronology that remains unique to this day. Aided by finds of subfossil oaks in the fluvial deposits of large rivers such as the Rhine, the Danube and the Main, he assembled an oak chronology that extends back without interruption to 8480 BCE.[14]

These chronologies, which have been expanded and extended further back by more recent research, form the basis for today's dendrochronology and dendroecology. Statistical tree ring analyses make it possible to calculate transfer functions that describe the relationship between climate and the growth performance of trees. For epochs and regions from which no written climate information has been passed down (prior to the eighth century), tree rings represent a high-resolution data source for climate reconstruction. Anatomical changes in wood also enable us to accurately date extreme weather conditions such as late or long-lasting winter freezes, especially damp or dry springs, floods and insect infestations to the year or season.[15]

3 Climate Information on Europe

Regionally focused climate reconstructions based on documentary records by various research groups now exist for Europe, such as for Germany,[16]

12 Douglass, Andrew E., "Some Topographic and Climatic Characters in the Annual Rings of the Yellow Pines and Sequoias of the Southwest," *Proceedings of the American Philosophical Society* 61 (1922): 117-122.

13 Huber, Bruno, and Veronika Giertz-Siebenlist, "Unsere tausendjährige Eichen-Jahrringchronologie durchschnittlich 57 (10-150)-fach belegt," in *Sitzungsberichte der Akademie der Wissenschaften, Mathematisch-Naturwissenschaftliche Klasse Abteilung 1*, 178 (1970), 37-42.

14 Friedrich, Michael et al., "The 12,460-Year Hohenheim Oak and Pine Tree-Ring Chronology from Central Europe: A Unique Annual Record for Radiocarbon Calibration and Paleoenvironment Reconstructions," *Radiocarbon* 46.3 (2004): 1111-1122.

15 Leuschner, Hans H., and Fritz H. Schweingruber, "Dendroökologische Klassifizierung und Auswertung häufig auftretender intraannueller holzanatomischer Merkmale bei Eichen und Kiefern," *Dendrochronologia* 14 (1996): 273-286.

16 Glaser, *Klimageschichte Mitteleuropas.*

Switzerland,[17] the Czech Republic,[18] Hungary,[19] the Netherlands,[20] England,[21] Scandinavia,[22] Spain,[23] Italy,[24] Greece,[25] as well as in the Levant,[26] in Turkey and in North Africa.[27] These are supplemented by countless individual regional studies.[28]

---

17 Pfister, *Wetternachhersage*; Wetter, Oliver et al., "The Largest Floods in the High Rhine Basin since 1268 Assessed from Documentary and Instrumental Evidence," *Hydrological Science Journal* 56.5 (2011): 733-758. doi:10.1080/02626667.2011.583613.

18 Brázdil, Rudolf et al., *Historické a současné povodně v České republice* [Historical and Recent Floods in the Czech Republic], *Historie počasí a podnebí v českých zemích* [History of Weather and Climate in the Czech Lands], vol. 7 (Brno, Prague: Masaryk University and Czech Hydrometeorological Institute, 2005), 370; Dobrovolný, Petr et al., "Monthly and Seasonal Temperature Reconstructions for Central Europe Derived from Documentary Evidence and Instrumental Records since AD 1500," *Climatic Change* 101.1-2 (2009): 69-107. doi:10.1007/s10584-009-9724-x.

19 Kiss et al., "An Experimental 392-Year Documentary-Based Multi-proxy (Vine and Grain) Reconstruction".

20 Engelen, Aryan F. van, Jan Buisman, and Frank Ijnsen, "Reconstruction of the Low Countries Temperature Series AD 764-1998," in *International Conference on Climate Change and Variability—Past, Present and Future*, ed. Mikami Takehiko (Tokyo: Tokyo Metropolitan University, 2000), 151-157; Engelen, Aryan F. van, Jan Buisman, and Frank Ijnsen, "A Millennium of Weather, Winds and Water in the Low Countries," in *History and Climate: Memories of the Future?*, ed. Philip D. Jones et al. (New York: Kluwer Academic/Plenum Publishers, 2001), 101-124.

21 Manley, Gordon, "Central England Temperatures: Monthly Means 1659 to 1973," *Quarterly Journal of the Royal Meteorological Society* 100 (1974): 389-405.

22 Leijonhufvud, Lotta et al., "Five Centuries of Winter/Spring Temperatures in Stockholm Reconstructed from Documentary Evidence and Instrumental Observations," *Climatic Change* 101.1-2 (2010): 109-141. doi:10.1007/s10584-009-9650-y.

23 Barriendos, Mariano, and Fernando S. Rodrigo, "Study of Historical Flood Events of Spanish Rivers Using Documentary Data," *Hydrological Sciences* 51.5 (2006): 765-783.

24 Camuffo, Dario, "History of the Long Series of Daily Air Temperature in Padova (1725-1998)," *Climatic Change* 53 (2004): 7-75. doi:10.1023/A:1014958506923; Camuffo, and Bertolin, "The Earliest Temperature Observations."

25 Grove, Jean M., *The Little Ice Age* (London, New York: Routledge, 1988).

26 Vogt, Steffen et al., "Assessing the Medieval Climate Anomaly in the Middle East: The Potential of Arabic Documentary Sources," PAGES *Newsletter* 19.1 (2011): 28-29.

27 Touchan, Ramzi et al., "May–June Precipitation Reconstruction of Southwestern Anatolia, Turkey during the Last 900 Years from Tree Rings," *Quaternary Research* 68.2 (2007): 196-202; doi:10.1016/j.yqres.2007.07.001; Touchan, Ramzi et al., "Spatiotemporal Drought Variability in Northwestern Africa Over the Last Nine Centuries," *Journal of Climate Dynamics* 37.1-2 (2010): 237-252. doi:10.1007/s00382-010-0804-4.

28 Munzar, Jan, "Weather Patterns in Czechoslovakia during the Years 1588-1598," in *European Climate Reconstructed from Documentary Data: Methods and Results*, ed. Burkhard Frenzel, Christian Pfister and Birgit Gläser, *Paläoklimaforschung*, vol. 7 (Stuttgart, New York: Fischer, 1992), 51-56; Munzar, Jan, "Early 17th Century Weather in Prague after

Multi-proxy studies compare and combine data originating from various sources in order to draw valid conclusions about the past climate. The regional differentiation of the proxy data frequently also enables interpolation over large distances.[29] For example, air pressure and precipitation fields for the years since 1500 were reconstructed from the time series available from across Europe.[30] Spatially integrated analysis allows assessing variations in large scale modes of natural climatic variability such as the North Atlantic Oscillation (NAO),[31] furthering our understanding of the dependencies between regional weather events and global circulation. At the mesoscale and over the last millennium in particular, historical sources offer the possibility of detailing the past climate with extremely high temporal resolution.

Selected data series for Europe are shown in Figure 3.1[32] and Figure 3.2[33]. Phases of coherent temperature development, i.e. 66 percent of the series

---

J. Kepler's Observations," in *Proceedings of the Regional Workshop on Climate Variability and Climatic Change Vulnerability and Adaptation*, ed. Ivana Nemešová (Prague: Institute of Atmospheric Physics Czech Academy of Sciences, 1996) 69-72.

29 Crowley, Thomas J., and Thomas S. Lowery, "How Warm Was the Medieval Warm Period?" *Ambio* 29.1 (2000): 51-54. doi:10.1579/0044-7447-29.1.51; Esper, Jan et al., "Climate: Past Ranges and Future Changes," *Quaternary Science Reviews* 24.20-21 (2005): 2164-2166. doi:10.1016/j.quascirev.2005.07.001.

30 Luterbacher, Jürg et al., "Reconstruction of Sea Level Pressure Fields over the Eastern North Atlantic and Europe Back to 1500," *Climate Dynamics* 18.7 (2002): 545-561. doi:10.1007/s00382-001-0196-6; Pauling, Andreas et al., "Five Hundred Years of Gridded High-Resolution Precipitation Reconstructions over Europe and the Connection to Large-Scale Circulation," *Climate Dynamics* 26 (2006): 387-405. doi:10.1007/s00382-005-0090-8.

31 Trouet, Valérie et al., "Persistent Positive North Atlantic Oscillation Mode Dominated the Medieval Climate Anomaly," *Science* 324.5923 (2009): 78-80. doi:10.1126/science.1166349.

32 Glaser, and Riemann, "A Thousand-Year Record of Climate Variations;" Pfister, Christian, and Urs Dietrich-Felber, "Euro-Climhist: A Data-Base on Past Weather and Climate in Europe and its Human Dimension," accessed November 7, 2013. <http://www.euroclimhist.com>; Leijonhufvud, et al. "Five Centuries of Winter/Spring Temperatures in Stockholm"; van Engelen, A.F., Buisman and Ijnsen, "Reconstruction of the Low Countries Temperature Series AD 764-1998"; Kilar, Piotr, "500 Year Poland Temperature and Precipitation Indices," in *IGBP PAGES/World Data Center for Paleoclimatology Data Contribution Series* (Boulder: National Climatic Data Center, 2010), 2010-2135. Also shown is the reconstructed North Atlantic Oscillation (NAO) index according to Cook et al. The data series are normalised 31-year moving averages. Cook, Edward R., "Multi-Proxy Reconstructions of the North Atlantic Oscillation (NAO) Index: A Critical Review and a New Well-Verified Winter NAO Index Reconstruction Back to AD 1400," in *The North Atlantic Oscillation: Climatic Significance and Environmental Impact*, ed. James W. Hurell et al. (Washington D.C.: American Geophysical Union, 2013), 63-79.

33 Wetter, Oliver, and Christian Pfister, "Spring–Summer Temperatures Reconstructed for Northern Switzerland and Southwestern Germany from Winter Rye Harvest Dates,

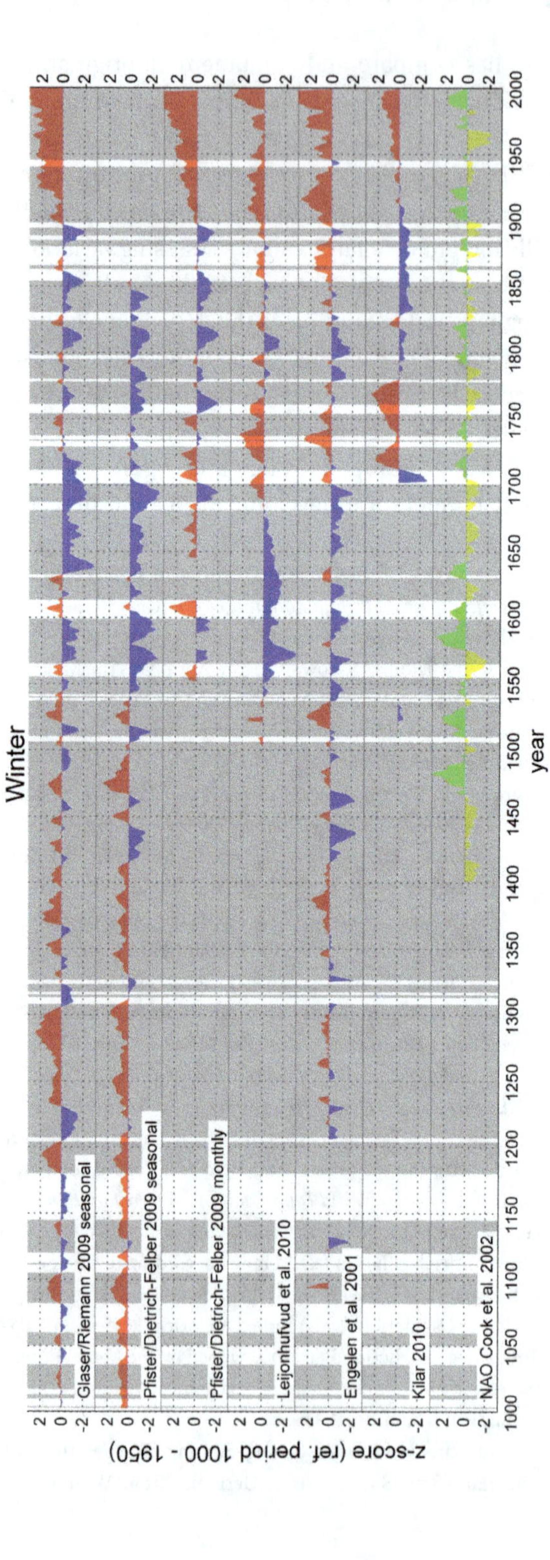

FIGURE 3.1 *Comparison of various winter temperature reconstructions*

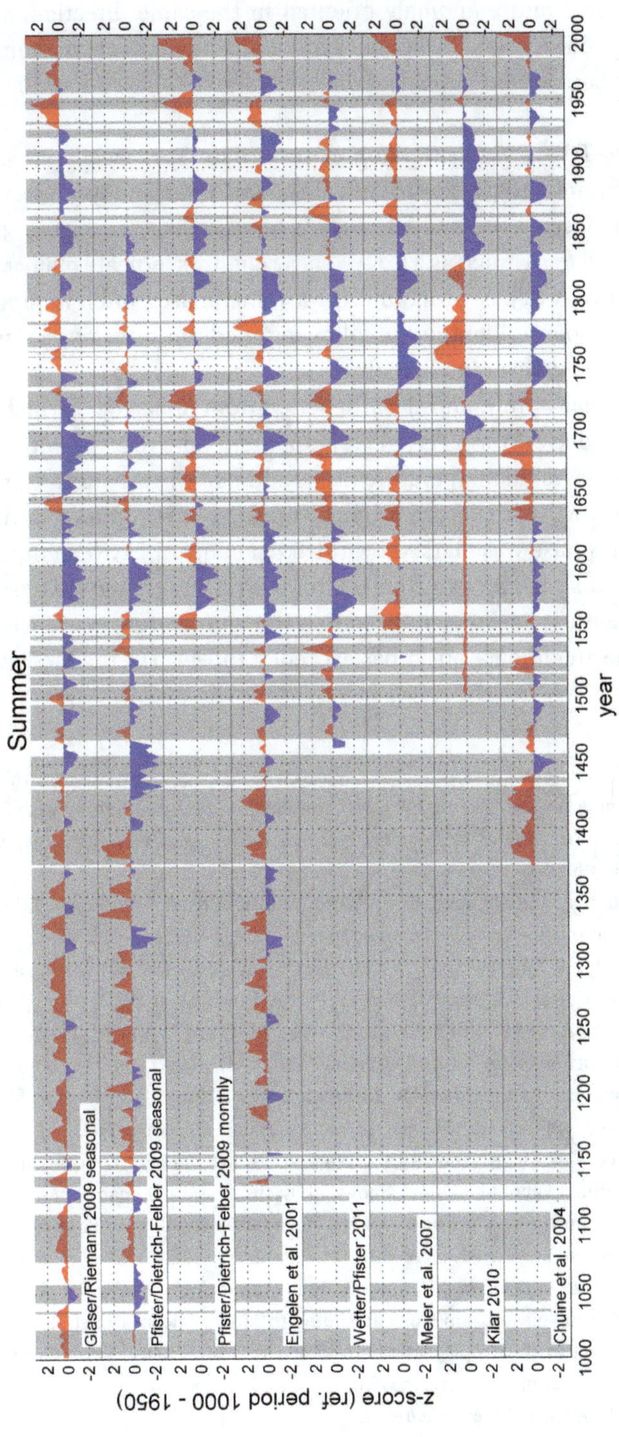

FIGURE 3.2 *Comparison of various summer temperature reconstructions (without the NAO)*

indicate a temperature anomaly oriented in the same direction, are shown with a grey background. Overall, very good agreement is demonstrated between the series, which clearly depict the climate phases described below.

3.1 *Long-Term Climate Development since the Year 1000*

The temperature progression of the last one thousand years reveals pronounced changes to the annual average temperature (cf. Figure 3.3[34]) that amount overall to as much as 2.5°C.[35] These changes indicate many rapid temperature drops as well as pronounced warming phases, often in rapid succession. It is possible to identify the following medium-term phases, which are also supported by the natural archive:

1. An especially pronounced warm period from 1100 to 1300, which could be called the "Medieval Warm Period". It is apparent here that the entire period is characterised by rapid temperature drops and rises. This pattern stands out in particular in the 14th century, when it occurred many times during a few decades. This phase was marked by a striking contrast between hot summers and cold to harsh winters. In comparison with the temperature development of the 20th century, it can be noticed that the temperature rise since the 1970s driven by the greenhouse effect has exceeded the level of the Medieval Warm Period.

1454-1970," *Climate of the Past* 7.4 (2011): 1307-1326. doi:10.5194/cp-7-1307-2011; Meier, Nicole et al., "Grape Harvest Dates as a Proxy for Swiss April to August Temperature Reconstructions Back to AD 1480," *Geophysical Research Letters* 34.20 (2007): L20705. doi:10.1029/2007GL031381.

34 Glaser, Rüdiger and Riemann, Dirk, "A Thousand-Year Record of Climate Variations." The eleven-year moving average is marked in red. Annual values are also depicted as of 1500. The uncertainty range is shown in grey. Also shown are the forcings: methane (green) and carbon dioxide (black) according to Etheridge, David M. et al., "Ice Core, Firn Air and Archived Air Atmospheric Methane Concentration Data," in *IGBP PAGES/World Data Center for Paleoclimatology Data Contribution Series #2002-039* (Boulder: National Climatic Data Center, 2002); and Etheridge, David M., et al., "Historical CO_2 Records from the Law Dome DE08, DE08-2, and DSS Ice Cores," in *Trends: A Compendium of Data on Global Change*, ed. Oak Ridge National Laboratory, U.S. Department of Energy (Oak Ridge: Oak Ridge National Laboratory, 1998). Shown above in blue is the change in solar irradiance resulting from the changed atmospheric chemistry according to Bard, Edouard, et al., "Reconstructed Solar Irradiance Data," IGBP PAGES/World Data Center for Paleoclimatology Data Contribution Series #2003-006 (Boulder: National Climatic Data Center, 2003) as well as due to volcanic emissions according to Mann, Michael E., et al., "Volcanic and Solar Forcing of the Tropical Pacific over the Past 1000 Years," *Journal of Climate* 18.3 (2005): 447-456. doi:10.1175/JCLI-3276.1.

35 Glaser, *Klimageschichte Mitteleuropas.*

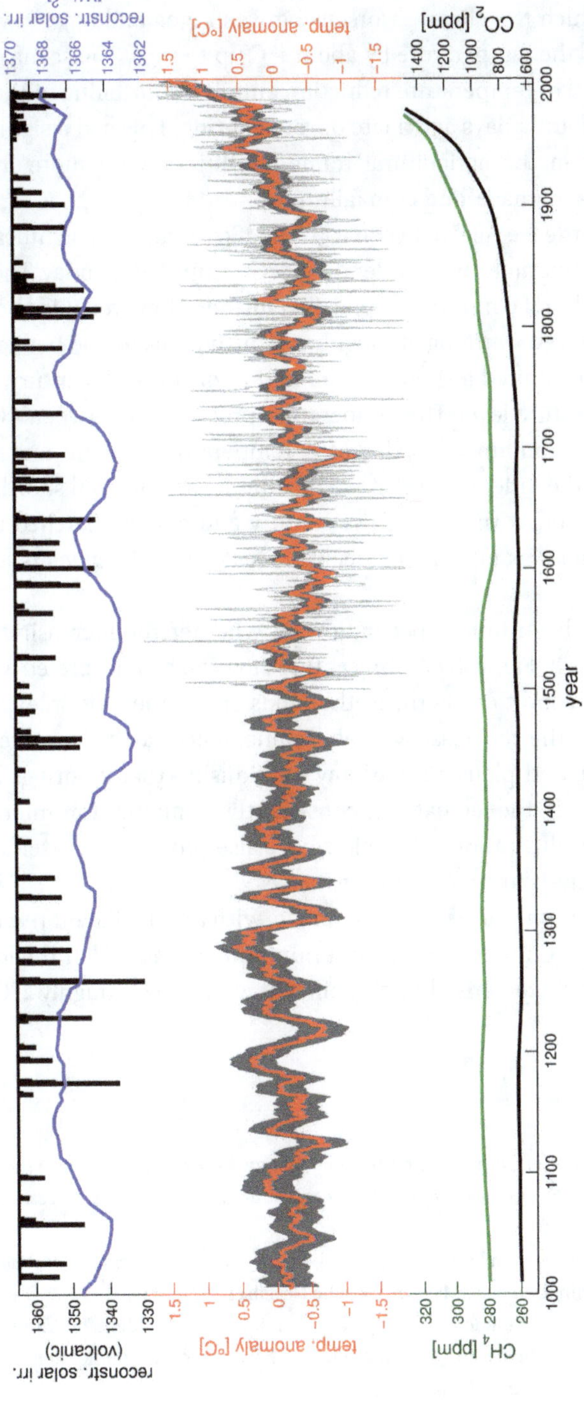

FIGURE 3.3 *Temperature reconstruction according to Glaser and Riemann for the last thousand years in Central Europe as an eleven-year moving average*

2. A transition period, from a global perspective often named the "AD 1300 Event", in which the climate worsens between 1300 and 1550,[36] with a steady reduction of the temperature by about 1°C up to 1550. The scenario indicated by this period of temperature reduction entails a shortening of the vegetation period by about 14 days and a late onset of spring. This had a significant negative impact on the agricultural frame conditions with more frequent bad harvests, frost damage, and crop failures.

3. The "Little Ice Age" between 1550 and 1850 as another significant negative temperature anomaly by an order of −1.5°C compared to today. The worsening of the agricultural conditions intensified further. However, it is also worth noting a pronounced warming of roughly 1°C during this period between 1700 and 1800, which exhibited a similar scope to that of the 20th century, even if at a lower temperature level. The temperature progression between 1800 and 1900 shows three pronounced cycles of a temperature decline and subsequent warming on the order of up to 1°C. The minima are situated shortly after 1800, around 1850 and at the end of the century and can be matched up to major glacier advances as long as the delayed reaction of the glaciers is taken into account.

There has been much speculation on whether the increasing number of winter images in Flemish painting at this time can be interpreted as a response to the colder conditions in the Netherlands or whether it is an expression of a genre shift. At the very least, a notable coincidence can be identified, although the existence and plausibility of any relationship that paintings and images may have with the climate is consistently a matter for interpretation.[37] Nevertheless, illustrations of glaciers have also proved to be useful subjects of scientific analysis in a number of instances.[38]

4. The modern warm period as of 1900, with the significant rise in temperature as of 1970. Compared to the last minimum at the end of the 19th century, this period is characterised by a significant warming of roughly 2°C.

---

36 For example, cf. Nunn, Patrick D., *Climate, Environment and Society in the Pacific during the Last Millenium. Developments in Earth and Environmental Sciences*, vol. 6 (London: Elsevier, 2007).

37 Wehry, Werner, and Franz Ossing, ed., *Wolken – Malerei – Klima in Geschichte und Gegenwart* (Berlin: Deutsche Meteorologische Gesellschaft, 1997).

38 Brunner, Kurt, "Klimaentwicklungen in alten Karten und Bildern," in *Natur und Mensch in Mitteleuropa im letzten Jahrtausend*, ed. Bayerische Akademie der Wissenschaften, (München: Friedrich Pfeil, 2007), 32.

4 Long-Term Climatic Development, Social Discourse, and Food Supply

Studies on the relationship and interactions between the climate and society have a long tradition and reflect in particular the fundamental developments of scientific disciplines. Issar and Zohar have visualised the shift from the (natural) deterministic paradigm of the 19th and early 20th century toward the anthropogenic paradigm as a pendulum that has been returning roughly to the middle point since the 1980s.[39] Environmental processes are once again 'on equal footing' with economic and political frameworks when it comes to evaluating the success and failure of societies. Emphasis is still regularly placed on one perspective or the other.[40] Vulnerability and resilience research has contributed in particular to a holistic understanding. For example, Kates can be held up as one of the protagonists who has applied himself to the relationship between climate and society.[41] On a historical time scale, a number of works have now been presented that generally address the impacts of exceptional climatic phases on societies.[42] The approaches refer to individual years ("1816—the year without summer"), short series of years, or longer phases such as the Little Ice Age, and focus largely on their agricultural–sociological consequences as well as their impact from a cultural–historical perspective.

4.1 *Conceptual Framework for the Assessment of Climate Vulnerability*

Risk and vulnerability approaches are particularly well suited for depicting the complex interactions between the environment and societies. For instance, the contribution of Working Group II of the Intergovernmental Panel on Climate Change (IPCC) to the Fifth Assessment Report of the IPCC comprehensively discusses climate risk and vulnerability.[43] Methodologically, the

39 Issar, Arie S., and Mattanyah Zohar, *Climate Change—Environment and History of the Near East* (Berlin, Heidelberg, New York: Springer, 2007).

40 Diamond, Jared M., *Collapse: How Societies Choose to Fail or Succeed* (New York: Viking Press, 2005); Nunn, *Climate, Environment and Society.*

41 [41] Kates, Robert W., "The Interaction of Climate and Society," in *Climate Impact Assessment. Studies of the Interaction of Climate and Society*, ed. Robert W. Kates, Jesse H. Ausubel and Mimi Berberian, *ICSU/SCOPE Report* 27 (Chichester: Wiley, 1985), 3-36.

42 Pfister, Christian, *Klimageschichte der Schweiz 1525-1860: Das Klima der Schweiz von 1525-1860 und seine Bedeutung in der Geschichte von Bevölkerung und Landwirtschaft* (Bern, Stuttgart: Haupt, 1988); Behringer, Wolfgang, Hartmut Lehmann, and Christian Pfister, ed. *Kulturelle Konsequenzen der Kleinen Eiszeit—Cultural Consequences of the Little Ice Age* (Göttingen: Vandenhoeck & Ruprecht, 2005); Nunn, *Climate, Environment and Society.*

43 Intergovernmental Panel on Climate Change (IPCC) *Climate Change 2014: Impacts, Adaptation, and Vulnerability* (Cambridge, New York: Cambridge University Press, 2014).

approach chosen here is based on vulnerability and resilience concepts detailed by Bohle, Gunderson et al., Redman, Turner et al., Watts and Bohle, and Wisner et al., and is diagrammed in Figure 3.4.[44] The following assumptions and causalities are taken as a starting point: Climatic stressors (hazards) such as temperature changes, ocean level rises, storm surges, droughts and floods have changed considerably in multidimensional ways in the last one thousand years.[45] Perturbations and interventions produce varying effects in dependence on the geo-ecological parameters and their exposure and sensitivity to these stressors. This biophysical vulnerability is coupled with regional socio-economic vulnerability. Based on various socio-economic settings and assets, specific options were available to individuals, groups and ultimately whole societies and led to specific response mechanisms. The various actors strove to maintain their own livelihood and security in order to retain or improve their standard of living by means of immediate, short-term changes in behaviour (coping), such as resowing after winter kills, as well as medium- to long-term strategies (adaptations), such as changing the crop rotations or the types of crops cultivated. In a historical context, food security was a key

44 Bohle, Hans-Georg, "Geographische Entwicklungsforschung," in *Geographie*, ed. Hans Gebhardt et al. (Heidelberg: Spektrum, 2007), 797-815; Gunderson, Lance H., and Crawford S. Holling, ed., *Panarchy: Understanding Transformations in Human and Natural Systems* (Washington, D.C.: Island Press, 2002); Redman, Charles L., "Resilience in Archaeology," *American Anthropology* 107.1 (2005): 70-77: doi:10.1525/aa.2005.107.1.070; Turner, Billie L. et al., "A Framework for Vulnerability Analysis in Sustainability Science," *Proceedings of the National Academy of Sciences* 100.14 (2003): 8074-8079. doi:10.1073/pnas.1231335100; Watts, Michael J., and Hans-Georg Bohle, "The Space of Vulnerability: The Causal Structure of Hunger and Famine," *Progress in Human Geography* 17.1 (1993): 43-67. doi:10.1177/030913259301700103; Wisner, Ben et al, *At Risk: Natural Hazards, People's Vulnerability and Disasters* (London: Routledge, 2004).

45 Brázdil, Rudolf et al., *Historické a současné povodně v České republice* [Historical and Recent Floods in the Czech Republic] (Brno, Prague: Masaryk University and Czech Hydrometeorological Institute, 2005); Brázdil, Rudolf, Zbigniew W. Kundzewicz, and Gerardo Benito, "Historical Hydrology for Studying Flood Risk in Europe," *Hydrological Sciences Journal* (*Special Issue: Historical Hydrology*) 51.5, (2006): 739-764. doi:10.1623/hysj.51.5.739; Glaser, *Klimageschichte Mitteleuropas*; Glaser, Rüdiger, Heiko Stangl, and Michel Lang, "Crues en Europe Centrale depuis l'an 1300 et dans leur contexte regional," *La Houille Blanche* 5 (2004): 43-49. doi:10.1051/lhb:200405005; Glaser, Rüdiger, Christoph Beck, and Heiko Stangl, "Zur Temperatur- und Hochwasserentwicklung der letzten 1000 Jahre in Deutschland," in DWD *Klimastatusbericht 2003*, ed. Deutscher Wetterdienst (Offenbach: Deutscher Wetterdienst, 2004), 55-67; Himmelsbach, Iso et al., "Flood Risk along the Upper Rhine since AD 1480," *Hydrology and Earth System Sciences Discussions* 12.1 (2015): 177-211, accessed July 10, 2015. doi:10.5194/hessd-12-177-2015.

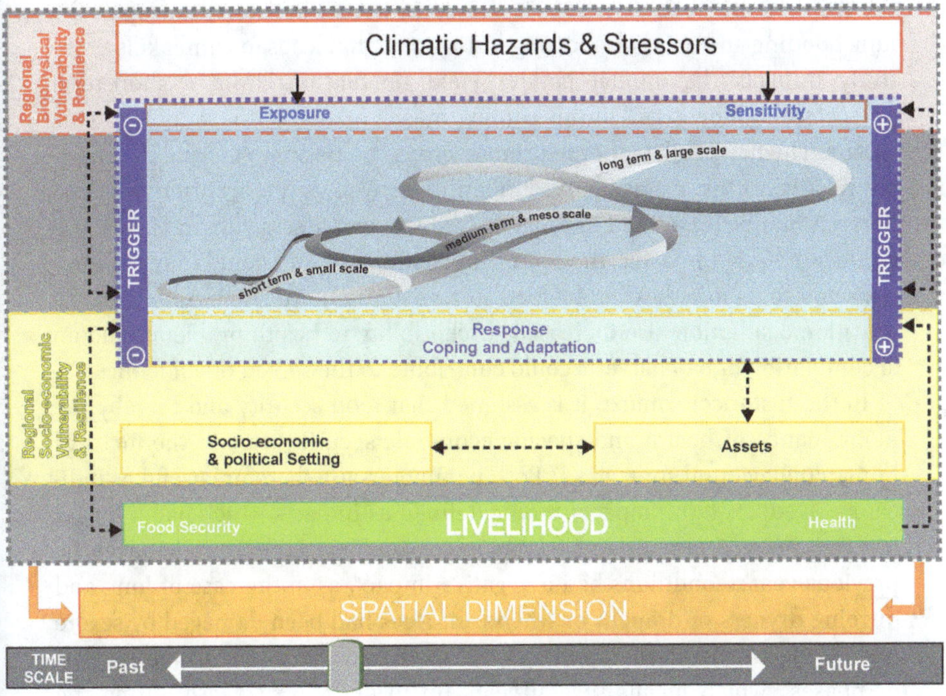

FIGURE 3.4 *Schematic representation of the conceptual framework* (AUTHORS' DESIGN)

priority, if not the most fundamental one of all.[46] The coupling of biophysical and socio-economic vulnerability is characterised by multi-scale response mechanisms and processes.

The shift from terp settlements to the construction of dikes as well as the fluctuations observed in the settling of low mountain ranges within the framework of inland colonisation and the subsequent desertion and reclamation of these areas can be listed as examples of especially fundamental indicators of these types of changes. Consideration must also be given to structural changes and shifts in agriculture.

4.2 *The Relationship between Climate Development and Food Security*

Harvest data and weather-related crop failures and the resulting hunger crises can be extracted from societal archives as additional material for the evaluation of short-term annual stresses. In the central European context 'severe and unfavourable weather' such as storms and, in particular, hail, single frost

46 Pfister, Christian, *Klimageschichte der Schweiz* 1525-1860.

events, extreme heat, dryness and droughts, excess moisture, downpouring rain, flooding and extreme and long-lasting cold that leads to winter kills negatively impacts the annual harvest yield fluctuations from a short-term perspective. "Short-term" in this context means related to the actual growing season. The harvest data for the primary crops can be analysed, and the shortfall relative to the average harvest quantity as well as the occurrence of bad harvests can be quantified in order to evaluate and assess vulnerability. Crop failures directly impacted the food supply situation and could only be compensated to a limited extent by food storage and imports. A worsening of the nutritional situation also increased susceptibility to health problems, and an accumulation of bad harvests could contribute to the spread of epidemics.

In the historical context, it is assumed that food security and thereby the maintenance of health and specific economic security were at the focus of human behavioural strategies.[47] Both parameters are therefore useful as indicators. Medium-term impacts can be defined as impacts which had a longer lasting influence onto the actual growing season. This is the case whenever the previous year's conditions had a negative impact, as in the case of long and ongoing dryness or drought or when fruit trees had been damaged by severe frost. Most cases of the medium scale are related to cold or dry or wet "periods or phases", which means several years up to decades, likewise during the Spörer- or Maunder-Minimum. A sequence of wet years, a period (of several years) characterized by long-lasting dryness and droughts, or increasing numbers of chilly winters are also examples of these mid-term variations. It is quite obvious that there cannot be in every case a clear distinction between these scales, but the medium-term category helps to structure the climatic stressors—and societal reaction and behaviour.

Long-term changes can be attributed to climatic change itself, especially to temperature and precipitation change. Long-term in this sense means changes in relation to a 31-year "normal period". The causal chain of such fundamental changes leads to sea level rise, for example, or to an significant extension or reduction of the vegetation period and growing season, and has had a long-term impact on the harvest situation and thereby on food security and food

47 Cf. Abel, Wilhelm, *Agrarkrisen und Agrarkonjunktur* (Hamburg, Berlin: Parey, 1978); Buszello, Horst, "Teuerung und Hungersnot am Ober- und Hochrhein im Spätmittelalter und in der frühen Neuzeit (circa 1300-1800)," in *Kriege, Krisen und Katastrophen am Oberrhein vom Mittelalter bis zur Frühen Neuzeit* (Schopfheim: Geschichtsverein Markgräflerland e.V, 2007), 32-71; Pfister, *Das Klima der Schweiz von 1525-1860*.

supply as well, as detailed by numerous studies in connection with historical climate change.[48]

Long-term responses and strategies consisted of agricultural–sociological changes, alterations to cultivation systems, shifts in cultivated crops and, above all, 'spatial variations' such as expanding the cultivated area and reclamation or desertion of areas. The great significance of population development notwithstanding, many long-term spatial consequences and mechanisms can be described in connection with climatic changes.

Within the regional context, large differences in harvest yields can be quantified in some cases. For example, the yield ratio of seeds sown to harvested material for the year 1803 can be found in the von Leiningen archive in Amorbach based on a complete survey carried out in 1803 (figure 3.5[49]). This survey shows that significantly lower yields were obtained from the poor soil conditions of the Odenwald woods with podzolic brown soils than in the neighbouring Bauland, where luvisols over loess are found. The somewhat higher elevation in the Odenwald and the overall lower temperature level negatively impacted the yields compared with the lower Bauland and Taubergrund region situated in the lee of the mountain range with a higher mean temperature level. In terms of vulnerability, the people in the Odenwald were more susceptible to climatic influences than those in Bauland.

To illustrate the long-term relationship between climate development and yields, high-resolution local harvest yield data extracted from accounting ledgers were aggregated over a large area and depicted as long-term yield data series. These data were then overlaid onto the temperature progression for the

48 Brückner, Eduard, "Der Einfluß der Klimaschwankungen auf die Ernteerträge und Getreidepreise in Europa," *Geographische Zeitschrift* 1.1/2 (1895): 39-51; 100-108; Glaser, Rüdiger, *Klimarekonstruktion für Mainfranken, Bauland und Odenwald anhand direkter und indirekter Witterungsdaten seit 1500, Paläoklimaforschung*, vol. 5 (Stuttgart, New York: Fischer, 1991); Hildebrandt, Helmut, and Martin Gudd, "Getreidebau, Missernten und Witterung im südwestlichen unteren Vogelsberg und dem angrenzenden Vorland während des 16. und frühen 17. Jahrhunderts," in *Archiv für hessische Geschichte*, vol. 49 (Darmstadt: Historischer Verein für Hessen e. V., 1991), 85-147; Pfister, *Klimageschichte der Schweiz*; Camenisch, Chantal, et al., "The early Spörer Minimum – a Period of Extraordinary Climate and Socio-Economic Changes in Western and Central Europe," *Climate of the Past Discussions* 12 (2016): 1-33, accessed April 4, 2017. <http://www.clim-past-discuss.net/cp-2016-7/cp-2016-7.pdf>.

49 Glaser, *Klimarekonstruktion für Mainfranken.* Poor soil conditions are found in the Odenwald woods with podzolic brown soils (green). Better conditions can be found on primarily luvisols on loess (brown).

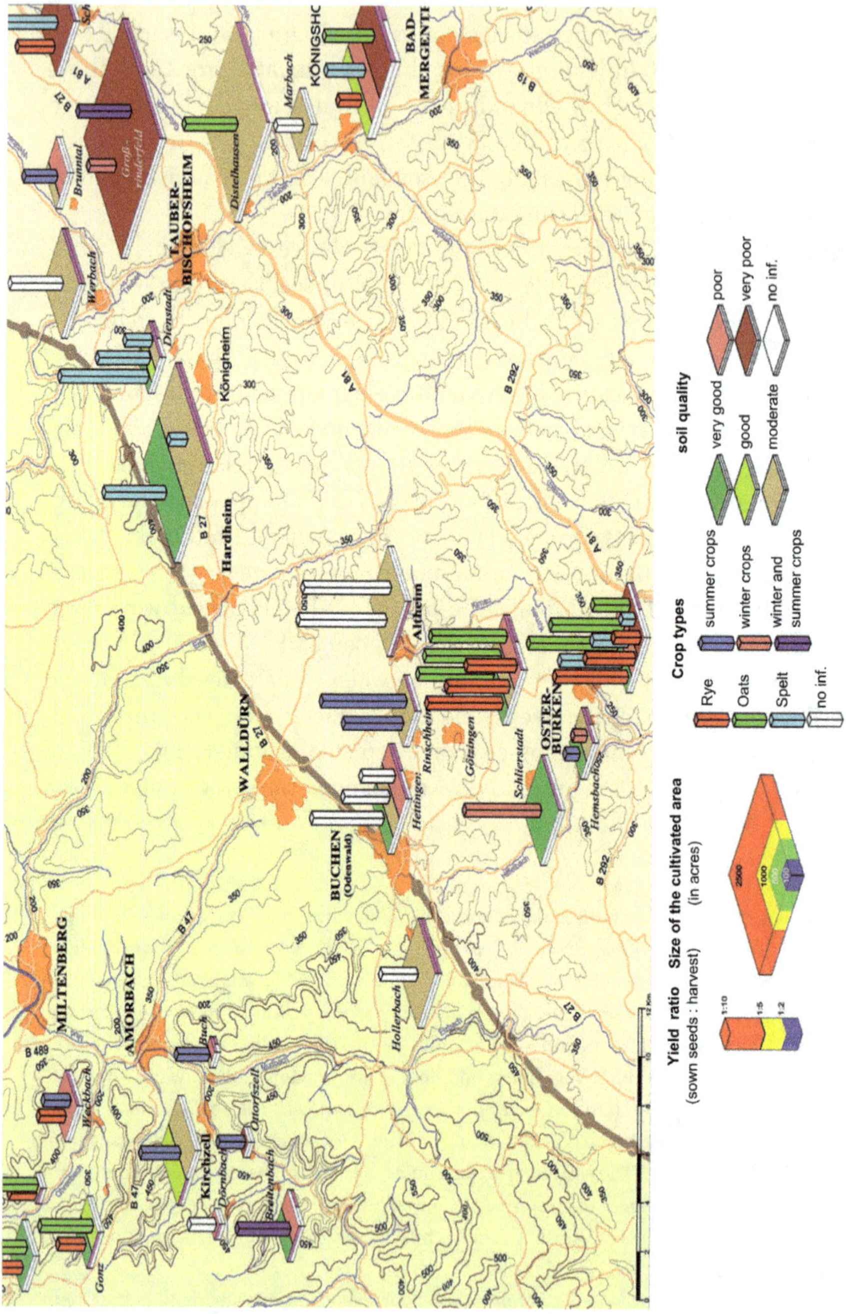

FIGURE 3.5 *Yield ratios in Odenwald and Bauland, 1803*

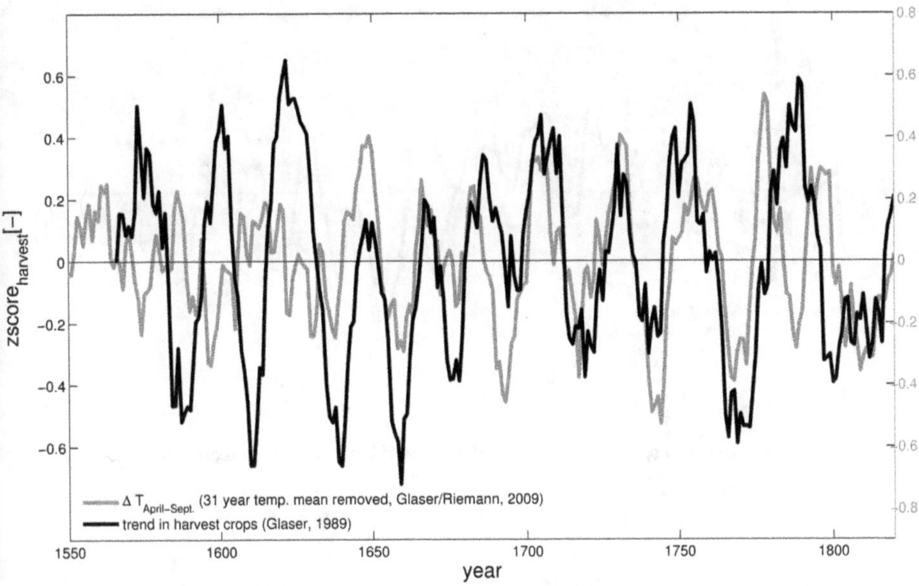

FIGURE 3.6 *Comparison of average temperatures in April–September (31-years detrended) together with the changes in harvest yields*

period of April–September, which is considered the main growing period (cf. figure 3.6[50]).

The graph clearly reveals phases of high congruency, i.e. consistent developments, as well as phases that are offset in time. Particularly apparent is the high correlation between the two series that set in after the Thirty Years' War, i.e. as of 1648, and remained stable until the period of industrialisation. Prior to the Thirty Years' War, the phases tend to be opposing or offset. These findings shall not be discussed further here but offered as a starting point for further research.

In order to further clarify the long-term course of the food supply situation as a base element of food security, the negative consequences of a shortage were considered. Price development, in particular inflation, and reports of famines were investigated as corresponding parameters. For better visualisation, these factors were overlaid with the long-term temperature development (cf. figure 3.7[51]).

50 Harvest data from: Glaser, *Klimarekonstruktion für Mainfranken.*

51 Glaser, and Riemann, "A Thousand-Year Record of Climate Variations"; Buszello, "Teuerung und Hungersnot"; Abel, *Agrarkrisen und Agrarkonjunktur.*

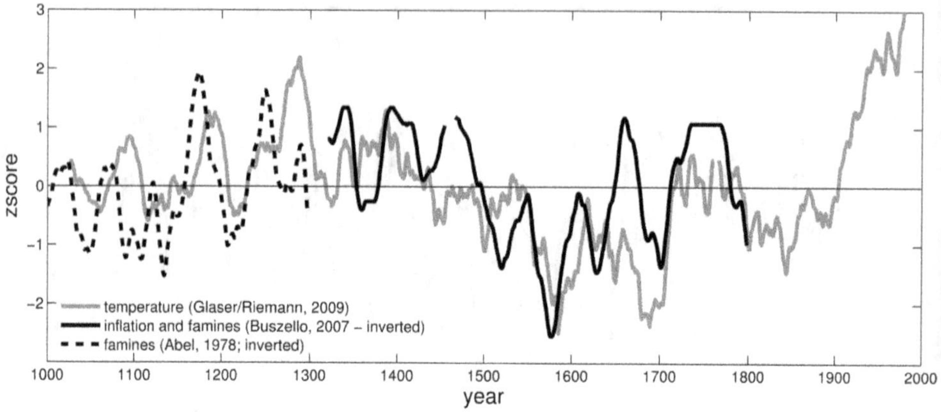

FIGURE 3.7 *Comparison of the frequency of inflation and famines with temperature developments*

Similar to the above comparison of temperature development and yields, extensive empirically verified agreement can be identified between these parameters as well. Over the last one thousand years, these phases of agreement predominate over the periods that exhibit a time offset or opposing trends. Based on the empirically evaluated parameters employed, it can be generally concluded that temperature developments have a wide-ranging impact on yields and, in this context, on pricing and the occurrence of inflation and famine. Without lapsing into the climate deterministic mode of thought, this can at least be postulated as an empirically justified working hypothesis in need of further specification within the regional context.

4.3 *Climate, Vulnerability, and Emigration 1815-1847: Three Case Studies for Short-Term Interactions*

The abovementioned concept of climate vulnerability can be applied on a smaller spatial and temporal (short-term) scale. The following three case studies investigate into the relation of climate and emigration from the Kingdom of Württemberg and the Grand Duchy of Baden in the first half of the 19th century.

Obviously, there is a high correlation between numbers of emigrants and the development of grain prices in the 19th century (cf. figure 3.8[52]), when

52 Shown here are official migration statistics for the Kingdom of Württemberg, taken from Hippel, Wolfgang von, *Auswanderung aus Südwestdeutschland: Studien zur württembergischen Auswanderung und Auswanderungspolitik im 18. und 19. Jahrhundert* (Stuttgart:

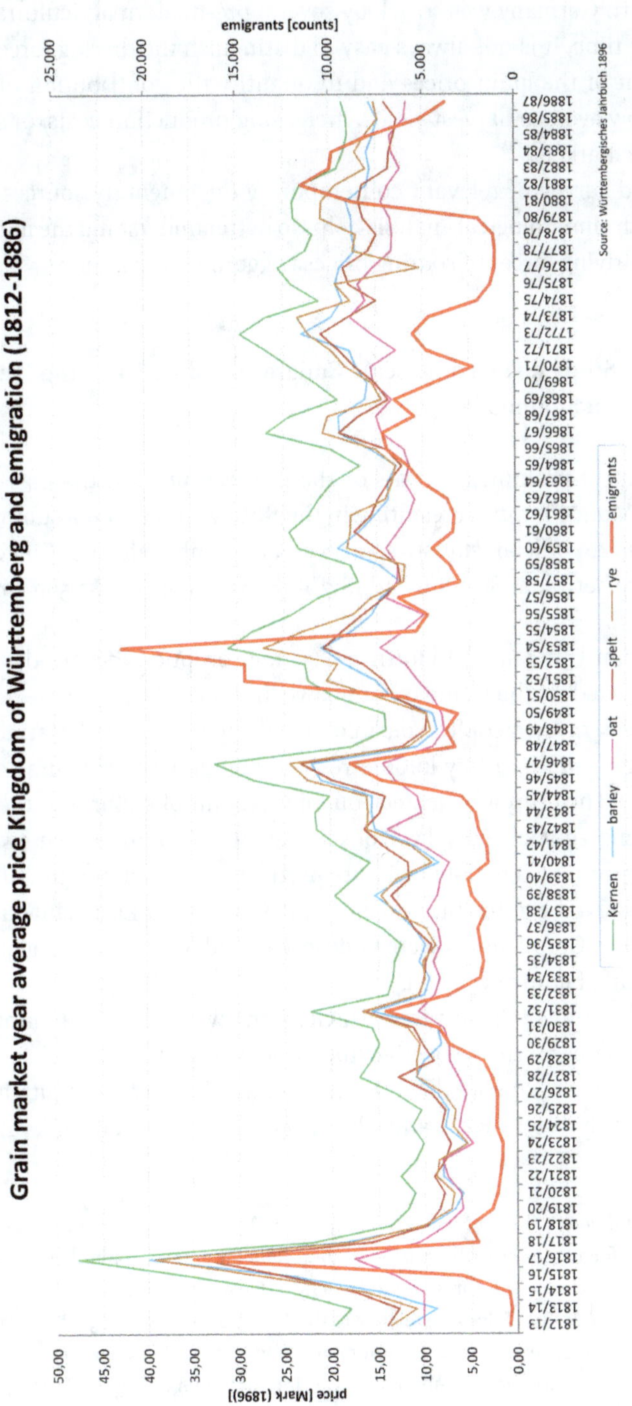

FIGURE 3.8 *Comparison of grain prices and number of emigrants 1812–1886*

south-western Germany was struck by several pre-modern agricultural crises. At the same time, it is not always easy to distinguish the driving forces of the development of the grain prices and to quantify the contribution of factors such as below average harvest yields, increasing production costs, or speculation in grain markets.

A detailed survey of relevant contemporary documentary sources in conjunction with climatological analysis helps to better understand the underlying causes and driving forces of soaring prices triggering migration.

5 Case Study 1: The Large-scale Emigration of 1816/17—the "Year without Summer"

One of the most prominent events of the 19th century was the massive 1815 eruption of Mount Tambora resulting in the 1816 "year without a summer".

According to official statistics, 22,630 emigrants left the Kingdom of Württemberg between 1815/1816 and 1817/1818, with a peak of 17,500 emigrants in 1816/1817.

Between April 5th and April 10/11th, 1815, huge eruptions occurred at Mount Tambora (located on the Island of Sumbawa, Indonesia) causing the deaths of approximately 71,000 people. Volcanic ashes expelled into the atmosphere reached heights of up to fifty kilometres and amounted to a volume of more than 150 km^3. The eruption ejected roughly sixty mt of sulfur into the stratosphere forming a global aerosol veil obscuring the sun for more than two years. This caused pronounced global climate perturbations. Temperature decrease has been observed e.g. for the Northeast of the USA, in East Canada and in Central Europe. Crop failures were widespread and as a consequence, subsistence crises and famine occurred.[53]

At those locations in south-western Germany which feature homogenised temperature series for the respective time period, it can be shown that the year 1816 was on average one of the coldest years of the century (Stuttgart 1816: 8.10°C, with only 1829: 7.83°C and 1879: 8.09°C being colder; Karlsruhe 1816:

Klett-Cotta, 1984); Güll, R., "Auswanderung aus Württemberg und Baden," Statistisches Monatsheft Baden-Württemberg 9 (2013): 41-48, and the annually published Württembergische Jahrbücher (WJB). For the Grand Duchy of Baden official statistics have been published only as of 1850. Prices are relative to the base year 1896 (WJB 1896, II.120-121).

53 Oppenheimer, Clive, "Climatic, Environmental and Human Consequences of the Largest Known Historic Eruption: Tambora Volcano (Indonesia) 1815," *Progress in Physical Geography* 27.2 (2013): 230-259. doi:10.1191/0309133303pp379ra.

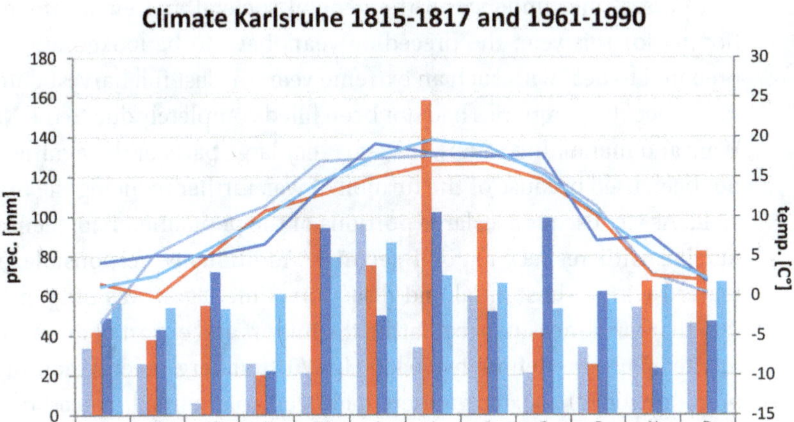

	J	F	M	A	M	J	J	A	S	O	N	D
1815 prec [mm]	34	26	6	26	21	96	34	60	21	34	54	46
1816 prec [mm]	42	38	55	20	96	75	158	96	41	25	67	82
1817 prec [mm]	49	43	72	22	94	50	102	52	99	62	23	47
1961-1990 prec [mm]	56,5	54,1	53,3	60,8	79,3	86,5	70,1	66,1	53,3	58,3	65,3	66,8
1815 temp [C°]	-3,2	5,3	8,8	11,3	16,9	17,3	17,6	18,0	15,9	11,2	2,7	0,4
1816 temp [C°]	1,5	-0,1	5,2	10,7	12,7	14,9	16,5	16,6	14,6	10,0	2,5	1,9
1817 temp [C°]	4,2	5,1	5,0	6,5	13,5	19,1	17,9	17,5	17,8	6,9	7,4	1,9
1961-1990 temp [C°]	1,2	2,5	6,0	9,9	14,3	17,5	19,6	18,8	15,4	10,4	5,3	2,2

FIGURE 3.9 *Temperature and precipitation for Karlsruhe, 1815-1817*

8.92°C, with only 1805, 1864, 1871 and 1879 being colder at 8.31°C; Bale 1816: 7.2°C, with 1805 and 1879 featuring the same lowest annual means in air temperature).[54]

The year 1816 is characterized not only by a below-average annual mean air temperature, but more importantly by the unfavourable distribution of temperature and precipitation during the growth period and the harvest period. Figure 3.9 shows this effect for air temperature and for the distribution and amount of precipitation in Karlsruhe.

For both countries, harvest yields for 1816 are described as "völlig missraten" (a bad failure).[55]

---

54 Data taken from KNMI-Explorer for station numbers: Stuttgart: WMO station code: 10739; Karlsruhe: WMO station code: 10727; Basel: Near WMO station code: 6642.1, accessed July 3, 2015. <http://climexp.knmi.nl/>.

55 For the Grand Duchy of Baden, cf. Stieffel, Philip, *Witterungskunde: Mit Rücksicht auf vermuthliche Witterung überhaupt und des Jahres 1842 insbesondere* (Karlsruhe: Verlag der

To assess and understand the general societal and environmental conditioning for this year, the preceding years have to be looked at. No one was prepared to deal with such an extreme year: the last full harvest dated back to 1812; since then granaries had not been filled completely due to the Napoleonic Wars and marauding troops. It seems that large parts of the agrarian land had not been used because of the turmoil of war, further reducing the total harvest yield. As a consequence, large portions of the population had been highly vulnerable with respect to food security. Additionally, responsible politicians reacted with a substantial and disastrous time lag: it was only in November 1816 that tariff measures regulating the import and export of grain were implemented. This was definitely too late: In Württemberg, several thousand bushels of grain worth 1,400,000 guilders had been sold abroad instead of supplying domestic markets. It was even too late for the government-initiated additional purchase of grain on foreign markets in October. It was not possible to deliver the urgently required goods into Württemberg or Baden before wintertime. Transport was severely blocked by fast freezing rivers with unfreezing taking place only late in the following spring. But it was not only the grain harvest that suffered from the adverse climate conditions. Forage production also failed and in addition, the wet and cold weather led to massive animal disease. Hence, farmers sold their livestock in large quantities before wintertime. Initially, this caused a substantial decline of prices in the meat market, which then turned into the opposite in the following spring.[56]

The municipalities, which had been in charge of the provisioning of the poor, were now forced to buy large amounts of grain from the royal granaries and, consequently, to increasingly run up debts. This led to tax debts amounting to 8,975,783 guilders in the Kingdom of Württemberg alone. Between November 1816 and January 1817, the government of Württemberg decreed regulations to set up so-called "Wohltätigkeitsvereine" (benevolent societies). The Grand Duchy of Baden followed in March 1817 with the founding of the "Allgemeiner Wohltätigkeitsverein für das Großherzogtum Baden zur Bewältigung der Auswirkungen des Krieges und der Mißernte" (Benevolent Society for the Grand Duchy of Baden to Overcome the Impacts of War and of Harvest Failures) under the auspices of the Grand Duchess Stephanie (1789-1860).

Chr. Fr. Müller'schen Hofbuchhandlung, 1842), 41; For the Kingdom of Württemberg, cf. Dürr, Friederich, *Heilbronner Chronik* (Heilbronn: Verlag Eugen Salzer, 1895), 322-324.

56 Gehlinger, G., "Überblick über die Entwicklung der Landwirtschaft in Württemberg seit der Mitte des 18. Jahrhunderts," *Württembergische Jahrbücher* 1897, I. 49-76.

In April 1817, large parts of the governmental staff were obliged to become compulsory members of the benevolent societies in Württemberg, and export tariffs were further increased.[57]

In Baden, income from tariffs had been re-allocated to the suffering municipalities in February 1817. In June 1817, both countries prohibited the sale of the coming grain harvest "auf dem Halm" (on the stem, i.e. sale of growing crops before the harvest), and in Württemberg, a census of all grain in stock took place, in the hope of triggering a decline in grain prices by demonstrating that there are sufficient provisions until the coming harvest.[58]

The development of grain prices in the year 1817 can also be shown with regard to the changing weight of bread (see figure 3.10). It was still common to sell certain types of bread by fixed prices. Consequently, in times of dearth and inflation, it was not the price that changed but the weight of the bread, and in case municipal committees intervened, to set the minimum weight of bread.

A comparison of Freiburg (Baden) and Riedlingen (Württemberg) clearly shows, that in Freiburg between February and September on average, residents could get more bread for the same amount of money, but the committee had to intervene already in May in order to avoid further decrease in bread weight. In contrast to the intervention in Freiburg, the committee in Riedlingen refrained from actively controlling the bread market. It acted only in mid-June, when a substantial increase in weight from thirty grams to more than sixty grams was decreed, at a point in time when there were sufficient indications that the coming harvest would not fail.

The example of the "year without a summer" shows that in Central Europe, at the beginning of the 19th century a combination of several stressors must have occurred which caused a severe crisis and triggered emigration. Harvest failure mainly caused by adverse climatic conditions alone should not have provoked a famine.[59] The preceding years of war had led to inadequately stocked granaries, there were great difficulties in getting substitution from foreign markets, and bad governance including delayed implementation of measures failed to provide relief. Another climate condition—the early

57 Concerning this and other measures taken in the Kingdom of Württemberg, cf. *Das Regierungs-Blatt für das Königreich Württemberg im Auszuge*, vol. 1 (1806-1816), and vol. 2 (1817-1824) (Stuttgart: 1840 and 1841).

58 For the Grand Duchy of Baden, cf. *Großherzoglich-Badisches Regierungsblatt*, vol. 14, and vol. 15 (Karlsruhe: 1809-1868).

59 For the Grand Duchy of Baden, cf. Heunisch, A.J. von, *Das Grossherzogthum Baden, historisch-geographisch-statistisch-topographisch beschrieben. Mit Beigaben von J. Bader* (Heidelberg: Verlag der Julius Groos'schen Universitätsbuchhandlung, 1857), 323; for the Kingdom of Württemberg cf. *Württembergische Jahrbücher*, vol. 1 (Stuttgart: 1818), 4-6.

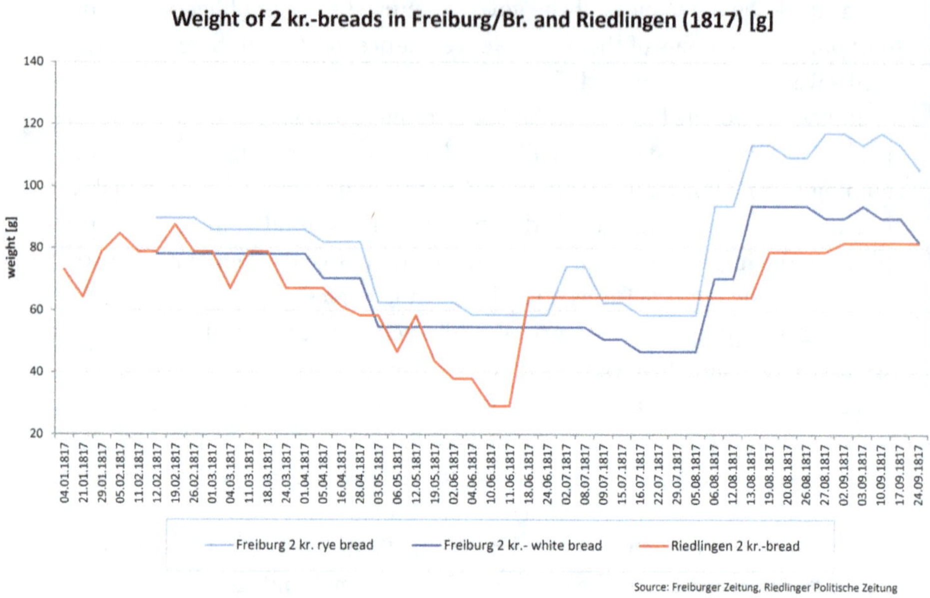

FIGURE 3.10 *Weight of 2-kr. bread in Freiburg/Br. and Riedlingen, 1817*

freezing of rivers—blocked the transportation system and further hampered the import of grain. In conclusion, the main reason for the crisis and famine of 1816 was the harvest failure in large parts of Europe due to the adverse climate conditions. One could argue that of all subsistence crises and famines triggering large-scale migrations in the 19th century, 1816 certainly is the one where weather and climate played a most decisive role.

6 Case Study 2: The Large-scale Emigration of 1829-1833

Between 1829 and 1833, the official number of emigrants from the Kingdom of Württemberg reached 18,650, with a peak of 7,066 emigrants in 1831/1832 (see figure 8). Regarding the climatic conditions, those years are not particularly anomalous, with the notable exception being the winter of 1829/1830. This winter featured 112 ice days in Stuttgart and 92 ice days in Karlsruhe between October 1829 and April 1830, and was one of the longest and coldest winters since 1788.[60]

---

60 *Württembergische Jahrbücher*, 1830, 2-4. On February 2nd, the cold in Stuttgart and Tübingen reached -32°C and in Tuttlingen even -35°C. Even in 1788, temperatures had not been as low.

The harvests of 1830 and 1831 in Baden were described as "mangelhaft" (deficient), whereas in Württemberg, the harvest was qualified as "mittelmäßig" (fair) and "noch ziemlich ergiebig" (still quite productive).[61] In 1832, an astonishing quick soaring of prices occurred, for which even contemporaries could not see any connection to harvest expectations and yields. Instead, speculation of wholesale and intermediary vendors was held responsible for the development in grain prices.[62] Nevertheless, the situation in Stuttgart got so much worse that soup kitchens had to be opened to relieve the poor.

Although the onset of this large-scale emigration coincides with one of the coldest winters in decades, and—at least in Baden—the harvest of the following years 1830 and 1831 had been insufficient, documentary sources point out that the main reason for the soaring grain prices triggering emigration was to be found in market speculations rather than in harvest failure.

7 Case Study 3: The Large-scale Emigration of 1846/1847

In the Kingdom of Württemberg, the number of emigrants had increased already since the year 1843/1844. According to official statistics, 16,244 persons emigrated between 1845 and 1847 (see figure 8).

Regarding the climatic conditions of these years, a generally wet summer in 1845, followed by an extensive period of rain in June and an exceptionally dry summer in 1846, are noticeable (see figure 3.11).

Harvest yields for the years 1843 to 1845 in the Kingdom of Württemberg for winter wheat are described as "nicht so ergiebig" (not so productive), but for spring wheat as "den Erwartungen entsprechend" (according to expectations).[63] In the Grand Duchy of Baden, harvest yields are labelled "mittlere Ernte" (average yield), in 1844 the harvest was "reich und gut" (rich and good) and in 1845 the weather had been "sehr fruchtbar" (very fertile). It was noted that the weather had been more favourable than in the preceding year, therefore "kleinmüthige Klagen" (fainthearted complaints) over the rain period in June could

61 Cf. Stieffel, *Witterungskunde*, 44-45; Pfaff, Karl, *Geschichte der Stadt Stuttgart nach Archival-Urkunden und anderen bewährten Quellen. 2. Teil* (Stuttgart: Sonnewald, 1846), 345-347. Note that the term "Eistag" (ice day) was used in the sense of what is nowadays commonly defined as "frost day", i.e. a day where air temperature drops below 0°C at least once a day, whereas "ice days" commonly are defined as days where air temperature never rises above 0°C during the whole day.

62 Pfaff, *Geschichte der Stadt Stuttgart*, 351. For the Grand Duchy of Baden, cf. *Freiburger Zeitung*, May 2, 1832 (Tagesausgabe).

63 Cf. *Württembergische Jahrbücher*, 1843-1845.

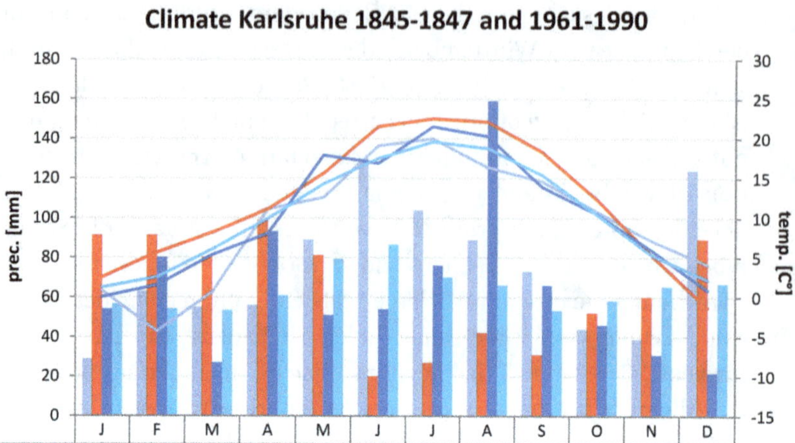

	J	F	M	A	M	J	J	A	S	O	N	D
1845 prec [mm]	29	63	55	56	89	130	104	89	73	44	39	124
1846 prec [mm]	91	91	80	99	81	20	27	42	31	52	60	89
1847 prec [mm]	54	80	27	93	51	54	76	159	66	46	31	22
1961-1990 prec [mm]	56,5	54,1	53,3	60,8	79,3	86,5	70,1	66,1	53,3	58,3	65,3	66,8
1845 temp [C°]	0,8	-4,3	0,6	11,1	12,6	19,1	20,0	16,4	14,6	10,8	7,1	4,1
1846 temp [C°]	2,5	5,6	8,1	11,1	15,6	21,5	22,5	22,1	18,3	12,3	5,3	-1,3
1847 temp [C°]	0,0	1,5	5,3	8,0	17,9	16,9	21,5	20,3	13,9	10,5	5,9	0,9
1961-1990 temp [C°]	1,2	2,5	6,0	9,9	14,3	17,5	19,6	18,8	15,4	10,4	5,3	2,2

FIGURE 3.11 *Temperature and precipitation for Karlsruhe, 1845-1847*

not be understood.[64] However, according to other reports for 1845, the grain harvest yield was less than on average.[65]

In 1846 in the Grand Duchy of Baden, overall harvest yields had been good only for oat and barley, the results for wheat and rye were considered poor, and average for spelt (the most important bread grain).[66] In Württemberg, both winter and spring wheat harvests resulted only in mediocre yields.[67] Mainly areas characterised by unfavourable edaphic and climatic conditions had been affected by insufficient yields, such as the Swabian Jura, the Odenwald, or the Black Forest.

During the year 1845, late blight—the potato disease that was the proximate cause for the great famine in Ireland—appeared for the first time both in

64 Großherzoglich-Badisches Landwirtschaftliches Wochenblatt, July/August 1844, Großherzoglich-Badisches Landwirtschaftliches Wochenblatt, July/August 1845.

65 Fleischmann, Sigmund, *Die Agrarkrisis von 1845-1855 mit besonderer Berücksichtigung von Baden* (Heidelberg: Karl Rössler, 1902), 9.

66 Ibid., 20.

67 *Württembergische Jahrbücher*, 1846.1, 41-43.

Württemberg and Baden.[68] The effects of late blight on the potato yield increased the pressure on grain prices from autumn 1845. Regarding tariffs on grain, both Baden and Württemberg had completely lost freedom of action after joining the Deutsche Zollverein (German Customs Union). For both countries, it was hardly possible to implement measures to restrict the export of grain, because the other members of the union denied such plans. Only a ban on the export of potatoes could be implemented.[69]

In the following spring, potatoes started to grow well, and as a consequence the export ban was released. However, the dry summer of 1846 generally was unfavourable to the growth of fruit. Additionally, the late blight spread into hitherto unaffected areas of Württemberg. And in Baden, again, those areas with unfavourable edaphic and climatic conditions such as the Odenwald and the Black Forest suffered from potato harvest failures.[70]

In the municipalities, which were still responsible for the relief of the poor, soup kitchens were opened already in the autumn of 1846.[71]

The grain price increase reached its peak between April and May 1847. From the beginning of May, riots caused by hunger occurred in the cities of Ulm, Tübingen, and Stuttgart (Kingdom of Württemberg), and also in Mannheim and Villingen (Grand Duchy of Baden).[72] At this point the governments of both countries again initiated a census of existing food stock in both countries to show that that there were sufficient provisions and to calm down the markets in order to level out prices at least until the next harvest.[73] These measures, together with further regulations, were successful in both countries.

The vast impact of the potato disease can only be understood considering the fact that in areas with partible inheritance, the increasing number of factory workers to a great extent also acted as part-time farmers. These subsistence-economy activities in horticulture and agriculture on small inherited plots played a crucial role to ensure the food security of those workers. The potato was widely distributed because the plant was well suited for such activities due to the relatively small expenditure of work and the relatively modest

68 Cf. Herrmann, Bernd, "Kartoffel, Tod und Teufel," in *"... mein Acker ist die Zeit": Aufsätze zur Umweltgeschichte* ed. Bernd Hermann (Göttingen: Universitätsverlag, 2011), 293-347.

69 Großherzoglich-Badisches Landwirtschaftliches Wochenblatt, July/August 1845, 13-15.

70 For these areas, dedicated collections of the benevolent societies took place from 1847 (see e.g. *Freiburger Zeitung*, May 14, 1847).

71 Cf. e.g. *Freiburger Zeitung*, May 18, 1847.

72 Dürr, *Heilbronner Chronik*, 343; Brüning, Rainer, and Peter Exner, ed., *Wege aus der Armut: Baden in der ersten Hälfte des 19. Jahrhunderts* (Karlsruhe: Generallandesarchiv, 2007), 13.

73 Freiburger Zeitung, May 7, 1847, 858; Regierungsblatt für das Königreich Württemberg, May 10, 1847, 175.

edaphic requirements. As a consequence, the potato harvest failure resulted in huge pressure on grain prices; not only in the cities but also in villages and rural areas.[74]

The large-scale emigration of the years 1846/1847 in Württemberg and Baden is closely connected to the fluctuations in grain prices. Unfavourable climatic conditions in 1845 (wet June) and 1846 (dry summer) lead to a modest grain harvest. The situation was exacerbated by the spreading of late blight and the following great losses in the potato harvest. The two effects in combination increased vulnerability towards climatic stressors in substantial parts of the population in both countries in urban and rural areas, and directly affected their food security.

8 Case Studies: Summary

In 1816/1817, only *one* severe harvest failure, which was clearly triggered by climatic stressors and subsequent adverse climatic conditions that effectively blocked the transport system for bulk commodities, was sufficient to cause famine and large-scale emigration. The crisis was worsened by an already increased vulnerability of major parts of society due to the preceding Napoleonic wars and bad governance during the crisis. Many adapted to the crisis by emigrating.

The beginning of the emigration during the years 1831/1832 is marked by the extremely long and cold winter of 1829/1830. Harvest yields for the Duchy of Baden were characterised as "insufficient", for the Kingdom of Württemberg at least as "fair". But the less-than-average yields stimulated market speculations, which led to soaring grain prices that severely affected food security in the more vulnerable parts of society. Again, emigration was an option that thousands chose.

For the emigration of the years 1846/1847, climatic stressors appear less prominent (wet early summer in 1845, dry summer in 1846). But the below-average harvest yields, in combination with another biophysical stressor—the late blight, which was severely affecting potato harvest yields—had a profound impact on food security of both the urban and rural population.

Summing up the case studies, one can conclude that the increase in the number of emigrants is closely related to the increase of grain prices in all

74 Cf. Megerle, Klaus, *Württemberg im Industrialisierungsprozeß Deutschlands: Ein Beitrag zur regionalen Differenzierung der Industrialisierung* (Stuttgart: Klett-Cotta, 1982), especially 215-217.

three cases. However, the reasons for the development of grain prices varied and the effects of the grain market on the food security of substantial portions of the population depended on specific combinations of climatic, biophysical, and socio-economic stressors. The relative importance of climatic stressors varies from case study to case study, with most prominent case in 1816/1817.

The stressors manifested themselves on different spatial scales in different ways, triggering different coping and adaption strategies. The three case studies clearly show that climate vulnerability was a very sensitive to context. With food security at risk, in the first half of the 19th century substantial parts of the more vulnerable part of the population in South-western Germany opted for emigration as an adaption strategy.

9 Summary

The climate developments over the last millennium can be reconstructed with increasing precision and differentiation on the basis of the combined study of natural and societal archives. Multi-proxy approaches aggregate the results and methods of various avenues of research in order to capitalise on the respective strengths.

Societal archives available for Europe represent an outstanding source of information. Historical climatology enables the quantitative reconstruction of climatic conditions and extremes over time, thereby significantly expanding the extent of knowledge concerning the climate. In addition to individual representations, climate time series for temperature and precipitation are now available. As of the year 1500, the pressure, temperature and precipitation data fields can be reconstructed for wide areas of Europe. These data are relevant to the current climate discussion in that it is possible to present, for example, the longer-term comparison data on climate extremes and their variations.

The results show that extreme climatic events occurred in all periods. The population was repeatedly surprised by heat waves and droughts, frost periods and heavy precipitation. In many regions, isolated floods significantly exceeded the "hundred-year flood" of the previous decade. A glance at the long data series makes the high variability apparent. In some river regions, our ancestors were subjected to a higher flood risk. This perspective casts quite a different light on many 'unheard of climate disasters' or 'self-inflicted floods'.

Climate developments can be placed in a broad social context within the framework of risk and vulnerability analyses. It is assumed that people focused on strategies to protect their livelihoods and primary needs, in other words on food security. In accordance with this perspective, the temperature data were compared with long-term harvest yield data and information on price developments as well as famines in order to empirically determine whether and to what extent a correlation exists at this mesoscale level. Such a correlation can be extensively confirmed: The long-term harvest yield developments as well as associated negative consequences such as hunger and price instability or inflation generally exhibit a high correlation with temperature developments over the last 1,000 years. On the other hand, phases of time offsets and opposing developments are also discernible, especially in the case of the harvest yields. Fundamentally, however, it is possible to infer a high climate-induced impact on the observed parameters of 'harvest yield, inflation and hunger'.

The extent to which this relationship is connected with regional and, in particular, short-term events can only be determined in a more extensive, higher-resolution study. This is especially true of the question of whether such relationships also stimulated waves of emigration and what part such movements played alongside other strategies for overcoming the associated difficulties. Emigration can be seen as an outlet of a multifold, complex stressor system. There is a clear underlying reaction chain, starting with unfavorable weather conditions, resulting in harvest failure and leading to high prices, which triggered emigration. But this reaction chain went along with bad governance, especially bad management of the crises, economic speculation, and the lack of perspective of the people, who suffered from hunger and diseases.

CHAPTER 4

History and Climate: The Crisis of the 1590s Reconsidered

*Geoffrey Parker**

Abstract

In 1985, a book entitled *The European Crisis of the 1590s: Essays in Comparative History*, edited by Peter Clark, examined the experience of ten individual regions of Western Europe, eight of them ruled either by Philip II or by his principal enemies: Elizabeth Tudor, Henry IV of France, and the Dutch. Although the individual authors noted specific disasters, most concluded that the 1590s were merely one of the cyclical crises that afflicted premodern societies. Since then, the publication of data on the global climate reveals that the 1590s saw some of the worst weather ever recorded in the northern hemisphere, a severe episode in the "Little Ice Age"—an era of major volcanic eruptions, reduced solar activity, and multiple El Niño events—linked with an increased frequency of extreme climatic events, plague, famine, and war.

1 Bringing Climate Back In

In 1985, Peter Clark published a pioneering volume of comparative history entitled *The European Crisis of the 1590s*. It included studies by eight historians of the European territories ruled by King Philip II (1556-1598) and his principal enemies—Elizabeth I of England, Henry IV of France, and the Dutch Republic—as well as 'diagnostic' chapters on four common denominators: "Demographic crisis", "Popular disorder", "The impact of war", and "The roles of the state and the town".[1] As often occurs in collections of essays, the various

* I thank Günhan Börekçi for extracting climatic data from Ottoman sources; Matthew Keith for doing the same with Chinese and Japanese sources; and Edward Tenace for drawing the chronicle of Jean Moreau to my attention. I also thank Rayne Allinson, Bruce Campbell, Dagomar Degroot, Steve Hindle, Emmanuel Le Roy Ladurie, Ruth MacKay, Christian Pfister and Sam White for sharing their insights with me.

1 Clark, Peter, ed., *The European Crisis of the 1590s: Essays in Comparative History* (London: George Allen & Unwin, 1985), based on a conference convened by Clark at the University of Leicester four years earlier.

 | DOI 10.1163/9789004356825_006

contributors reached somewhat different conclusions. Thus, although in his "Introduction" Clark asserted that "the 1590s cannot be dismissed simply as one of the periodic mini-crises" of early modern times, in his "Epilogue" John H. Elliott disagreed. "Readers of this volume will be conscious of a vein of scepticism running through it," he averred. "A number of contributors have obvious doubts about the applicability of the notion of crisis to the decade of the 1590s", and "the impression conveyed" by their essays "is less bleak than their general theme might suggest". Indeed, Elliott wondered, "Is there really any case for singling out the 1590s as a decade in which conditions were such as to warrant a special degree of attention?"[2]

Elliott's scepticism reflected two important gaps in the volume's evidentiary base. First, it included only limited parts of Europe (readers will find virtually nothing about Scotland, Ireland, Scandinavia, or anywhere east of the Elbe) and it ignored the experience of other parts of the globe. Second, although mean temperatures in the later 16th century fell by perhaps 2°C, while droughts, floods, and frosts repeatedly decimated or destroyed harvests, the volume lacked a 'diagnostic' chapter on climate.[3] In 1999, the journal *Climatic Change* rectified this omission with a special issue entitled *Climatic Variability in Sixteenth-Century Europe*, containing a dozen articles, many of them documenting the global cooling of the 1590s. When the experience of the territories ruled by Philip II and his principal enemies is set within the broader framework of what historians of climate call the Little Ice Age, the last decade of the 16th century certainly warrants "a special degree of attention".[4]

---

2 Clark, *The European Crisis*, 18; Elliott, John H., "Yet another Crisis?," in *The European Crisis of the 1590s: Essays in Comparative History*, ed. Peter Clark (London: George Allen & Unwin, 1985), 301, 305.

3 Admittedly, Clark conceded that for much of Europe during the 1590s "the most obvious general source of difficulty was bad weather" (Clark, *The European Crisis*, 8-9), while several contributors noted extreme climatic conditions in the regions: ibid., 74-75 (the Dutch Republic), 126-127 (Southern France), 158 (Northern Italy), 199 (Sicily), and 214 (Spain).

4 Pfister, Christian, Rudolf Brázdil, and Rüdiger Glaser, ed., *Climatic Variability in Sixteenth-Century Europe and its Social Dimension* (Dordrecht: Kluwer Academic Publishers, 1999), included all items in the special issue of *Climatic Change* 43 (1999): Pfister, Christian, and Rudolf Brádzil, "Climatic Variability in Sixteenth-Century Europe and its Social Dimension: A Synthesis," *Climatic Change* 43 (1999): 5-53. On the nature of the data see Bradley, Raymond S., *Paleoclimatology: Reconstructing Climates of the Quaternary* (San Diego: Academic Press, 1999), 441-470; Pfister, Christian, Rudolf Brázdil, and Manuel Barriendos, "Reconstructing Past Climate and Natural Disasters in Europe Using Documentary Evidence," *Pages News* 10.3 (2002): 6-8; Pfister, Christian, "Weeping in the Snow: The Second Period of Little Ice Age-Type Impacts, 1570-1630," in *Kulturelle Konsequenzen der "Kleinen Eiszeit"*, ed. Wolfgang Behringer, Harmut Lehmann and Christian Pfister (Göttingen: Vandenhoeck & Ruprecht, 2005), 31-86;

Christian Pfister, the foremost living historian of climate, disagreed: he noted an unusual run of cool summers in Europe at the end of the 16th century, caused "by subnormal pressure over large parts of Central Europe and by a weak and remote Azores anticyclone". He continued: "At present, such situations usually last for a week, in severe cases up to a month. In the late sixteenth century, summers were predominantly cool and rainy for no less than thirteen years."[5] Specifically, Central Europe suffered an "uninterrupted sequence of eight cool summers from 1591 to 1598" that "may have been unique in the last 500 years". Some Alpine glaciers reached their most advanced positions in historical times around 1600: the Lower Grindelwald glacier (the most closely studied Alpine glacier because of its proximity to human settlements) advanced no less than a kilometre between 1580 and 1602.[6] Between 1591 and 1596 the Eastern Mediterranean suffered the longest and most intense drought recorded in the past six centuries, as well as some of the coldest winters. In 1604, at Florence, "the winter set in so fast that the River Arno froze over so hard" that people first walked and then skated from one side to the other.[7]

---

and Brázdil, Rudolf, and Christian Pfister, "Historical Climatology in Europe—the State of the Art," *Climatic Change* 70 (2005): 363-430.

5 Pfister, Christian, "Climatic Extremes, Recurrent Crises and Witch Hunts: Strategies of European Societies in Coping with Exogenous Shocks in the Late Sixteenth and Early Seventeenth Centuries," *The Medieval History Journal* 10.1-2 (2007): 52.

6 Briffa, Keith R. et al., "European Tree-Rings and Climate in the 16th Century," *Climatic Change*, 43 (1999): 166; Landsteiner, Erich, "The Crisis of Wine Production in Late 16th-Century Central Europe: Climatic Causes and Economic Consequences," *Climatic Change* 43 (1999): 325; Pfister, and Brádzil, "Climatic Variability," 24, 30; and Zumbühl, Heinz J., *Die Schwankungen der Grindelwaldgletscher in den historischen Bild- und Schriftquellen des 12. bis 19. Jahrhunderts* (Basel: Birkhäuser Verlag AG, 1980). According to Glaser, Rüdiger, *Klimarekonstruktion für Mainfranken, Bauland und Odenwald* (Stuttgart: Fischer, 1991), 111 (figure 30b), the 1590s saw the coolest summers recorded between 1500 and 1800 in central Germany.

7 Anonymous, *Relazione delle feste fatte in Fiorenza sopra il ghiaccio del fivme d'Arno l'vltimo dì di dicembre MDCIV* (Florence, 1604); Touchan, Ramzi, "Reconstruction of Spring/Summer Precipitation for the Eastern Mediterranean from Tree-Ring Widths and its Connection to Large-Scale Atmospheric Circulation," *Climate Dynamics* 25 (2005): 75-98; Touchan, Ramzi, "Preliminary Reconstructions of Spring Precipitation in South-Western Turkey from Tree-Ring Width," *International Journal of Climatology* 23 (2003): 157-171; Luterbacher, Jurg, and Elena Xoplaki, "500-Year Winter Temperature and Precipitation Variability over the Mediterranean Area and its Connection to the Large-Scale Atmospheric Circulation," in *Mediterranean Climate: Variability and Trends*, ed. Hans-Jürgen Boehle (Heidelberg: Springer Verlag, 2003), 133-153; and, more recently, Roberts, Neil et al., "Palaeolimnological Evidence for an East–West Climate See-Saw in the Mediterranean Since AD 900," *Global and Planetary Change* 84-85 (2012): 23-34; and Kaniewski, David, Elise Van Campo, and Harvey Weiss, "Drought Is a

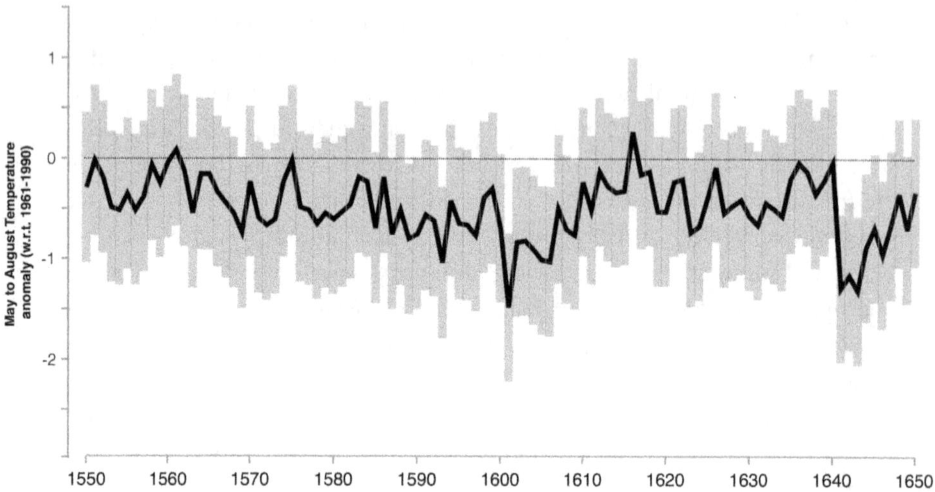

FIGURE 4.1 *Estimated northern hemisphere summer temperature anomalies, 1550-1650*

A recent "ensemble reconstruction" of the global climate over the past millennium revealed a unique alignment of cooler temperatures in both the northern and southern hemispheres between 1594 and 1677. Sam White has concluded that between 1580 and 1610, "the Northern hemisphere cooled at its fastest rate in at least the past 500 years". The 'unsmoothed' estimates of annual temperatures derived from tree-ring data from around the world north of 20°N show cool growing seasons in 1591-1593, 1595-1597 and particularly 1601-1602—the coldest summer recorded during the past six centuries.[8]

These climatic abnormalities reflected three natural phenomena: a reduction in solar energy; a spate of major volcanic eruptions; and a greater frequency of El Niño episodes. Reduced deposits of those trace elements in the Arctic icecap that normally correlate with the number of sunspots (and therefore with the solar energy received on earth), suggest a 'sunspot minimum' in the 1590s. The same icecap evidence also reveals major volcanic eruptions in low latitudes in 1588-1590, 1595-1597, and 1600-1603; while the record of dust

Recurring Challenge in the Middle East," *Proceedings of the National Academy of Sciences* 109 (2012): 3862-3867.

8 Neukom, Raphael, et al, "Inter-hemispheric Temperature Variability over the Past Millennium," *Nature Climate Change* 4 (May 2014): 362-367; White, Sam, *A Cold Welcome. The Little Ice Age and Europe's Encounter with North America* (Cambridge: Harvard University Press, 2017), chapter 1; Briffa, Keith R., and Timothy J. Osborn, "Blowing Hot and Cold," *Science* 295 (2002): 2227-2228, attached data.

deposited by storms at Summit, Greenland, likewise shows high deposits in 1588, 1589, 1593, 1595, 1597, 1602, and 1603. According to Pfister, "considering the number and density, this series of [volcanic] explosions seems to be unique within the last millennium".[9] Each eruption hurled sulphur dioxide into the stratosphere, where it deflected some of the sun's radiation back into space, further reducing temperatures in all areas of the earth beneath the dust clouds (which included the most densely populated areas of the globe).

The linkage between volcanic activity and global cooling is particularly clear after the eruption of Huaynaputina in Peru in February and March 1600, "one of the largest eruptions of historical time," which "buried at least ten villages, claimed about 1500 lives, and affected the overall economy of south Peru". Its 'global footprint' includes sulphate acid spikes in both Greenland and Antarctica; narrow tree rings in both Northern Europe and North America; descriptions and drawings of the "reddish and faint" sun dimmed by a constant haze (in Iceland the summer sunlight in 1601 was so faint that it cast no shadows); and vivid written complaints about extreme weather.[10] To take a single example of the latter, according to English soldiers on campaign in Ireland, some days during the winter of 1600-1601 were so "dark and bitter" that "looking, [we] could not see"; "in all the time of my service in this land, I have not

9 Pfister, "Weeping in the Snow," 82; Pustil'nik, Lev, and Gregory Yom Din, "On Possible Influence of Space Weather on Agricultural Markets: Necessary Conditions and Probable Scenarios," *Sun and Geosphere* 4 (2009): 35-43, figure 5. See also: Mayewski, Paul. A. et al., "Major Features and Forcing of High-Latitude Northern Hemisphere Atmospheric Circulation Using a 110,000-Year-Long Glaciochemical Series," *Journal of Geophysical Research* 102 (1997): 26345-26366; Tingley, Martin P., and Peter Huybers, "Recent Temperature Extremes at High Northern Latitudes Unprecedented in Past 600 Years" *Nature* 496 (2013): 201-205; and Sigl, Michael, et al., "Timing and Climate Forcing of Volcanic Eruptions for the Past 2,500 Years", *Nature* 533 (2015): 543-549. Greenland data was graciously supplied by John Brooke. Astronomical observations before the invention of the telescope in 1610 were insufficiently accurate to confirm the absence of sunspots as they do during the longer period of global cooling 1643-1715 known as the "Maunder Minimum".

10 Thouret, Jean-Claude, et al-. "Reconstruction of the AD 1600 Huaynaputina Eruption Based on the Correlation of Geologic Evidence with Early Spanish Chronicles", *Journal of Volcanology and Geothermal Research* 115 (2002): 529-570; Buisman, Jan, *Duizend jaar weer, wind en water in de Lage Landen IV: 1575-1675* (Franeker: Van Wijnen, 2000), 200, reproduced a striking drawing of the hazy red sun seen in Friesland in January 1601; Guaman Poma de Ayala, Felipe, *El primer nueva corónica, y buen gobierno por D. Filippe Guaman Poma de Aiala* (Royal Library, Copenhagen, GkS 2232 4to), 1061, shows Arequipa under a cloud of ash so thick that "for thirty days one could not see sun, moon or stars." See also Atwell, William S., "Volcanism and Short-Term Climatic Change in East Asian and World History, c. 1200-1699," *Journal of World History* 12 (2001): 56-57 and 72-83.

seen such tempestuous weather so long together"; and "the abundance of snow which is fallen" made campaigning almost impossible.[11]

The same period also saw an abnormal frequency of El Niño (or ENSO: El Niño—Southern Oscillation) events. Nowadays, an El Niño episode occurs on average once every five years, but "strong" or "very strong" El Niño episodes occurred in 1591, 1594, and 1596, with "moderate" events in 1597 and 1601—more than twice as often as today.[12] This was important because, in normal summers, easterly winds blowing from equatorial America to Australia and Southeast Asia prevail, and so heavy rains fall on South and East Asia, nurturing the harvest; whereas in El Niño years, westerly winds blowing from equatorial Asia to America prevail, releasing the monsoon rains eastward towards Central and South America. El Niño thus produces catastrophe simultaneously on both sides of the Pacific, bringing drought to one and floods to the other. In addition, El Niño's 'global footprint' also normally includes the Caribbean, which almost always suffers floods; Ethiopia and northwest India, which usually endure droughts; and Northeast America and northwest Europe, which frequently experience harsh winters. We observe this same 'footprint' in the 1590s—but twice as often as today.[13]

2 Contemporary Perceptions of Climate Change

Although few contemporaries knew about sunspot minima, volcanic ash or El Niño, they could not ignore global cooling. In southern Sweden, Pastor Petrus Magni noted in his diary that 1591 had been a "black year in which the grass did not turn green at all", and that in 1596 and again in 1597 prolonged torrential rain turned the fields into lakes, destroying both hay and grain so that cattle died for lack of fodder; while there was "so dreadful a hunger that the greater

11 Atkinson, Ernest G., ed., *Calendar of State Papers Relating to Ireland of the Reign of Elizabeth, 1600-1601* (London: Public Record Office, 1905), 106, Mountjoy to Fenton, 26 Dec. 1600; 139, Fenton to Cecil, 6 Jan 1601; and 147, Covert to Cecil, 28 Mar. 1601 (all dates 'old style'.)

12 Gergis, Joëlle L., and Anthony M. Flower, "A History of ENSO Events since A.D. 1525: Implications for Future Climate Change," *Climatic Change* 92 (2009): 370-372, provides a chronology of El Niño events.

13 The 'global footprint' is admirably discussed in Grove, Richard H., and John Chappell, ed., *El Niño: History and Crisis* (Cambridge: The White Horse Press, 2000), chapter 1, and Caviedes, Cesar, *El Niño in History: Storming across the Centuries* (Gainesville: University Press of Florida, 2001), 198.

part of the people had to [eat] bread made of bark".[14] Across the Baltic in Northern Germany, Martin Behm was one of several Lutheran pastors who composed special hymns that reproached God for "holding back the sunshine and sending heavy rain".[15] His colleague Pastor Daniel Schaller complained in 1595 that

> There is no real prolonged sunshine, no settled winter or summer; the earth's crops and produce do not ripen, and are no longer as wholesome as they were in former times. [...] The water is no longer as rich in fish as it used to be; the forests and fields are no longer so full of animals and livestock; the air is no longer so full of birds.[16]

"The fields and farms have grown tired of bearing crops," Schaller concluded, and he (like several other Lutheran writers) predicted the imminent end of the world: apocalyptic pamphlets appeared in Germany in unprecedented numbers during the decade.[17]

The 1590s also saw a notable rise in accusations blaming the "unnatural weather" on witchcraft, with a close correlation between persecutions and climatic deterioration. A chronicler in the archdiocese of Trier, in West Germany, commented that the almost constant "lack of grain, the rigours of climate, and crop failures" in the later 16th century led "everybody to think that the continuous crop failure was caused by witches" and to demand their "eradication".

14 Utterström, Gustav, "Climatic Fluctuations and Population Problems in Early Modern History," *Scandinavian Economic History Review* 3 (1955): 3-47, provides a wealth of detail from contemporary sources, including Pastor Petrus Magni, living on Kalmar Sound.

15 Hille, Martin, "Mensch und Klima in der frühen Neuzeit: Die Anfänge regelmässiger Wetterbeobachtung, 'Kleine Eiszeit', und ihre Wahrnehmung bei Renward Cysat (1545-1613)," *Archiv für Kulturgeschichte* 83 (2001): 89, quotes this and four other Lutheran hymns about the weather composed around 1600, and still included in Evangelical Hymn Books.

16 Lehmann, Hartmut, "Frömmigkeitsgeschichtliche Auswirkungen der 'Kleinen Eiszeit'," in *Volksreligiosität in der modernen Sozialgeschichte*, ed. Wolfgang Schieder (Göttingen: Vandenhoeck & Ruprecht, 1986), 35, quoting Schaller, Daniel, *Herold. Ausgesandt in allen Landen öffentlich zu verkündigen und auszurufen. Dass diese Welt mit ihrem Wesen bald vergehen werde, und der Jüngste Gerichtstag gar nahe vor der Tür sei* (Magdeburg, 1595).

17 Lehmann, "Frömmigkeitsgeschichtliche Auswirkungen," 38, quoting Schaller, *Herold*. See also Lehmann, Hartmut, "Endzeiterwartung im Luthertum im späten 16. und im frühen 17. Jahrhundert," in *Die lutherische Konfessionalisierung in Deutschland: Wissenschaftliches Symposion des Vereins für Reformationsgeschichte 1988*, ed. Hans-Christof Rublack (Gütersloh: Gerd Mohn, 1992), 545-558 (including a spirited discussion); and Leppin, Volker, *Antichrist und Jüngster Tag: Das Profil apokalyptischer Flugschriftenpublizistik im deutschen Luthertum 1548-1618* (Heidelberg: Gütersloher Verlagshaus, 1999), 32-34, 53.

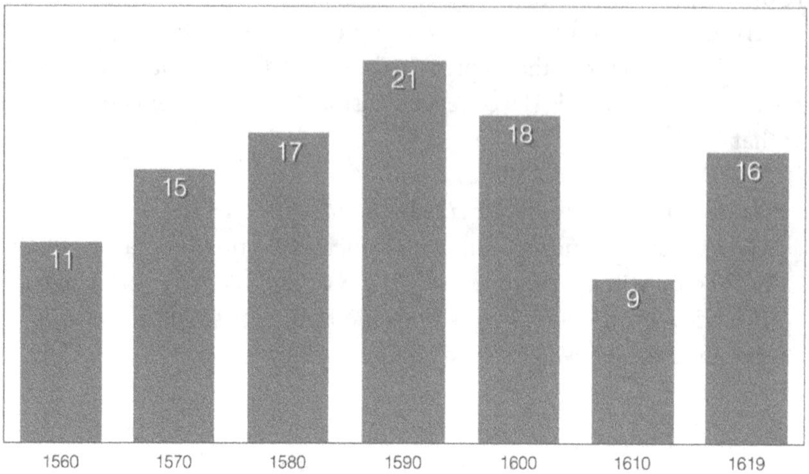

FIGURE 4.2 *The rhythm of German broadsheets predicting the "End of the World", 1550-1620*

The authorities of Trier burnt over three hundred people for witchcraft. Throughout Western Europe, learned treatises about witchcraft appeared in print.[18]

In 1600, Felipe Guaman Poma de Ayala, with both Spanish and Inca parents, began to travel around Peru to collect material for a chronicle that would chart the changes in the world around him "in my lifetime." In a particularly bitter passage he noted the frequent earthquakes, volcanic eruptions, tsunamis and epidemics, as well as the "frost that killed the maize and potatoes", and "the great fall of snow and hail that fell from the sky and covered all the mountains to the depth of two metres and in some places two metres, killing many people and livestock." He ironically hailed them all as "the miracles of God."[19] A decade later Renward Cysat, botanist, archivist, and town historian of Lucerne, Switzerland, added a new section to his *Chronica* entitled "The Seasons of the Year", beginning with a "Warning to the Reader". "I would not wish the reader to judge my compilation of the various seasons and years of our times as a wasteful, useless or childish work," Cysat wrote, "because the past few years have seen such a strange and wondrous succession of changing weather." He had decided to record his observations concerning climate change

18 Details from Pfister, "Climatic Extremes", and Behringer, Wolfgang, *Witches and Witch-Hunts: A Global History* (Cambridge: Polity Press, 2004), 93-98, 102-104.

19 Guaman Poma de Ayala, *El primer nueva corónica*, 94-95.

> As a service and a favour to future generations because unfortunately, on account of our sins, for some time now the years have shown themselves to be more rigorous and severe, and in the recent past we have seen deterioration amongst living things, not only among mankind and the animal world but also of the earth's crops and produce; as well as extraordinary alterations of the elements, stars and winds.[20]

Overpopulation and war intensified the impact of this "more rigorous and severe" weather. Partly thanks to a half-century of largely favourable weather, many parts of the northern hemisphere boasted more people than ever before: by 1590, China had perhaps 150 million inhabitants, India 116 million, and Europe 100 million.[21] In some areas, population densities reached dizzying levels. The Lower Yangzi Valley boasted a population of about twenty million, an average of almost 1,200 persons per square mile (today the overall population density of the modern Netherlands, the most densely settled part of Europe, is 1,000 persons per square mile). In Anatolia, heartland of the Ottoman Empire, in the later 16th century the number of farmsteads tripled in some areas; permanent settlements grew up on marginal land or land previously uninhabited; and woodland steadily disappeared under the plough. Rural settlements reached a density unequalled until the 20th century.[22] In 1609 an English preacher looked back nostalgically to the time, a century before, when

> [T]his Kingdome was not so populous as now it is. Ciuell warres at home, and forreine wars abroad, did cut off the ouer-spreading branches of our people. Our country then yeelded vnto all that were in it a surplussage of all necessities: it yeelded preferment in due correspondencie, for al degrees & sorts of men. The commons of our Country lay free and open

20 Cysat, Renward, *Collectanea Chronica und denkwürdige Sachen pro Chronica Lucernensi et Helvetiae*, vol. 4.2, ed. Josef Schmid (Lucerne: Diebold-Schilling-Verlag, 1969), 898. Cysat's section "Stationes Annorum" included extreme weather events recorded since 1435.

21 Although the figures for the total population of Europe and China are generally accepted, the figure for India is not. Guha, Sumit, *Health and Population in South Asia from Earliest Times to the Present* (London, Hurst, 2001), 31-34, provides the most careful estimate of the Indian population circa 1600—16 million, somewhat lower than that of previous scholars—but it remains speculative.

22 Yong, Xue, "Agrarian Urbanization: Social and Economic Changes in Late Imperial Jiangnan" (PhD diss., Yale University, 2006), 239-240; and Ozel, Oktay, "Population Changes in Ottoman Anatolia during the Sixteenth and Seventeenth Centuries: The 'Demographic Crisis' reconsidered," *International Journal of Middle Eastern Studies* 36 (2004): 187-188.

> for the poore Commons to inioy, for there was roome enough in the land for euery man.

But now, he lamented, "God hath prospered vs with the blessings of the wombe, & with the blessings of the brests, the sword deuoureth not abroad, neither is there any feare in our streets at home; so that we are now... a great people, and the lande is too narrow for vs."[23]

Moreover, although climate change made it harder for governments to mobilise and deploy their resources, many of them fought major wars during the 1590s: Japan invaded Korea, provoking a massive Chinese military response; Poland fought both Sweden and Russia; the Ottoman empire campaigned against a Christian coalition in Hungary; above all, Philip of Spain conducted operations on land and sea against England, France and the Dutch Republic.

3 The Crisis of the 1590s and the Monarchy of Philip II

Queen Elizabeth of England had no doubt about the role of the climate in human affairs. In a celebrated contemporary portrait of her, done shortly after 1588, George Gower showed storms destroying the Spanish Armada (Figure 4.3); while in a Proclamation of 1597 the queen gloated that "it is manyfestly seen that the king of Spayne" had been thwarted in all his military and naval enterprises "not only by her Majestie's forces on the sea, but specially by the ordynance of Almightie God with tormentes of wynde and forcible tempestes at tymes and places unlooked for".[24]

Some contributors to *The European Crisis of the 1590s* agreed with Gloriana. Timothy Davies noted that in Sicily "The meagre harvest of 1590 was followed by a long spell of drought until March 1591 when torrential rain fell throughout the spring months"; while James Casey observed that "The 1590s were a decade of extremely bad weather" in Spain: the "chronicles and tithe records for Valencia and Murcia in the period 1580-1630" revealed "an exceptional run of wet years between 1589 and 1598," while "1599 was the single worst year in modern Spanish history".[25] Data from Castile, the economic and military heart of

---

23 Gray, Robert, *A Good Speed to Virginia* (London, 1609), Sigs. B2 - B2v.

24 Collier, J.P., ed., *The Egerton Papers* (London: Camden Society, 1840), 259-263, Proclamation of September 29, 1597.

25 Davies, Timothy B., "Sicily," in *The European Crisis of the 1590s: Essays in Comparative History*, ed. Peter Clark (London: George Allen & Unwin, 1985), 199; Casey, James G., "Spain: A Failed Transition," in *The European Crisis of the 1590s: Essays in Comparative History*, ed.

FIGURE 4.3
George Gower's "Armada Portrait" of Elizabeth

the Monarchy, likewise suggest a catastrophic fall in both production and population in the late 16th century. To take a single example: at Pozuelo de Aravaca (now Pozuelo de Alarcón, a suburb of Madrid, which boasts excellent records) the value of annual agricultural production (cereal, wine and sheep) fell from over 12,000 ducats in 1576 to under 6,500 in 1590-1595, and its population fell from 263 households in 1591 to 244 in 1596. The plague epidemic in 1599 removed one parent from 31 households and both parents from nine others. Striking figures, indeed: but were they typical?[26]

A census of Castile commissioned in 1591 provides (despite some ambiguities and omissions) a unique snapshot of the kingdom just before the onset of the extreme climatic events, with data on 12,000 separate communities and perhaps 6.5 million inhabitants.[27] Almost three-quarters of the Castilian popu-

---

Peter Clark (London: George Allen & Unwin, 1985), 211. See also Davidson, Nicholas, S., "Northern Italy in the 1590s," in *The European Crisis of the 1590s: Essays in Comparative History*, ed. Peter Clark (London: George Allen & Unwin, 1985) 157-176; and Stella, Domenico, *Crisis and Continuity: The Economy of Spanish Lombardy in the Seventeenth Century* (Cambridge: Harvard University Press, 1979), 35-36.

26 Barbazza, Marie-Catherine, La société paysanne en Nouvelle Castille: Famille, mariage et transmission de biens à Pozuelo de Aravaca (1580-1640) (Madrid: Casa de Velázquez, 2000), 30-31, 38, 83.

27 García España, Emilio, and Annie Molinié-Bertrand, ed., *Censo de Castilla, 1591: Vecindario y estudio analítico*, 2 Vols. (Madrid: Instituto Nacional de Estadística, 1980-1986), provide a facsimile and analysis of the Censo. Molinié-Bertrand, Annie, *Au Siècle d'Or: L'Espagne et ses hommes: La population du royaume de Castille au XVIe siècle* (Paris: Economica, 1985), 17-21, offers a concise evaluation of the Censo and other demographic surveys of Castile

lation lived in settlements with 1,000 or fewer households, with the rest in towns (albeit only eleven exceeded 5,000 households). But these data concealed striking regional disparities. First, 9,500 communities (most of them small) lay north of the Guadarrama Mountains and only 2,500 settlements lay to the south. Second, paradoxically, although the population of the kingdom as a whole had grown by at least fifty percent between the 1520s and 1591, that growth was far stronger in New Castile, which registered an increase of 85 percent, as against only twenty percent in Old Castile, which contained many deserted and declining villages. Thus, the Census listed 18 of the 209 settlements in the Tierra of Ledesma (Salamanca) as "depopulated", while 71 more had fewer than four households. Only one other region reported a notable population loss: the kingdom of Granada, where the ethnic cleansing that followed the revolt of the Moriscos two decades before had left numerous abandoned or moribund villages. Finally, the census revealed a high proportion of both "poor households" and single women who headed households. The two categories often overlapped, and the majority depended on charity.[28] They would suffer first and most when the climate began to deteriorate.

Increased volcanic and El Niño activity affected the climate of Castile even before the enumerators began their work. Parts of the south suffered from repeated floods. The *Andalusian Seasonal and Annual Rainfall Index, 1501-1997* records an unparalleled sequence of extremely wet years between 1590 and 1593 and again between 1595 and 1597. In May 1598, the town council of Cádiz feared famine because of "the meagre grain harvest in this area this year".[29] Further north, the problem was not too much rain but too little: a series of tree rings from Navacerrada, high in the Guadarrama Mountains that divide Old

in the 1590s. On Granada, see Castillo, Alvaro, "El 'Servicio de Millones' y la población del reino de Granada en 1591," *Saitabi* 11 (1961): 61-91. The Censo was commissioned in order to achieve a more equitable distribution of the new Millones tax voted by the Cortes in the wake of the failure of the Armada.

28 García España and Molinié-Bertrand, *Censo de Castilla*, vol. 2, 177-178, on Ledesma. Growth rates taken from Molinié-Bertrand, *Au Siècle d'Or*, 308-309; and Brumont, François, *Paysans de Vieille Castille aux XVIe et XVIIe siècles* (Madrid: Casa de Velázquez, 1993), 355. On the 'viudas pobres', see Molinié-Bertrand, *Au Siècle d'Or*, 341-348, and Barbazza, *La société paysanne*, 82-114.

29 Complutense University of Madrid, "Andalusian Seasonal and Annual Rainfall Index, 1501-1997," accessed November 28, 2014. <http://pendientedemigracion.ucm.es/info/reclido/es/basesdatos/Rainfallindex.txt>. The next wettest period, in 1640-1643, saw less than half the precipitation of 1590-1597. 1592, 1597 and 1642 were by far the three wettest years in the five centuries covered by the Index; Archivo Municipal, Cádiz, Ms. 10,001, "Libro de Acuerdos 1596-1599," folio 102, Acuerdo of May 15, 1598.

and New Castile reveal that the lowest annual precipitation ever recorded in the past millennium occurred during the 1590s.[30]

Drought also afflicted the Tierra de Campos around Palencia, north of Navacerrada, which normally produced and exported a surplus of grain. Many communities reported serious harvest shortfalls in 1593, 1595 and above all 1598, with burials exceeding baptisms each year from 1595 to 1599. The population of some villages fell by one-third and more.[31] Urban records tell the same story. In Palencia itself, in October 1595 the municipal granary had to provide seed-corn to "the farmers on the city's lands so that they can sow" the next harvest; 1597 proved sterile; and in 1598 the city's magistrates discussed "the setbacks of the year, including the recent destruction of the vines by the great frosts, which we believe have affected the whole region" and "the great shortage of grain we expect next year, because of the drought and evident sterility of this year". Their options dwindled after "God Our Lord was pleased to send hail on the 19th of June which beat down on the fields around this city with the greatest force ever seen". In August, the council discussed the "great danger posed to this city through the multitude of poor and migrant people who converge on us", and they resolved to expel all "destitute people" not born in Palencia and instructed the guards on the city gates to admit no one. A modern analysis of surviving parish registers in Castile revealed the same pattern almost everywhere: a steep rise in burials, and a sharp fall in baptisms and marriages.[32]

The fugitives from starvation were joined by others trying to escape an epidemic of bubonic plague that began in the port-cities of Cantabria in 1596 and inexorably moved south. In Mateo Alemán's bitter novel *Guzmán de Alfarache*, as the hungry narrator travelled from Cazorla to Madrid in 1598 he observed grimly that

30 Manrique, Emilio, and Ángel Fernández-Cancio, "Extreme Climatic Events in Dendroclimatic Reconstructions from Spain," *Climatic Change* 44 (2000): 128 (graphs), and 133-135 ('frequency of out-of-range values').

31 Brumont, *Paysans de Vieille Castille*, 219-222, 229-230 and graphs at 449-460.

32 Archivo Municipal, Palencia, "Libro de Acuerdos, 1595-1600," folios 186-187, acuerdos of October 27 and 30, 1595; and folios 320v, 327, 345v, acuerdos of June 12 and 22, and August 11, 1598; Pérez Moreda, Vicente, *Las crisis de mortalidad en la España interior: Siglos XVI–XIX* (Madrid: Siglo XXI, 1980), 270-272 (parish register data) and 259 (map). More details in Marcos Martín, Alberto, *Economía, sociedad, pobreza en Castilla: Palencia 1500-1814*, 2 Vols. (Palencia: Diputación Provincial, 1985); and González Muñoz, María del Carmen, *La población de Talavera, siglos XVI-XX: Estudio socio-geográfico* (Toledo: Diputación Provincial, 1974), 125, 176-177.

> People gave little charity, and no wonder because the year was generally sterile; and if it was bad in Andalucía, it was worse as one entered the kingdom of Toledo; and the further inland, the worse the scarcity. That was when I heard: "God save you from the plague descending from Castile and the hunger rising from Andalucía.[33]

Some of the sermons delivered at the memorial services for Philip II late in 1598 complained about the desperate state in which he had left Spain. Fray Lorenzo de Ayala, preaching in Valladolid, the late king's birthplace, noted:

> Our Catholic King died after a drought that lasted almost nine months without a break, revealing that the earth had declared itself bankrupt—just like an unsuccessful merchant. At the same time, the price of everything in Castile increased, as supplies ran short, coinciding with the collapse of public health throughout the kingdom and opening the door to plague in many areas. These disasters [...] were harbingers of the greatest catastrophe Spain has ever suffered since our Patriarch Tubal, grandson of Noah, settled here.[34]

López de Ayala's apocalyptic fears soon materialised. A reconstruction of the demographic disasters that afflicted Castile between 1575 and 1855 suggested that 1598 and 1599 were "the two worst years in the entire [Iberian] peninsula", with double the intensity of any other mortality crisis. Valladolid, for example, had perhaps 36,000 inhabitants when López de Ayala delivered his sermon but within the year plague had killed over 6,000 of them (figure 4.5[35]). In all, "the plague descending from Castile" and "the hunger rising from Andalucía" probably carried off ten percent of the kingdom's population: 600,000 souls.[36]

Although Philip II died before the crisis peaked, he already knew that Spain faced a "catastrophe". In 1591, the year of the census, a message arrived from his

33 Alemán, Mateo, *Guzmán de Alfarache*, (Madrid, 1599, Lisbon, 1604), part 2, ii.

34 Íñiguez de Lequerica, Juan, ed., *Sermones funerales, en las honras del rey nuestro señor don Felipe II* (Madrid, 1599), fos 85v–86, López de Ayala's sermon on November 15, 1598 ("el mayor de los daños").

35 Source: Bennassar, Bartolomé, Valladolid au siècle d'Or: *Une ville de Castille et sa campagne au XVIe siècle* (Paris, The Hague: Mouton, 1967).

36 Pérez Moreda, *Las crisis de mortalidad*, 257-293; Bennassar, Bartolomé, *Recherches sur les grandes épidémies dans le nord de l'Espagne à la fin du XVIe siècle: Problèmes de documentation et de méthode* (Paris: SEVPEN, 1969), 19.

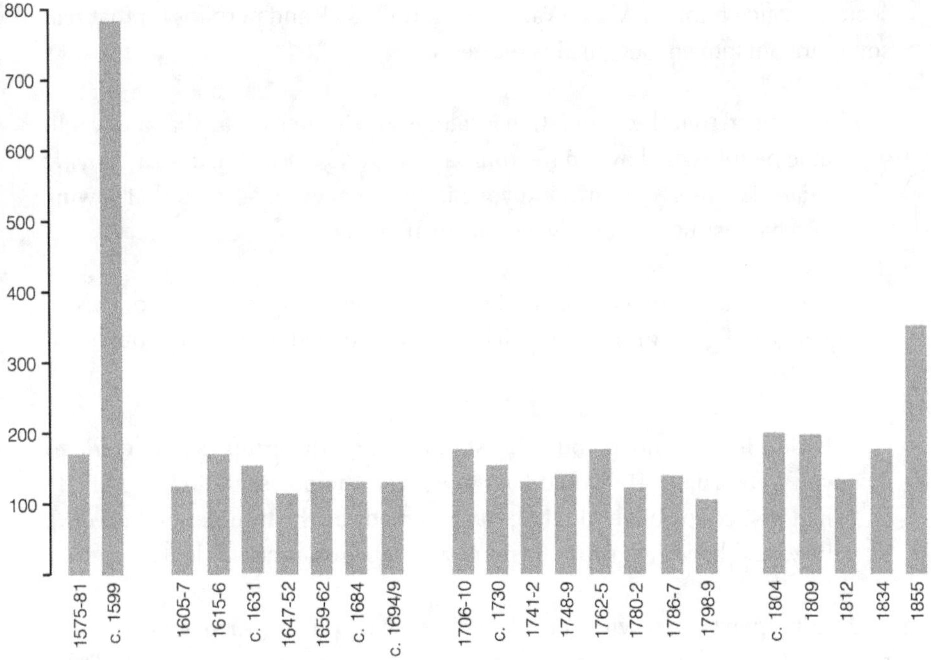

FIGURE 4.4 *Intensity of mortality crises in Castile, 1575-1855*

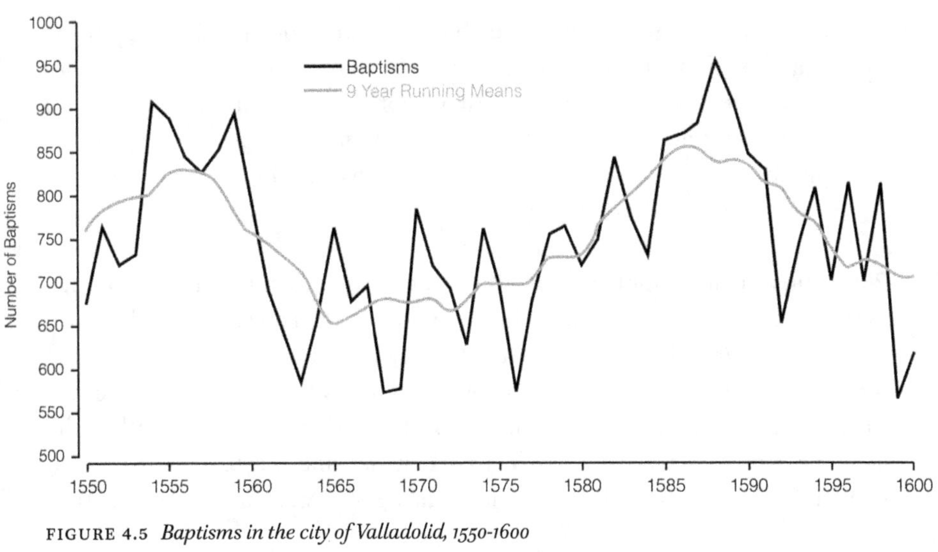

FIGURE 4.5 *Baptisms in the city of Valladolid, 1550-1600*

secretary and chaplain, Mateo Vázquez, mortally sick and perhaps for that reason more outspoken, begging his master to

> Take heed that the population is falling, and in such a way that many reliable people who have come from various parts of the kingdom are saying that it is a marvel to meet anyone in the smaller villages, so that sowing and harvesting are rapidly coming to an end.

Vázquez therefore argued that Philip must cease spending the resources of Castile so prodigally on foreign wars, and instead find alternative sources of revenue.

> If God had intended Your Majesty to heal all the cripples who come to you to be cured, He would have given you the power to do it; and if He had wished to oblige Your Majesty to remedy all the troubles of the world, he would have given you the money and the resources to do it.

If the king persevered with the same expensive policies, Vázquez continued relentlessly, "Everything may collapse at once for lack of money". Philip addressed this passionate complaint with remarkable equanimity—although of course he rejected it. "I know you are moved by the great zeal you have for my service to say what you did," he gently chided his minister,

> But you must also understand that these are not matters that can be abandoned by a person who is as conscientious about his responsibilities as you know me to be, because they depress me and matter to me more than to anyone. Taken together, they involve far more problems than people think [...]. Moreover, these issues involve religion, which must take precedence over everything.[37]

Philip thus did not dispute the material evidence of impending disaster; rather, his faith-based political vision led him to ignore it. He likewise ignored a warning from a different quarter, received one week later. The king had just ordered the cities of Castile to raise and arm 60,000 soldiers for national defence, but the magistrates of Guadalajara refused "because of the dreadful weather, the great dearth" and the high taxes "that exhaust and waste the assets and strength" of the country. "If Your Majesty does not remedy and overcome [our

---

37 Instituto de Valencia de Don Juan, Madrid, envío 51/1, Vázquez to Philip and rescript, February 8, 1591.

difficulties]," they wrote, "we foresee total ruin and collapse."[38] Their prediction soon came true: a few weeks later anti-tax riots broke out in Madrid, Seville, Ávila and other cities of Castile; and although the king's agents soon managed to restore order there, suppressing the rebellion that began in the neighbouring kingdom of Aragon in May 1591 required six months, a large army, and an outlay of 1,370,000 ducats.[39]

The king therefore convened the Cortes of Castile and asked them to vote yet more taxes to fund his foreign wars. The assembly lasted twice as long as any of its predecessors because the deputies dug in their heels, "beseeching Your Majesty most insistently that, before anything else, you will order a reduction in war expenditure, both in the Netherlands and elsewhere". One deputy stated outright that "although the wars with the Dutch, England and France are holy and just, we must beg Your Majesty that they may cease", while another urged "Your Majesty to abandon all these wars, making the best terms that you can".[40]

Once again, the king refused to abandon his faith-driven foreign policy. He instructed his ministers to remind the deputies "that these wars are necessary and unavoidable unless Spain is to suffer the miseries that afflict other parts of Christendom". Therefore,

> They should and must place their trust in me, in the love I have for these kingdoms, and in the long experience I have in governing them, [and accept] that I shall always do what is in their best interests. Speak to them at length in this vein and advise them that they are never, on any pretext, to come to me with such a suggestion again.

38 Olivari, Micheli, *Entre el trono y la opinión: La vida política castellana en los siglos XVI y XVII* (Valladolid: Junta de Castilla y León, 2004), 130, concejo of Guadalajara to Philip, February 15, 1591.

39 Lorenzo Cadarso, Pedro L., *Los conflictos sociales en Castilla (siglos XVI–XVII)* (Madrid: Siglo XXI, 1996), 232-233, reports that several communities in Castile escaped direct taxation until the 1590s but saw their obligations rise rapidly thereafter; Gracía Rivas, Manuel, *La "invasión" de Aragón en 1591: Una solución militar a las alteraciones del reino* (Zaragoza: Diputación General de Aragón, 1992), 277-280, gives the cost of the campaign. On the revolts of 1591, see Parker, Geoffrey, *Felipe II: La biografía definitiva* (Barcelona: Planeta, 2010), chapter 18.

40 *Actas de las Cortes de Castilla*, vol. 16, 166-167, Juan Vázquez de Salazar to Philip, April 28, 1593, vol. 12, 456 (procurador of Sevilla, May 19, 1593); vol. 16, 170 (procurador of Burgos, summarized by Juan Vázquez, May 6, 1593).

Philip left nothing to chance. He offered some deputies bribes in return for a favourable vote; he intimidated others with summary arrests and house searches; and he sent a message to "forewarn the theologians of Madrid" so "that if the deputies of the Cortes should turn to them for advice, they may include in their opinions a full account of my obligations, and the reasons why we must find a remedy for the needs that face us".[41]

Instead of winning over "the troublemakers" (*los dificultosos*, as the government termed its critics), the king's intransigence fostered in the Cortes, perhaps for the first time, an "organised opposition with a coherent political agenda": an agenda that I.A.A. Thompson has termed "Castile first".[42] Gerónimo de Salamanca, a leading *dificultoso*, articulated the agenda most clearly. He was willing to vote new taxes, he declared, but only if the entire proceeds went towards defending Spain:

> He wished to serve His Majesty [by voting taxes] as much as anyone else in the Cortes, and if they were destined to send one fleet in the Atlantic and another in the Mediterranean, to defend them and drive away the [Dutch and English] pirates and to fortify Spain's frontiers, he would urge the Cortes to give Your Majesty as much and more than anything proposed so far.

Another *dificultoso* concurred: voting taxes to continue the disastrous wars in France and the Netherlands was pointless.

> It is important for God's service that these kingdoms [Spain] do not expend all their strength in these [foreign] wars, because we face other enemies who are no less dangerous, but rather more so—namely the Turks and Moors—and the coasts, which form our frontier with them, are undefended.[43]

Some royal ministers became equally pessimistic. In 1596 the marquis of Poza, president of the Council of Finance, described Spain's impossible financial

41 *Actas de las Cortes de Castilla*, vol. 16, 169, 173, and 195-197, Philip to Juan Vázquez, Apr 28, May 6, and July 23, 1593; and Jago, Charles, "Taxation and Political Culture in Castile 1590-1640," in *Spain, Europe and the Atlantic World: Essays in Honour of John H. Elliott*, ed. Richard L. Kagan and Geoffrey Parker (Cambridge: Cambridge University Press, 1995), 52.

42 Thompson, Irving A.A., "Oposición política y juicio del gobierno en las Cortes de 1592-98," *Studia Historica, Historia Moderna*, 17 (1997): 37, 43 n.18, and 48-49.

43 Ibid., 48-49, quoting Salamanca and don Pedro Tello.

and strategic position to Don Cristóbal de Moura, the king's principal adviser, just as bluntly as the "troublemakers" of the Cortes.

> His Majesty must see that it is impossible to carry on as we are, because although we have already spent all His Majesty's revenues until the year 1599, his expenses continue and even increase, so that even if his revenues were unencumbered, we could not carry on. To do this, one only needs to know what I know: that His Majesty must either reduce his expenditure voluntarily until things improve, or else find a way to achieve the impossible.[44]

The capture and sack of Cádiz by Anglo-Dutch forces in July 1596, the month after Poza wrote his plea, made it impossible for the king "to reduce his expenditure". Instead he used the humiliation to browbeat the Cortes into voting more taxes; but they did not suffice. Since, in Moura's colourful phrase, "we are drowning tied back-to-back", Philip signed the fourth Decree of Bankruptcy of his reign, confiscating the capital of all outstanding loans and suspending all interest payments to his creditors. Castile had become the world's first serial defaulter on sovereign debt and Philip had become, in the evocative phrase of a recent book, "the borrower from hell".[45]

In July 1597, the widespread misery led the royal confessor, Fray Diego de Yepes, to send a hard-hitting paper of advice to Philip. "Although Your Majesty's ailments prevent me from saying some things to you in person", he began, "I cannot refrain from saying them in writing."

> I grieve to see Your Majesty grow weaker each day, with business of state held up, and the prince so isolated from it. Many fear that unless Your Majesty informs and initiates him, and instructs him in your presence, that the government and the administration of justice will falter.

After providing some specific examples, Yepes attempted naked spiritual blackmail.

44 British Library, London, *Additional Ms* 28,378/69-73v, Poza to Moura, June 9, 1596.

45 British Library, London, *Additional Ms* 28,378/128-31, Poza to Moura and rescript, July 28/31, 1596; Drelichman, Mauricio, and Hans-Joachim Voth, *Lending to the Borrower from Hell: Debt, Taxes, and Default in the Age of Philip II* (Princeton: Princeton University Press, 2014).

> Your Majesty should consider the toil, vigilance and care that you have invested in the government of these kingdoms to maintain peace and justice, and consider what may be lost while you are ill, and what the consequences may be when these kingdoms see that Your Majesty no longer watches over them. Since God entrusted the conservation of the Catholic faith and the expansion of the Christian religion to Your Majesty, and since they both depend on the sound government, justice and prosperity of these kingdoms, I hope you will be pleased to arrange matters so that on the Day of Judgement (which is not far off), you can appear in the presence of God confident that you have done everything possible.[46]

Ten months later, since nothing had improved, Yepes sent an equally stark warning to don Cristóbal de Moura that the intolerable burden imposed on the king's subjects by rising taxes at a time of dearth "will bring down our world".

> Your Lordship can see where this leads. The poor cry out, but His Majesty does not hear them, nor does he want to hear me. I simply do not know what to do. May God in His mercy keep me in His hand, and protect Your Lordship, whom the world blames for all this. [...] I beg Your Lordship to look into this carefully, and advise His Majesty, because I have had no luck in writing or speaking to him.

The king changed nothing before his death, six months later, and the crisis did indeed almost bring down "our world".[47]

4 The Enemies of Philip II and the Crisis

Two factors sheltered Philip and his subjects from the full effects of "the greatest catastrophe Spain has ever suffered since our Patriarch Tubal, grandson of Noah, settled here". First, despite serious delays caused by English and Dutch efforts to intercept them, the huge fleets sailing from America transported silver bullion in record quantities throughout the 1590s: official cargoes worth over eighty million ducats reached Spain, of which over 25 million went straight into the royal coffers. This windfall revenue, arriving precisely as

46 Hispanic Society of America, *Altamira Papers* 18/IV/3c, Yepes to Philip, July 1, 1597, holograph.

47 Hispanic Society of America, *Altamira Papers* 12/I/1, # 13, Yepes to Moura, May 17, 1598, holograph.

treasury receipts from other sources plunged, forestalled worse defaults by the Spanish treasury.[48] Second, the same double curse of weather and war also affected Spain's principal enemies. Between 1591 and 1596, the lands of the Ottoman sultan around the Eastern Mediterranean endured the longest and most intense drought recorded in the past six centuries (ch. 1). At the same time, the central government escalated its demands for money and supplies for its armies fighting in Hungary—a decision that, in the felicitous phrase of Sam White, turned "a meteorological disaster into a human crisis".

> Population pressure had squeezed supply, and the growth of cities and the military had raised demand for basic commodities. Moreover an unprecedented drought was bringing famine to the same core provinces presently called upon to meet the exactions of war—and not just any war, but a drawn-out war of sieges and counter-sieges.[49]

Dry summers prevented navigation on the Danube at Budapest while an Ottoman chronicler with the troops compared the torrential autumn rains with "Noah's flood". According to another chronicler, "The winter, the so-called 'merciless soldier', pressed with its full strength. Terrible storms and snow-storms occurred; all hands and feet were crippled". Observers in Moldavia and Wallachia also reported unusually harsh winters in 1594-1595 and again in 1597-1598, when the major rivers (including the Danube) froze over and "people are at one in saying there has not been a winter like it in fifty or sixty years". In Istanbul, snowstorms during the winter of 1594-1595 interrupted the food supply and many citizens starved; in summer 1596, "the palace had never had such a shortage of food"; while in 1600, the price of bread quadrupled and the city suffered from "extreme dearth".[50]

---

48 Elliott, John H., *Imperial Spain, 1469-1716* (London: Arnold, 1963), 175; Chaunu, Huguette, and Pierre Chaunu, *Séville et l'Atlantique (1594-1650). VI.1: Tables statistiques* (Paris: SEVPEN, 1956), tables on 186-187, 256-257, 280-281, 296-297, 312-313, and 316-317 show the longer voyages of the *flotas*, which often numbered over hundred ships, in the 1590s; the table on 474 shows the value of the cargoes.

49 White, Sam, *The Climate of Rebellion in the Early Modern Ottoman Empire* (Cambridge: Cambridge University Press, 2011), 141, 144.

50 Data from: Yılmazer, Ziya, ed., *Topçular Kâtibi 'Abdülkādir (Kadrî) Efendi Tarihi* [The Chronicle of Abdulkadir Efendi, Secretary of the Artillery Corps] (Ankara: Türk Tarih Kurumu, 2003), vol. 1, 51, 90, 205-207, 218-223, 300, 330, 347; İpşirli, Mehmet., ed., *Tarih-i Selâniki* [The Chronicle of Selâniki] (Istanbul: İstanbul Üniversitesi Edebiyat Fakültesi Yayınları, 1989), vol. 1, 425, 431, and vol. 2, 444, 593, 615, 849, 853; Baykal, Bekir. S., ed., *Peçevi Tarihi* [The Chroncle of Pecevi] (Ankara: Kültür ve Turizm Bakanlığı, 1982), vol. 2, 143, 202

In many parts of Anatolia, from the 1580s tax rolls showed a rural proletariat of "landless men" who either married late or never married at all—presumably because they could no longer make a living from farming. Others left their villages either to become religious students or else to join the army, a local bandit gang, or the armed entourages maintained by provincial officials. The intense and prolonged drought of the 1590s, resulting in repeated harvest failures, created what the Ottoman historian Mustafa Akdag called "The Great Flight" from the countryside; worse, the synergy between weather and war provoked the worst episode of domestic violence ever to afflict the Ottoman Empire: the Çelali Rebellion. From 1596 until 1610, "riding a wave of popular desperation and discontent, mercenary leaders gathered rebel armies that plundered the provinces and defied the imperial government", producing a contraction of population and agriculture in the core provinces of the empire. According to a report to the sultan in 1603,

> The leading men have fled. The poor and the taxpayers have gone to the mountains [...] abandoning their wives and children. No cattle remain. The whole world is in confusion. Those who can use a gun have joined the rebels.

Clearly the sultan was in no position to attack Spain.[51] Adverse weather also afflicted France. Emmanuel Le Roy Ladurie's *Histoire humaine et comparée du climat* presented abundant evidence, in a chapter entitled "Froids et fraîcheurs du fin de siècle: les 1590's". For example, between 1591 and 1597, grapes throughout France ripened two, three or more weeks later than normal.[52] Troops also devastated many areas in the bitter civil war that began in 1589 and ended in 1598, while some cities endured prolonged blockades. The cost of a *sétier* of wheat in Paris rarely exceeded ten livres in the 1580s, but exceeded 19 livres in

---

225; Teodoreanu, Elena, "Preliminary Observations on the Little Ice Age in Romania," *Present Environment and Sustainable Development* 5 (2011): 187-194; and White, *The Climate of Rebellion*, 153-155, 175 and 178.

51 Yılmazer, *Topçular Kâtibi 'Abdülkādir (Kadrî) Efendi Tarihi*, 1:344-345, quoting a report to the sultan from rebellious sipahis in Istanbul, 4 January 1603. On the links between climate change, the Great Flight (*Büyük Kaçgunluk*) and the Çelali rebellions, see White, *The Climate of Rebellion*, chap. 7 (quotation from p. 185). On the Çelali rebellions, see Griswold, William J., *The Great Anatolian Rebellion, 1000-1020/1591-1611* (Berlin: K. Schwarz Verlag, 1983).

52 Le Roy Ladurie, Emanuel, *Historie humaine et comparée du climat I: Canicules et glaciers (XIIIe–XVIIIe siècle)* (Paris: Fayard, 2004), 239-242; and *II: Disettes et révolutions (1740-1860)* (Paris: Fayard, 2006), 522-539.

the famine years of 1592, 1593 and 1597, and forty livres when the city came under siege in 1591.[53] Further south, serious peasant revolts (the *Tards-Avisés* or *Croquants*) paralysed Périgord and Limousin between 1593 and 1595; while the East of France experienced a "crise catastrophique" in 1592-1593 and 1596-1597 caused by a combination of harvest failure and war. In the southeast, Notary Eustache Piémont recorded several famines in his "Memoirs". He blamed them in part on plague and military operations, but the main culprit was always the hostile climate that ruined the harvest: heavy precipitation (prolonged snow followed by endless rain) in 1590, 1591, 1598 and 1599; drought sandwiched between snow and rain in 1595.[54] In the West, Canon Jean Moreau identified four scourges in his Breton chronicle: war, famine, plague, and wolves. Although other Frenchmen believed that the "great dearth and famine" had caused most damage, because they produced "such a great scarcity that even the richest peasants died of hunger", Moreau claimed that the wolves did the worst damage. "War did not afflict a place for ever", he noted (and people could often avoid danger through flight), while "famine only lasted one year or two", and a plague epidemic scarcely six months. By contrast, "the wolves continued their furious attacks" for almost a decade. According to Moreau, the events "that we have seen with our own eyes, heard with our own ears, and experienced" during the civil war caused "the death of more than one million people" as well as creating desolation and insecurity—admittedly "not always at the same time throughout the kingdom but always somewhere, and sometimes everywhere".[55] By the time the civil war ended, and France made peace with Spain in 1598, the combination of famine, plague, war and wolves had reduced agricultural and

53 Baulant, Michèle, and Jean Meuvret, *Prix des céréales de la mercuriale de Paris I: 1520-1620* (Paris: SEVPEN, 1960), 243.

54 Greengrass, Mark, "The Later Wars of Religion in the French Midi," in *The European Crisis of the 1590s: Essays in Comparative History*, ed. Peter Clark (London: George Allen & Unwin, 1985), 105-134 (above all the map of wars and revolts in the kingdom at p. 107 and the tabulation of Piémont's climatic data at pp. 126-127). Other data from: Benedict, Philip, "Civil War and Natural Disaster in Northern France," in *The European Crisis of the 1590s: Essays in Comparative History*, ed. Peter Clark (London: George Allen & Unwin, 1985), 84-105; Cabourdin, Guy, *Terre et hommes en Lorraine, 1550-1635: Toulois et comté de Vaudémont*, 2 Vols. (Nancy: Annales de l'Est, 1977), 365, 392; and Bercé, Yves, *Histoire des croquants* (Paris: PUF, 1986), 89-100.

55 Waquet, Henri, ed. *Mémoires du Chanoine Jean Moreau sur les Guerres de la Ligue en Bretagne* (Quimper: Archives historiques de Bretagne, 1960), 4, 2. Other pessimistic eyewitnesses are quoted in Croix, Alain, *Nantes et le pays nantais au XVIe siècle: Étude démographique* (Paris: SEVPEN, 1974), 165-166, and Croix, Alain, *Bretagne aux XVIe et XVIIe siècles: La vie, la mort, la foi*, 2 Vols. (Paris: Maloine 1981), 279-281.

industrial production to the lowest levels recorded in three centuries, and diminished the population of the kingdom by perhaps one-fifth.[56]

Adverse weather also afflicted England in what Patrick Collinson has called "the nasty nineties". In 1596, Queen Elizabeth's chief minister received a report that

> The estate of the poore is verie miserable, for their neade is great, and by reason of this extraordinary and excessive wett, worke is verie harde to be gotten, whereby their earninges towardes their maintenance are too little, and therewith allsoe charity genrally is verie colde.

The writer predicted that "[t]his yeare wyl be the hardeste yeare for the poore people that hathe happened in anie man's memory".[57] William Shakespeare agreed. In *A Midsummer Night's Dream*, probably written and first performed that same year, Titania lamented that

> [...] the winds, piping to us in vain,
> As in revenge, have suck'd up from the sea
> Contagious fogs; which falling in the land
> Have every pelting river made so proud
> That they have overborne their continents:
> The ox hath therefore stretch'd his yoke in vain,
> The ploughman lost his sweat, and the green corn
> Hath rotted ere his youth attain'd a beard [...]
> And through this distemperature we see
> The seasons alter: hoary-headed frosts
> Far in the fresh lap of the crimson rose,
> And on old Hiems' thin and icy crown
> An odorous chaplet of sweet summer buds
> Is, as in mockery, set: the spring, the summer,
> The chiding autumn, angry winter, change
> Their wonted liveries, and the mazed world,

56 Dupâquier, Jacques, ed., *Histoire de la population française I: Des origines à la Renaissance* (Paris: PUF, 1988), 60, estimated the population of France (1520 borders) at 14.3 million in 1580 but only 11.7 million in 1600.

57 Collinson, Patrick, "Ecclesiastical Vitriol: Religious Satire in the 1590s and the Invention of Puritanism," in *The Reign of Elizabeth I: Court and Culture in the Last Decade*, ed. John A. Guy (Cambridge: Cambridge University Press, 1995), 170; British Library, *Lansdowne Ms* 81/152v, Sir Henry Cock to Lord Bughley, July 21, 1596, a year of high volcanic activity and a strong El Niño.

By their increase, now knows not which is which.[58]

The following year the curator of the London College of Physicians' herb garden noted that he had once been able to plant "the blacke winter cherrie [that] is brought out of Spaine and Italy, or other hot regions [...] in my garden where they haue borne floures"; but now they "haue perished before the fruite could growe to maturitie, by reason of those vnseasonable yeeres, 1594. 95. and 96".[59]

Dr John King, a celebrated preacher, observed in a sermon that the "abundance of the rains that fell" was worse "than the agedest man of our land is able to recount", because "our July hath been like to a February; our June even as an April; so that the air must needs be corrupted". In another sermon, in 1596, "concerning a dearth for three years now successively", Dr King complained about the "famine God has brought upon our land".

> One year there hath been hunger; the second there was dearth; and a third there was great cleanness of teeth. And see, whether the Lord doth not threaten us much more, by sending such unseasonable weather and storms of rain among us; which if we will observe, and compare it with that which is past, we may say that the course of nature is very much inverted. Our years are turned upside down. Our summers are no summers; our harvests are no harvests; our seedtimes are no seedtimes. For a great space of time, scant any day hath been seen that it hath not rained upon us.[60]

Conditions remained "*inverted*" for one more year, and near Oxford a vicar recorded in his Parish Register that 1597 had been "a sorroful time for the poore of the land; God grant that such a darth and famyne may never be sene again".[61]

---

58 Shakespeare, William, *A Midsummer Night's Dream*, Act II, Scene I, Lines 88-95, 106-114.

59 Gerard, John, *The Herball, or General Generall Historie of Plantes* (London: 1597), 271 (book II, chapter 52, 'Of Winter Cherries'). Gerard's garden lay between Chancery Lane and Fetter Lane in central London.

60 Strype, John, ed., *Annals of the Reformation and Establishment of Religion and Other Various Occurences in the Church of England, during Queen Elizabeth's Happy Reign* (Oxford: Clarendon Press, 1824), vol. 4, 293-295, sermon by Dr John King, dated by Strype to 1594. Abbot, George, *An Exposition Upon the Prophet Ionah* (London: 1600), 365-366, a lecture delivered in Oxford in 1596, used virtually the same language, so that either one author copied the other, or else both copied a third, now lost.

61 Walter, John, "A 'Rising of the People'? The Oxfordshire Rising of 1596," *Past and Present* 107 (1985): 95, quoting the vicar of Wendlebury.

The failure of four consecutive harvests produced (in the words of historian M.J. Power) "the most sustained and severe inflation of prices in the sixteenth and seventeenth centuries and culminated in the lowest real wages in English history in 1597"—the year in which English Privy Council blamed the "late dearth of all kind of graine and of butter and cheeses and other victuells in the moste parte of this realme" on the "unseasonable wether". Pamphleteers likewise attributed the "dearth of graine", which had caused the ranks of the poor to swell "in all countreys", to "unseasonable weather, especially in harvest time."[62] Almost one-fifth of English parishes experienced a major subsistence crisis between 1596 and 1598: baptisms and marriages plunged while burials and migration soared.

Thanks to its ability to import grain, London suffered less, but a plague epidemic in 1592-1593 carried away almost ten percent of the capital's population and another in 1603 killed almost twenty percent: over 25,000 people.[63]

In England, as in Spain, dearth coincided with an increase in public disorder. Criminal prosecutions in the Courts rose sharply, especially for thefts involving foodstuffs, and in 1596 the magistrates of London created temporary "street cages" where they locked up vagrants and vagabonds.[64] In the West Country, one magistrate warned the central government about "the fury of the inferior multitude, if they should happen to break out into sudden outcry for want of relief, as (without good circumspection) many suspect they may and will do"; while another predicted the outbreak of organised disorder because

62 Power, Michael J., "London and the Control of the 'Crisis' of the 1590s," *History* 70 (1985): 371; Dasent, John R., *Acts of the Privy Council of England: New Series XXVII, 1597* (London: HMSO, 1908), 359, letter to all local magistrates, August 27, 1597; Anonymous [Henry Arthington], *Provision for the Poore, Now in Penurie, Out of the Store-House of God's Plenty* (London: 1597), sig C1v—although he reminded his readers that all "ill weather proceeds directly from the justice of God".

63 Data from: Hindle, Steve. "Dearth, Fasting and Alms: The Campaign for General Hospitality in late Elizabethan England." *Past and Present* 172 (2001): 44-86; Appleby, Andrew B., *Famine in Tudor and Stuart England* (Liverpool: Liverpool University Press), 1978, 134-141; Slack, Paul, *The Impact of Plague in Tudor and Stuart England* (London: Routledge & Kegan Paul, 1985), 58-78 (based on records from over three hundred parishes and 14 cities) and 145-151; and Power, Michael J., "A 'Crisis' Reconsidered: Social and Demographic Dislocation in London in the 1590s," *London Journal* 12 (1986): 134-145.

64 Power, "London and the Control of the 'Crisis'", 380, noted the "street cages"; Lawson, Peter, "Property Crime and Hard Times in England, 1559-1624," *Law & History Review* 4.1 (1986), table 4 at p. 110, showed that during the dearth of 1597-1598 prosecutions almost doubled in Hertfordshire, Surrey and Sussex, and more than doubled in Kent. Wrightson, Keith, and John Walter, "Dearth and the Social Order in Early Modern England," *Past and Present* 71 (1976): 24-25, found that prosecutions in Essex for theft more than doubled.

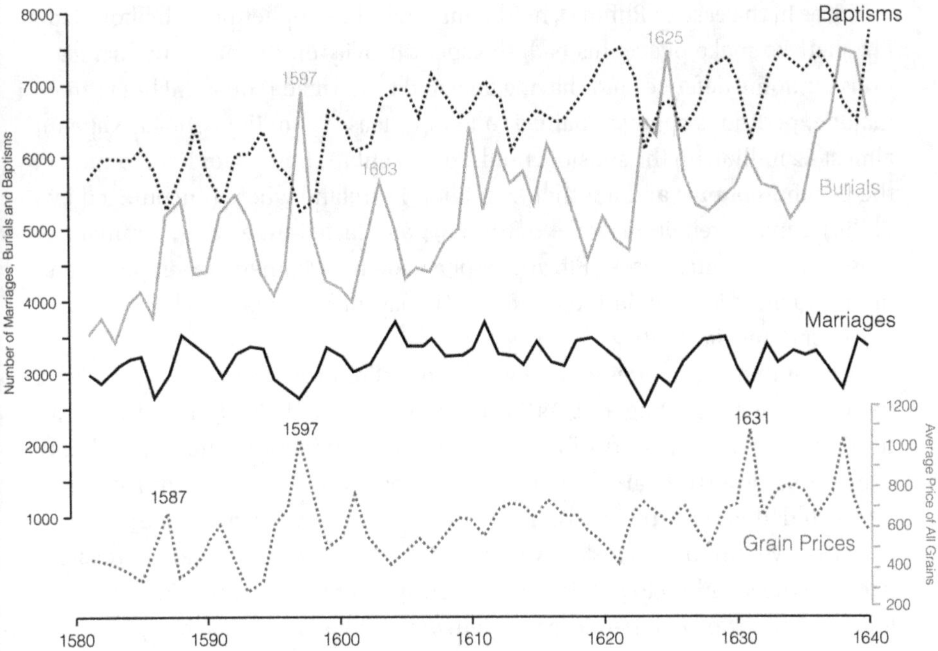

FIGURE 4.6 *Grain prices, baptisms, burials, and marriages in England, 1581-1640*

"the great scarsytye and dearth" had abated respect and fear, and many people claimed "that they weare as good be slayne in the markett place [stealing grain] as starve yn their owne howses".[65] No sooner had these concerns receded than the queen's Favourite, the earl of Essex, rebelled, provoking Elizabeth to fill a holograph letter to her cousin and likely successor, James VI of Scotland, with despair at "this wretched faithless time, in which the subjects have license to deprive both life and land from prince's rule". The queen concluded in apocalyptic terms that "These latter days of the world are too weak to retain so sound bodies as may carry good minds, but rather all inclined to what may be worst thought and wickedliest done".[66]

65 Historical Manuscripts Commission, *The Manuscripts of the Duke of Somerset, Fifteenth Report,* Appendix VII (London: HMSO, 1897), 20, Earl of Bath to Privy Council, Devon, October 25, 1596; British Library, *Lansdowne Ms* 83/49, Edward Hext to Lord Burghley, Somerset, February 21, 1597.

66 British Library, *Additional Ms.* 23,240/148r–149r, Elizabeth to James VI, spring 1601. I thank Rayne Allinson for bringing this document to my attention, for providing a transcript and for helping me to establish the probable date.

Yet, as in the case of Philip II, neither natural catastrophes nor rebellions led Elizabeth to make peace. Instead, she spent heavily on the navy, on fortifications for home defence, and (having defeated the Armada) on launching three major expeditions against Spain: a total of at least £2 million. She also spent almost £2 million on the subsidies and troops sent to Henry IV of France and to the Dutch Republic; and another £2 million in Ireland where (encouraged by Philip) a major rebellion broke out in 1594 and lasted nine years. To finance these various enterprises, Elizabeth persuaded Parliament to impose "an unprecedented level of national taxation" to pay for her wars, which continued until her dying day in 1603.[67]

The combination of dearth and war in the 1590s nevertheless had one positive consequence for England. As Steve Hindle has noted, the decade proved "a turning point in the history of English poor relief" because the numerous laws concerning poverty, charity and poor relief passed by Parliament in 1597 and 1601 would "dominate the lives of many English men, women and children over the next two hundred years".[68] Above all, the Poor Law of 1601 required each parish in the kingdom to provide certain "entitlements" (as we would call them today) at times of need—old age; widowhood; illness and disability; unemployment; dearth—but only to those normally resident within the parish. Funding for the system came from a tax on income from local property, administered by "Overseers" chosen from the local property owners who presented their accounts annually for audit by the local magistrates. The devolution of public welfare to parish level proved a stroke of genius. The involvement of magistrates (backed up, if necessary, by the royal courts) ensured that all the rich contributed; while the limitation of benefits to local residents meant that, in a crisis, the poor would stay in their parishes where their 'entitlements' were guaranteed, instead of seeking relief in the nearest town and overloading its resources (as happened in other countries). Although the system was by no means perfect, the climate-induced crisis of the 1590s nevertheless gave rise to

67 Naval spending is from Parker, Geoffrey, "The *Dreadnought* Revolution of Tudor England," *Mariner's Mirror* 82 (1996): table at p. 289; other figures are from Outhwaite, Brian, "Dearth, the English Crown, and the 'Crisis of the 1590s'," in *The European Crisis of the 1590s: Essays in Comparative History*, ed. Peter Clark (London: George Allen & Unwin, 1985), 23-43; quotation from Power, "London and the Control of the 'Crisis'," 382.

68 Hindle, "Dearth, Fasting and Alms," 48; Dean, David, *Law-making and Society in Late Elizabethan England: The Parliament of England, 1584-1601* (Cambridge Cambridge University Press, 1996), 287.

the first 'welfare state' in Europe, if not in the world; and in doing so, it provided an essential precondition for the first Industrial Revolution.[69]

The Dutch Republic also registered gains as well as losses from waging war in a period of climate change. Like other parts of the Northern hemisphere, the Netherlands experienced a succession of cool summers and bitter winters in the 1590s when, according to the English traveller Fynes Moryson,

> Rivers, lakes and pools, [...] are almost all winter frosen over with a thicke ice, so as they will beare some hundreths of young men and women, sliding upon them with pattins [skates] according to their custome. [Even the Zuider Zee] though it be large and deepe, [...] is many times in winter so frosen over, as victualers erect tents in the middest of it, having beere and wine and fier made upon iron furnaces, to refresh such as pass upon sledges, or sliding upon iron pattens from one shoare to the other.

According to a chronicler, 1594-1595 was "the coldest winter that we have ever experienced. The frosts started three weeks before Christmas, and lasted at least a month; and then after two or three days it started to freeze hard again for another six weeks". When the snow and ice melted in March, the rivers burst their banks; and then in April another three feet of snow fell, producing more floods when they melted. The Netherlands saw winter snow twice as often in the 1590s as in the decades before and after, as well as two complete harvest failures (in 1590 through drought, and in 1594 through late frost and then excessive rain).[70] In addition, the war with Spain brought devastation to many communities along the Eastern and southern borders of the Republic but Philip's reduced ability to deploy his resources in the Netherlands allowed the small army commanded by Maurice of Nassau to force one unpaid Spanish garrison after another to surrender. By 1598, the area in revolt against Spain had almost doubled.[71]

---

69 Solar, Peter M. "Poor Relief and English Economic Development before the Industrial Revolution," *Economic History Review* 48 (1995): 4-6.

70 Moryson, Fynes, An Itinerary Containing his Ten Yeeres Travell: Through the Twelve Dominions of Germany, *Bohmerland, Sweitzerland, Netherland, Denmarke, Poland, Italy, Turky, France, England, Scotland & Ireland,* 4 Vols. (Glasgow, 1908), 4-53; Buisman, Duizend jaar weer, 149, quoting the "Dagboek" of Jan de Pottre. Ibid., 104-213, documents the climate of the Netherlands and its neighbours year by year from 1587 to 1602: see, for example, p. 155 on the areas flooded in spring 1595.

71 Noordegraaf, Leo, "Dearth, Famine and Social Policy in the Dutch Republic at the End of the Sixteenth Century," In *The European Crisis of the 1590s: Essays in Comparative History,* ed. Peter Clark (London: George Allen & Unwin, 1985), 67-83.

5 A Global Crisis in the 1590s

Many other parts of the Northern hemisphere experienced climatic adversity. In North America, the "epic drought" that began in the 1540s and culminated in the last two decades of the century "stands out as the most severe sustained drought in this region for the past 1500 years", with precipitation failing for multiple summers and winters on end.[72] In Europe, hostilities between Sweden and Poland both exacerbated and prolonged the impact of natural disasters. In Sweden, when the Lutheran Duke Charles of Södermannland organised opposition to his Catholic nephew King Sigismund, the latter (also king of Poland) invaded Sweden and the scorched earth campaign of 1598-1599 caused extensive damage. Popular revolts broke out both in Finland, which (except for 1599) experienced a major crop failure every year between 1595 and 1601, and in the Polish-Lithuanian Commonwealth.[73] Dutch attempts to find a "Northeast passage" to China all failed, in part because of the extreme cold. An account of the expedition of 1594-1595 made the point explicitly: the native inhabitants of the islands beyond the North Cape told their frozen Dutch visitors that "there had never been such a hard winter and unstable summer".[74]

In Russia, according to an English visitor, the "tyrannous practice" of the tsars "has so troubled that country, and filled it so full of grudge and mortal hatred ever since, that it will not be quenched (as it seems now) till it burn again into a civil flame". Two catastrophic winters separated by a cool, wet summer (all probably caused by the eruption of Huaynaputina), destroyed the harvests of 1601 and 1602, and the little seed-corn planted over the following winter failed to ripen, producing a third year of famine and initiating a period later known in Russian history as "the great trouble" (*smuta*), in which both Poles and Swedes fanned the "civil flame" ignited by various rebel factions.[75]

72 Stahle, David W, et al., "Tree-Ring Data Document 16th-Century Megadrought over North America," *EOS* 81.12 (2000): 121-125; Stahle, David W., et al., "Tree-Ring Reconstructed Megadroughts over North America since AD 1300," *Climatic Change* 83 (2007): 133-149; White, Sam, "Cold, Drought, and Disaster: The Little Ice Age and the Spanish Conquest of New Mexico," *New Mexico Historical Review* 89 (2014): 425-458.

73 Myllyntanus, Timo, "Summer Frost as a Natural Hazard with Fatal Consequences in Preindustrial Finland," in *Natural Disasters, Cultural Responses: Case Studies towards a Global Environmental History*, ed. Christoph Mauch and Christian Pfister (Lanham: Rowman & Littlefield, 2009), 83; Mauelshagen, Franz, *Klimageschichte der Neuzeit* (Darmstadt: Wissenschaftiche Buchgesellschaft, 2010), 99, notes the popular revolts.

74 Degroot, Dagomar, "Testing the Limits of Climate History: The Quest for a Northeast Passage during the Little Ice Age, 1594-1597," *Journal of Interdisciplinary History* 45 (2015): 459-484.

75 Bond, Edward, ed., *Russia at the Close of the Sixteenth Century* (London: Hakluyt Society, 1856), 34, 45, 163, 206, quoting Giles Fletcher. On the "forbidden years", when taxpayers

East Asia also suffered. According to a recent study, due to the combination of civil war and famine, "the late sixteenth century saw one of the most disastrous periods recorded in the Vietnamese annals". Never before "had there been so many refugees mentioned repeatedly in such a short time", and in the course of 1594 "a third of the population died of starvation".[76] In China, torrential rains between 1592 and 1596 caused both the Yellow and the Yangzi Rivers repeatedly to burst their banks, causing widespread floods that washed away homes, drowned livestock, submerged the wheat crop so it could not be harvested, and prevented the sowing of millet. Henan Province was hit particularly hard. In 1592-1593, 44 of the 96 surviving county gazetteers recorded famine; and in 1594 (the worst year for Vietnam) a provincial official, Yang Dongming, composed perhaps the most vivid account of an early modern famine ever written, later printed under the title *Album of the Famished* (*Jimin tushuo*), with 13 harrowing sketches of the consequences of dearth.[77] One portrayed a mother selling her children to a rich neighbour in return for a single pot of rice, while other mothers watch and wait to do the same (figure 4.8[78]). In another, five members of a family of seven had hanged themselves from the branches of the trees growing in the garden of the local magistrate who had failed in his duty to "nourish the people".

Yang boldly warned the Wanli emperor (1573-1620) that "The ruler depends on the people for wealth and status. If he wants to protect wealth and status he cannot allow his people to starve to death."

> If orders to suspend the land tax are not issued, the local authorities will continue to collect them strictly. If funds are not made available from the

lost their freedom of movement and peasants sold themselves into slavery to survive, see Dunning, Chester S.L., *Russia's First Civil War: The Time of Troubles and the Founding of the Romanov Dynasty* (University Park: Pennsylvania State Press, 2001), 68-69 and 97-99.

76 Li, Tana, *Nguyên Cochinchina: Southern Vietnam in the Seventeenth and Eighteenth Centuries* (Ithaca: Cornell University Press, 1998), 26 and 162, citing the *Đại Việt sử ký toàn thư* [Complete Annals of Đại Việt].

77 Satō Taketoshi 佐藤武敏, ed., *Chūgoku saigaishi nenpyō* 中国災害史年表 [Chronology of the History of Natural Disasters in China] (Tokyo: Kokusho Kankokai, 1993), 274-275 (based on *Mingshi* [History of the Ming]); Song Zhenghai 宋正海, *Zhongguo gudai zhongda ziran zaihai he yichang nianbiao zongji* 中国古代重大自然灾害和异常年表总集 [A Comprehensive Chronology of Significant Natural Disasters and Unusual Occurences in Ancient China] (Guangzhou: Guangdong jiaoyu chubanshe, 1992), categories 4-8, 9 and 5-1 through 5-12 (based on Local Gazetteers); and Des Forges, Roger V., *Cultural Centrality and Political Change in Chinese History: Northeast Henan in the Fall of the Ming* (Stanford: Stanford University Press, 2003), 35.

78 Des Forges, *Cultural Centrality*, 40-41. Image reprinted by courtesy of Roger Des Forges.

FIGURE 4.7 "The Hanging Tree", from Yang Dongming, *An Album of the Famished*

> royal treasury, people below will cease to look up to the throne with hope. If people are urged to pay their taxes when they cannot, there will be fines that will push the people into revolt.

It was better, Yang affirmed, for the emperor to spend ten percent of his resources at once on relief than to spend ninety percent later on repression.[79]

Yang's *Album* with its graphic images stirred the Wanli emperor to action. He made substantial donations both from his personal fortune and the central treasury, and he also encouraged his relatives and ministers to contribute to relief efforts. He diverted 150,000 bushels of rice from government reserves to Henan and sent a senior official, Zhong Huamin, who had already distinguished himself in distributing relief in the neighbouring province of Shandong, to coordinate welfare in Henan. Zhong immediately set up soup kitchens, medicines and refuges, pardoned prisoners who agreed to provide grain, and provided money to redeem those who had sold themselves or their children through desperation. He also reformed and stocked the public

79 Loc.cit., quoting Yang Dongming, *An Album of the Famished* (with more information at pp. 35-51 and 76-77.)

granaries so that future harvest shortfalls would not produce a similar catastrophe.[80]

Unfortunately, Wanli failed to respond effectively to the similar crises caused by subsequent extreme climate events: more floods in 1596; severe winters in 1594-1595 and 1597-1598; widespread drought in 1599 and 1600; and a combination of natural disasters in 1601-1602.[81] Instead, he deployed his resources to fight Japan, where Toyotomi Hideyoshi, having defeated the last of his domestic enemies, resolved to become ruler of China as well. He called on Korea, a Chinese vassal, to grant his troops safe passage. The Korean Court unwisely assumed that Hideyoshi was bluffing and so was completely unprepared when 150,000 Japanese troops invaded in 1592. Initially, superior Japanese tactics and the systematic use of atrocity prevailed; but eventually the pressure of adverse weather (floods and famine in 1593; famine again in 1594), combined with tenacious guerrilla resistance, Korean superiority at sea and Chinese military intervention, forced them to withdraw. The parties agreed to an armistice, but in 1597 Hideyoshi's troops invaded again until the same combination of weather, Korean guerrilla and naval action, and Chinese intervention drove them out for good.[82]

Although Korea and its Chinese allies won the war, victory came at a high cost. During the 1590s the Korean peninsula lost perhaps one-fifth of its total population and four-fifths of its arable land. In China, according to the latest study of the First Great East Asian War, the Wanli emperor

> Sent a total of 167,000 troops on two different occasions to Korea's defense and spent millions of ounces of silver. All along there were those in the government who favored negotiating a truce, but in the end the conscience of Wanli prevailed and the Ming lived up to their tributary responsibilities. [...] As Wanli declared: "The imperial court will not let losses get in the way and we won't rest until the bandits are extirpated and our vassal state is at peace.[83]

---

80 See details in ibid., 44-54, based on Zhong's report (Wanli donated 100,000 *shi*, which I take to be 148,000 bushels.)

81 See Satō, *Chūgoku saigaishi nenpyō*, 275-276; and Song, *Zhongguo gudai zhongda ziran zahai*, categories 4 and 5.

82 We should not dismiss Hideyoshi's plans—well described in Berry, Mary, E., *Hideyoshi* (Cambridge: Harvard University Press, 1982), 206-217—as a chimera: four decades later the Manchu, with far fewer resources, would accomplish his goals.

83 Swope, Kenneth W., *A Dragon's Head and a Serpent's Tail: Ming China and the First Great East Asian War, 1592-1598* (Norman: University of Oklahoma Press, 2009), 297 (quotation), with figures from p. 286. Swope offers the first comprehensive account of the military

Wanli—just like Elizabeth I and Philip II—placed foreign policy goals above domestic problems, using his resources to fight wars rather than to relieve "the Famished".

6 A Fatal Synergy

Deliverance came only after temperatures began to rise again, shortly after the volcanic dust from Huaynaputina settled. In Sweden, the same Pastor Magni who complained of floods and famines in the 1590s hailed the harvests of 1604, 1605, and 1606 successively as "so prolific" that "never has such a thing been [seen] in living memory". In 1607, his cup overflowed: "No more fruitful year, with such splendid and heavy grains and others crops, has there been in living memory." In Japan, too, the improved climate led to bountiful harvests, until in 1614 a chronicler wrote euphorically:

> What a marvellous age! Even peasants like me enjoy tranquillity and happiness, and there are wonderful things to be heard and seen. [...] [We] dwell in the land of bliss. If this is not a [Buddhist Paradise], then how is it that I and other men could meet with such great fortune?[84]

This revival reflected not only a milder climate but also more "responsible" policies by humans. In both Northern Italy and Northwest Spain, the catastrophic failure of the grain harvests in the 1590s encouraged farmers to experiment with maize, which proved more resistant to drought.[85] In addition, in almost all parts of Western Europe, governments at last embraced policies of retrenchment. Thus, having made the concessions necessary to end both the civil war and the war with Spain in 1598, Henry IV ruled France in peace for over a decade. This allowed him to reduce taxes while building roads and bridges, stimulate industry and commerce, and patronise manuals on how to

---

campaigns and their aftermath, but he largely ignores the adverse economic and climatic context.

84 Utterström, "Climatic Fluctuations," 64, quoting Pastor Petrus Magni; Elison, George, "The Cross and the Sword: Patterns of Momoyama History," in *Warlords, Artists and Commoners: Japan in the Sixteenth Century*, ed. George Elison and Bardwell L. Smith, 55-85 (Honolulu: University of Hawai'i Press, 1981), 55, quoting Miura Jōshin in 1614.

85 Società italiana di demografia storica, *La popolazione italiana nel Seicento* (Bologna: CLUEB, 1999), 333. The same partial adaptation occurred in Northeast Spain: Marcos Martín, Alberto, *España en los siglos XVI, XVII y XVIII: Economía y sociedad* (Barcelona: Crítica, 2000), 479-482.

improve agriculture (Olivier de Serres filled his 1,000-page *Théâtre d'Agriculture* with details and illustrations on how to farm more efficiently: seven editions appeared between 1600 and 1619). The Turkish sultan made peace in Hungary in 1606 and used his army to suppress the Çelali rebellion. Since King James of Scotland had no quarrel with Spain, as soon as he succeeded Elizabeth on the English throne he ordered an immediate ceasefire; while Philip III, for his part, made all the painful concessions required to conclude a permanent settlement in 1604. Five years later he did the same with the Dutch, and for a brief period the Spanish Monarchy was at peace.

In his excellent 1980 monograph *War and Rural Life in the Early Modern Low Countries*, Myron Gutmann drew on the detailed records of the Maasland (currently part of Belgium and the Netherlands) to attribute the catastrophe that struck the region in the mid-17th century to a "fatal synergy" between war and the weather. "Seldom did the effects of bad weather last more than two years without some recovery," Gutmann noted, while "epidemics and bad weather might produce a year or two of difficulties, and a few years of recovery; but they could not turn around [a] region's generally favourable economic and demographic climate". By contrast, he continued, "people suffered most when war combined with bad weather sufficiently severe to reduce agricultural production".[86] The same fatal synergy occurred in the 1590s: not only a sustained episode of climatic instability caused by solar cooling and high volcanic and El Niño activity, creating a series of extreme weather events that destroyed several harvests in succession, but also a spate of major wars, both between and within states. These greatly increased the demands on societies at the precise moment when they had fewer resources to spare than before.

7 A Turning Point in World History?

Although the entire Northern hemisphere apparently saw disaster throughout the 1590s, the decade only marked a decisive turning point for two areas: East Asia and the Mediterranean. The devastation caused by climate and war weakened not only Korea but also China. The Wanli emperor managed to finance his first campaign against the Japanese invaders (1593) with reserves built up in

86 Gutmann, Myron P., *War and Rural Life in the Early Modern Low Countries* (Princeton: Princeton University Press, 1980), 199. Gutmann was one of the first modern historians to devote serious attention to the impact of the weather on a specific region: see his data on "extreme events" in the Maasland at pp. 81-82 and 211-230.

the years of prosperity, but the cost of the second campaign (which lasted until 1601, when the emperor finally judged it prudent to demobilise his troops) severely overstrained the treasury at a time of nationwide harvest failure. This led to the neglect of other security issues, especially the control of Manchuria, allowing a minor leader called Nurhaci to unite into a single confederation the various tribes that inhabited the steppes of central Asia. By 1600 Nurhaci commanded at least 15,000 warriors, organised into Banners (*gūsa*, the Manchu term for a "large military division"), each one commanded by a member of Nurhaci's family, and in 1618 they occupied most of China north of the Great Wall. A decade later, Banner troops forced Korea to acknowledge Manchu suzerainty, and in 1644 they captured Beijing and founded a new Imperial dynasty that replaced the Ming.

Likewise, neither the Spanish nor the Ottoman Empire entirely recovered from the crisis of the 1590s. Since many of the people living around the Mediterranean relied on a single crop of wheat or barley, agriculture in the region was highly vulnerable to floods and droughts. Moreover, Mediterranean commerce remained a very efficient vector for communicating plague. Consequently famines and epidemics were worse, and population loss much higher, in the Anatolian, Balkan, Italian and Spanish peninsulas. Finally, although both empires benefitted from windfall flows of gold and silver from distant territories (America and Egypt), both also leaned heavily on the rural communities of the core territories (Castile and Anatolia) for extra taxation during military mobilisation. These core territories were among those most vulnerable to climate fluctuations, so that extreme weather events could suddenly undercut money and supplies for ambitious foreign wars.

It is no coincidence, as William S. Atwell has noted, that the 1590s saw an outpouring of "crisis and advice literature" in all three empires: *nasihatname* ("letters of advice") literature in the Ottoman Empire; *jingshi jimin* (vaguely, "manage the state's affairs, provide for the people") writers in China; and the *arbitristas* ("reformers") in Spain. All three complained of very similar problems, but none offered realistic remedies.[87] In short, although consideration of just Western European evidence in the later sixteenth century in isolation led some historians to question the scale and importance of the crisis, once we take into account the global context, and include climate as a protagonist in

87 Atwell, William S., "Ming Observers of Ming Decline: Some Chinese Views on the '17th Century Crisis' in Comparative Perspective," *Journal of the Royal Asiatic Society of Great Britain and Ireland* 120 (1988): 316-348. See also Parker, Geoffrey, *Global Crisis: War, Climate Change and Catastrophe in the Seventeenth Century* (London, New Haven: Yale University Press, 2013).

human affairs, an overwhelming case exists "for singling out the 1590s as a decade in which conditions were such as to warrant a special degree of attention".[88]

88 Elliott, "Yet another Crisis?," 305.

CHAPTER 5

The "Black Swan" of 1540: Aspects of a European Megadrought

*Christian Pfister**

Abstract

This paper provides coherent evidence on what Europe might expect in the case of a worst-case heat and drought event. The record-breaking heat in 1540 was an analogous case to the 2003 event, albeit more intense, longer lasting and affecting a larger area—extending from France to Poland and from Italy to Germany, also including Spain and Morocco. Both in Switzerland and in Poland, precipitation in spring, summer, and autumn was below twentieth century averages. Discharge deficits of ninety percent were assessed for major rivers. Due to the extreme soil dessication, maximum temperatures in early August probably rose above 40°C. By then, forest and settlement fires were ravaging throughout continental Europe.

Premodern societies were surprisingly resilient to extreme conditions, notwithstanding the widespread dysentery, cattle mortality, and forest fires. The majority of the impact of a 1540-like event on present societies would be caused by the resulting severe water shortage and its cascading across interlinked systems. In particular, fossil and nuclear energy production, which depend on a sufficient amount of cooling water, would be significantly affected. Such a shortage might entail longer-lasting power blackouts with disruptive impacts on societies and economies.

•••

* Acknowledgements are due to the Oeschger Centre for Climatic Change Research at the University of Bern for providing a workplace and logistic support, Dr. Ursula Bieber, University of Salzburg for providing evidence on Russia, Dr. Antonio Contino, University of Palermo, for providing evidence on Italy, Prof. Dr. Adriaan de Kraker, Free University, Amsterdam for providing information about dysentery in 1540, Dr. Franz Mauelshagen, Potsdam, for the critical and constructive lecture on the text, Tamara Widmer, Institute of History, University of Bern, for bibliographical and clerical work, and Prof. Dr. Alfred Johny Wüest, Swiss Federal Institute of Aquatic Science and Technology (EAWAG), Switzerland for providing information on green algae in Mediterranean rivers.

 | DOI 10.1163/9789004356825_007

> What is surprising is not the magnitude of our forecast errors, but our absence of awareness of it.
>
> NASSIM TALEB[1]

1 Black Swans and Hot Summers

A "black swan" is an outlier outside the realm of normal expectations carrying an extreme impact. Nassim Taleb, who coined the metaphor with regard to the global financial crisis, notes in this context, that most people expect all swans to be white because that's what their experience tells them; a black swan is by definition a surprise, at least outside a zoo, where black swans from Australia may be kept. Our minds are designed to retain, for efficient storage, past information that fits into a compressed narrative, but we are blind with respect to randomness, particularly to large deviations.[2] Taleb distinguishes between two different kinds of black swans. The *narrated* black swan, present in the current discourse, is overestimated, while the *implausible*[3] black swan, about which nobody talks, is underestimated and may thus come as a surprise.[4] To face the black swan, "you need a story to displace the story".[5]

With regard to climatic change, the implausible black swan is as a low-probability, high-impact event of an unimaginable magnitude. Studying the past is the best way to derive credible scenarios for such extremes, because what has already happened in the past may happen again. A recent World Bank study expects that extreme heat waves, that without global warming would be expected to occur once in several hundred years, will be experienced during almost all summer months in many regions.[6] Most climate scientists would argue that the well-known European heat wave in summer 2003 meets this

1 Taleb, Nassim N., *The Black Swan: The Impact of the Highly Improbable* (London: Allen Lane, 2007), XXV.

2 Taleb, *The Black Swan*, XX.

3 This term was coined by the author.

4 Taleb, *The Black Swan*, XX.

5 Ibid., XXXI.

6 World Bank, ed., "The Turn Down Heat: Why a 4°C Warmer World Must be Avoided," Report for the World Bank by the Potsdam Institute for Climate Impact Research and Climate Analytics, published November 2012, accessed November 20, 2014. <http://www.worldbank.org/en/topic/climatechange/publication/turn-down-the-heat-climate-extremes-regional-impacts-resilience>.

criterion. In this context, they would probably refer to the study by Isabel Chuine and co-authors published in the renowned journal *Nature*. The authors conclude from a series of grape harvest dates in Dijon (France) that the heat from April to August was probably "higher than in any other year since 1370".[7] Their findings are supported by Ulf Büntgen and co-authors who concluded from a series of tree-ring maximum latewood density measurements made in the inneralpine dry valley of Lötschental (Canton Valais, Switzerland) that the summer of 2003 was the warmest since AD 755.[8]

The summer of 2003 thus corresponds to the narrated black swan. However, was it really the worst case documented in the past centuries? García-Herrera and co-authors concede in their review article that "several summers, such as 1420 or 1540 and a few others, were as warm as or even warmer than 2003".[9]

Two recent papers demonstrate that the obsession of many scientists with summer 2003 obscured the view of the event in 1540. Not only was the heat wave in 1540 much longer, more extended and more severe than in 2003;[10] to make things worse, the associated drought persisted for eleven months, affecting most of continental Europe.[11] Several authors claimed that the drought of 1540 was outstanding (Pfister, Glaser et al., Brazdil et al. and Dobrovolný et al.).[12] Carlo Casty and co-authors concluded that the years 1540, 1921, and 2003

7 Chuine, Isabelle et al., "Historical Phenology: Grape Ripening as a Past Climate Indicator: Summer Temperature Variations Are Reconstructed from Harvest Dates since 1370," *Nature* 432 (2004): 289-290.

8 Büntgen, Ulf et al., "Summer Temperature Variation in the European Alps: A.D. 755-2004," *Journal of Climate* 19 (2006): 5606-5623.

9 García-Herrera, Ricardo et al., "A Review of the European Summer Heat Wave of 2003," *Critical Reviews in Environmental Science and Technology* 40.4 (2010): 267-306, accessed April 4, 2017 doi:10.1080/10643380802238137.

10 Wetter, Oliver, and Christian Pfister, "An Underestimated Record Breaking Event—Why Summer 1540 Was Likely Warmer than 2003," *Climate of the Past* 9 (2013): 1-16, accessed April 6, 2017. doi:10.5194/cp-9-41-2013; Wetter, Oliver et al., "The Year-Long Unprecedented European Heat and Drought of 1540—A Worst Case," *Climatic Change* 125 (2014): 349-363, accessed April 6, 2017. doi:10.1007/s10584-014-1184-2.

11 Wetter et al., "The Year-Long Unprecedented European Heat and Drought," 351.

12 Pfister, Christian, *Das Klima der Schweiz von 1525 bis 1860 und seine Bedeutung in der Geschichte von Bevölkerung und Landwirtschaft*, 2 Vols. (Bern: Paul Haupt, 1984); Pfister, Christian, *Wetternachhersage: 500 Jahre Klimavariationen und Naturkatastrophen 1496-1995* (Bern: Haupt, 1999); Glaser, Rüdiger et al., "Seasonal Temperature and Precipitation Fluctuations in Selected Parts of Europe During the Sixteenth Century," *Climatic Change* 43.1 (1999): 169-200, accessed April 5, 2017. doi:10.1023/A:1005542200040; Brázdil, Rudolf, "Fir Tree-Ring Reconstruction of March-July Precipitation in Southern Moravia (Czech Republic), 1376-1996," *Climate Research* 20.3 (2002): 223-239, accessed November 5, 2013.

were very likely the driest in the Greater Alpine area in the context of the last five hundred years,[13] while the year 1540 is not mentioned by Pauling and co-authors among the driest seasons in their gridded multi-proxy-reconstruction for Europe.[14]

The evidence for 1540 taken from documentary data offers a broad spectrum of evidence on past weather and climate and their societal impacts.[15] It is described in a sample of 312 first-hand documentary weather reports originating from continental Europe. Most chroniclers were mainly concerned about precipitation and its human and environmental impacts, which are thus more extensively discussed than temperature.[16]

The present study is organised as follows: The second section summarises the course of the European summer heat wave of 2003 discussing the pattern of temperature anomalies and drought from multiple perspectives and addressing the role of the main contributing factors involved in the occurrence of this event. The third section describes how contemporary chroniclers and diarists witnessed the course and the consequences of the 1540 event distinguishing between meteorological, hydrological, agricultural and socio-economic drought. It is then explained how spring-summer temperatures and seasonal precipitation were assessed for 1540. The fourth section compares the socio-economic impacts of the extreme events in 1540 and 2003, and discusses the vulnerability of present-day societies in the face of a recurrent 1540-like megadrought.[17] The final section summarises the results and the main lessons that can be drawn from the study.

---

doi:10.3354/cr020223; Dobrovolný, Petr et al., "Precipitation Reconstruction for the Czech Lands, AD 1501-2010," *International Journal of Climatology* 2 (2014), accessed April 6, 2017, doi:10.1002/joc.3957.

13 Casty, Carlo, et al., "Temperature and Precipitation Variability in the European Alps since 1500," *International Journal of Climatology* 25 (2005): 1825-1880.

14 Pauling, Andreas et al., "Five Hundred Years of Gridded High-Resolution Precipitation Reconstructions over Europe and the Connection to Large Scale Circulation," *Climate Dynamics* 26.4 (2006): 387-405, accessed April 6, 2017. doi: 10.1007/s00382-005-0090-8.

15 Brázdil, Rudolf et al., "European Climate of the Past 500 Years: New Challenges for Historical Climatology," *Climatic Change* 101 (2010): 7-40, accessed April 6, 2017. doi: 10.1007/s10584-009-9783-z.

16 Pfister, Christian, and Rudolf Brázdil, "Climatic Variability in Sixteenth-Century Europe and its Social Dimension: A Synthesis," *Climatic Change* 43.1 (1999): 5-53, accessed April 6, 2017. doi: 10.1023/A:1005585931899.

17 The term "megadrought" is used here to denote an extreme drought extending over several seasons. In the US southwest, where droughts are common, the term is used for persistent period of drought conditions lasting decades to centuries. (Cf. Coats, Sloan et al.,

2 The Heat Wave in Summer 2003

The atmospheric circulation in summer of 2003 was characterised by a northward displacement of the Azores anticyclone which shielded the continent from the Atlantic depressions. The shift led to consecutive episodes of intense anti-cyclonic anomalies, particularly for the months of June and August. The extensively heated continental surface induced strong meridional (South-North) air flow components which intensified the already elevated temperatures.[18]

The summer of 2003 set new standards for heatwaves and heat-periods within large parts of Europe. The largest anomalies (+4 to +5°C) were measured in a strip extending from Southern France over the Alps to the Adriatic Sea, while other parts of Central Europe still experienced anomalies of +2 to +4°C. It was the hottest summer since the beginning of instrumental measurements in most of Germany[19] with average maximum temperatures up to 30.4°C (Karlsruhe).[20] In the centre of the anomaly (Switzerland) the summer average temperature was 4 to 5.5°C higher than in the 1961-1990 reference period, surpassing the warmest summer since the beginning of systematic instrumental measurements (1864) by 2 to 3°C. Likewise, the highest average summer temperature anomaly in France was 5.3°C in Lons-le-Saunier, situated in the plain west of the Jura Mountains.[21] Maximum temperatures exceeding 30°C were measured in Basel on 51 days.[22] Two distinct periods of exceptional heat occurred during the season: the first in June and the second in the first half of August. In June, the duration of the heat was exceptional, rather than the maximum temperatures. During the August heatwave, France, Germany, Switzerland and the UK experienced record-breaking maximum temperatures.[23] In Portugal, the historical record of absolute extreme temperatures was surpassed in early August with anomalies of up to 11°C in the central sector of the country.[24] Precipitation in spring and summer 2003 was substantially below average

---

"Megadroughts in Southwestern North America in ECHO-G Millennial Simulations and their Comparison to Proxy Drought Reconstructions," *Journal of Climate* 26 (2013): 7636.

18 García-Herrera et al., "A Review," 277-279.

19 Bundesanstalt für Gewässerkunde, *Niedrigwasserperiode 2003 in Deutschland: Ursachen – Wirkungen – Folgen*, Mitteilungen, Nr. 27 (Koblenz, 2006).

20 Wahl, Laurent et al., "Les canicules de l'été 2003: Un événement météorologique exceptionnel dans le quart nord-est de la France," *Revue géographique de l'est* 45.2 (2005): 67-77.

21 Ibid., 70.

22 Ibid., 69.

23 García-Herrera et al., "A Review," 269.

24 Ibid., 269.

within a zone extending from the Pyrenees over Southern France, Southern Germany, Switzerland, Austria and Northern Italy to the Northern Balkans.[25] Monthly precipitations in France, Switzerland and in Germany were below average from February to October with a maximum deficit of fifty percent in June.[26] The smallest deficits were measured in May, July and September.[27] In France, they were between 20 and 55 percent in summer. However, the summers of 1921, 1964, 1976 were much drier in this country.[28] Likewise, spring and summer in Switzerland were not particularly dry considering that spring precipitation in 2003, since the beginning of regular measurement in 1864, ranks 14th in Geneva and 43rd in Zürich. The values for summer are 16th in Zürich and 51st in Geneva.[29] The authors of a study by the Swiss Office for the Environment and the National Weather Service MeteoSwiss concluded that the summer half year in 1921, 1929, 1944 was drier than 2003, to a considerable extent also in 1947 and 1976, and that 1540 was probably even drier.[30] In summer the drought also affected the Czech Republic, Slovakia and parts of Romania. Annual precipitation deficits in Austria were twenty to thirty percent with substantial regional differences.[31] The conditions under which temperatures rise to record breaking levels were investigated after the 2003 event.[32] There is consensus that soil moisture-temperature interactions were a

---

25 Bundesanstalt für Gewässerkunde, *Niedrigwasserperiode 2003*, 21.

26 Wahl et al., "Les canicules," 10; Bundesamt für Umwelt, Land und Landschaft (BUWAL), "Hitzesommer 2003: Synthesebericht 2005," accessed March 28, 2017. <https://web-beta.archive.org/web/20160802130808>/<http://www.occc.ch/products/heatwave03/PDF_D/Hitzesommer03_d.pdf>.

27 Bundesanstalt für Gewässerkunde, *Niedrigwasserperiode 2003*, 21

28 Wahl et al., "Les canicules," 69.

29 Values from Begert, Michael, Thomas Schlegel, and Walter Kirchhofer, "Homogeneous Temperature and Precipitation Series of Switzerland from 1864 to 2000," *International Journal of Climatology* 25.1 (2005): 65-80. doi:10.1002/joc.1118. Updated from Meteo Schweiz.

30 BUWAL, Auswirkungen des Hitzesommers 2003 auf die Gewässer. Schriftenreihe Umwelt 369 (Bern 2004), 20.

31 Eybl, Jutta et al., *Trockenheit in Österreich im Jahr 2003: Ein hydrologischer Situationsbericht* (Wien: Bundesministerium für Land und Forstwirtschaft, Umwelt und Wasserwirtschaft Abteilung VII/3, 2004), 1/11, accessed November 20, 2014. <https://www.google.de/search?q=Trockenheit+in+%C3%96sterreich+im+Jahr+2003%3A+Ein+hydrologische r+Situationsbericht+&ie=utf-8&oe=utf-8&client=firefox-b-ab&gfe_rd=cr&ei=cfvlWKCF PLLVXqW6tPAB>.

32 See the review article by Seneviratne, Sonia I. et al., "Investigating Soil Moisture—Climate Interactions in a Changing Climate: A Review," *Earth-Science Reviews* 99.3-4 (2010): 125-161. doi: 10.1016/j.earscirev.2010.02.004.

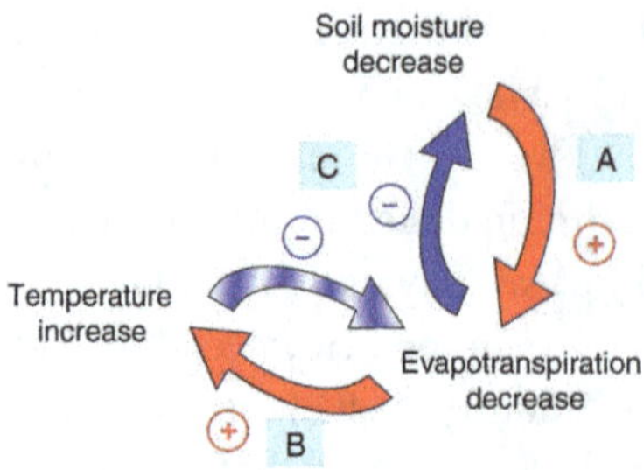

FIGURE 5.1
Processes contributing to soil moisture–temperature coupling and feedback loop

key driver in the sequence of events that led to the exceptional heat wave in August, 2003. In temperate climates, a considerable part of incoming shortwave radiation is generally used for evapotranspiration, i.e. for driving the so-called latent heat flux. The remaining sensible heat flux impacts air temperature. In case of an initial strong soil moisture deficit, which may occur after a dry spring, the share of sensible heat increases with the higher position of the sun in early summer, leading to higher air temperatures. This leads to a higher evaporative demand, and thus to a potential increase in evapotranspiration despite the dry conditions, possibly leading to a further decrease in soil moisture (figure 5.1[33]).

Sensitivity analyses suggest that, given climatologic mean soil moisture and similar continental-scale circulation, the 2003 summer (June, July, August) surface temperature anomalies would have been reduced by around forty percent. Thus, in the absence of soil moisture feedbacks, summer 2003 would still have been warm, but it would not have been such a devastating event as it turned out to be.[34]

Large rivers dropped to low levels. The discharge deficit of major German rivers in the hydrological summer half-year (April to September) was more than fifty percent for the river Meuse, situated in the Western part of the country, and for the river Odra situated in the North-eastern part, while it was somewhat lower for the Elbe (−47 percent) and the Rhine (−37 percent), fed by melt-water from the Alpine firns and glaciers.[35] However, the year 2003 does not rank among the smallest annual values in many very long German discharge-series.[36]

---

33 Seneviratne et al., "Investigating Soil Moisture."

34 Fischer, Erich M. et al., "Soil Moisture–Atmosphere Interactions during the 2003 European Summer Heat Wave," *Journal of Climate* 20 (2007): 5097. doi:10.1175/JCLI4288.1.

35 Bundesanstalt für Gewässerkunde, *Niedrigwasserperiode 2003*, 77.

36 Ibid., 94-95.

Lake Constance, the largest lake in Central Europe, dropped to the lowest monthly average level in August and September registered since 1866. Likewise, the level of Lake Maggiore (Southern Switzerland/Northern Italy) was extremely low in July and August. The strong insulation and the extreme temperatures warmed the watercourses up massively. Water temperatures of 26°C were measured in the High Rhine at the climax of the heat wave on 12th August leading to the death of 52,000 graylings (*Thymallus thymallus L*).[37] Temperatures in the German catchment area of the Rhine were even up to 28°C.[38] It was estimated that Alpine glaciers lost five to ten percent of their total mass. The freezing limit rose above 4,500 metres above sea level for ten days, which resulted in anomalous thawing and degradation of mountain permafrost and thus a destabilisation of Alpine rock walls.[39]

Large forest fires were not registered in continental Europe, in contrast to the Mediterranean region. The 2003 fire season in Portugal represents an outlier year considering the outstanding total burned area of roughly five percent of the Portuguese territory.[40]

3 The Year 1540—an Evidence-based Worst Case Scenario

The first part of this section presents a set of documentary narratives describing the mega-drought of 1540. The focus is put on Northern Switzerland, Eastern France and Southern Germany, from where the most detailed weather descriptions are available. Probably this region was—like in 2003—situated in the centre of the anomaly. It may then become clear from these narratives why both grape-harvest dates and tree rings failed to adequately indicate the record-breaking heat wave in 1540. Subsequently it is outlined how seasonal temperature and precipitation were assessed and in what sense these results are physically coherent. With regard to dating it needs to borne in mind that the Julian style was still generally in use in 1540. The Gregorian reform was only introduced step by step after 1582. Julian dates are converted to Gregorian style

37 ProClim-Forum, and Plattform of the Swiss Academy of Sciences, "Hitzesommer 2003," 8, accessed March 29, 2017. <http://proclimweb.scnat.ch/products/heatwave03/PDF_D/Hitzesommer03_d.pdf>.

38 Bundesanstalt für Gewässerkunde, *Niedrigwasserperiode 2003*, 168.

39 García-Herrera et al., "A Review," 295.

40 Trigo, Ricardo M. et al., "Atmospheric Conditions Associated with the Exceptional Fire Season of 2003 in Portugal," *International Journal of Climatology* 26.13 (2006): 1741-1757. doi:10.1002/joc.1333.

by adding ten days, which is, of course, not possible, when descriptions refer to an entire month.

Documentary sources, being understood as physical units of man-made information on weather and climate, may contain two different kinds of data, namely, narrative accounts on (unusual) weather spells and weather-induced disasters on the one hand, and references to (bio) physical proxy evidence such as vegetation advances or delays in the boreal summer half year and the presence or absence of frost, ice and snow-cover on the other hand. Narrative accounts put biophysical observations within their meteorological context by shedding light on the interplay of different weather elements, such as heat, frost, wetness and drought and their impact on ecosystems and societies, which is particularly important for the analysis of record-breaking extreme events. With regard to 1540, the analysis of the megadrought mainly draws on coherent narrative accounts, while the assessment of the related heat wave draws on biophysical evidence.

Another important distinction is related to the agents who kept the sources distinguishing between individuals and institutions. Individuals were mainly motivated by concerns about the human impacts of extreme events. In doing so, chroniclers referred to (bio) physical proxy evidence as quasi-objective indicators that could be compared over time.

Institutions such as monasteries and town administrations provided public services within given territorial structures. Their office holders laid down year-by-year records on weather-dependent resources and activities, such as the time of grape and grain harvests, often for several centuries. Institutional evidence thus has the advantage, in comparison to narrative accounts, of providing continuous, quantitative and largely homogenous series of biophysical proxy data, similar to evidence from archives of nature.[41] For the present analysis, institutional proxy data in the form of grape harvest dates provide the framework for the assessment of temperatures. Prior to the French revolution, vine-growers were not free to harvest at will. They had to wait for an order by the municipality to begin the harvest.[42] In the weeks preceding the harvest, the vineyards were banned, i.e. guarded day and night. The main reason for the

41 Pfister, Christian et al., "Documentary Evidence as Climate Proxies," "White Paper" written for the Proxy Uncertainty Workshop in Trieste, 9-11 June, 2008, accessed November 6, 2013. <http://www.pages.unibe.ch/index.php/products/2029-proxy-data-uncertainties-white-paper-documentary-evidence-as-climate-proxies>.

42 Garnier, Emmanuel et al., "Grapevine Harvest Dates in Besançon (France) between 1525 and 1847: Social Outcomes or Climatic Evidence?" *Climatic Change* 104.3-4 (2011): 703-727. doi:10.1007/s10584-010-9810-0.

vintage ban was preventing theft or clandestine harvesting before the owners of the vineyards and the beneficiates of tithe (i.e. taxes) could monitor the correct delivery of their dues. The date on which the grape harvest was enabled was noted down in the protocols of the community and communicated to all parties concerned.[43]

Subsequently, some narrative descriptions of the megadrought are presented.[44] Severe drought was reported from 312 chronicles representing an area of two to three million km² ranging from France to Poland including the London Basin and from Tuscany to the Northern border of Germany (figure 5.2[45]). Marc Stefanon and co-authors worked out a set of six typical European heat wave patterns for the period 1950 to 2009. The spatial location and dimension of the 1540 pattern agrees best with the pattern for "Western Europe" except that the drought in 1540 extended further east to Poland and farther south to the entire Po valley[46] (figure 5.2[47]).

Based on a network of 165 tree-ring series from the Alps, Italy, Spain, Morocco, Algeria, Greece, and Turkey, Nicault and co-authors reconstructed the widely-used Palmer Drought Severity Index for the larger Mediterranean land area over the last five hundred years.[48] Their reconstruction supports the documentary-based outstanding drought in the Alps, Italy and France. In addition, it points to severe drought in Southern Spain, Morocco, Algeria, and parts of Libya (figure 5.3[49]).

Summer 1540, as the reconstruction by Jacobeit and co-authors (figure 5.4[50]) illustrates, was characterised by a persistent diagonal South-west to North-East

---

43 Wetter and Pfister, "An Underestimated Record Breaking Event," 43.

44 Observations taken from Pfister, *Das Klima der Schweiz,* 138-140; Glaser et al., "Seasonal Temperature," 193.

45 Wetter et al. "The Year-Long Unprecedented European Heat and Drought of 1540," 352.

46 Stefanon, Marc, Fabio D'Andrea, and Philippe Drobinski, "Heatwave Classification over Europe and the Mediterranean Region," *Environmental Research Letters* 7 (2012): 1-9, doi:10.1088/1748-9326/7/1/014023.

47 Nicault, Antoine et al. "Mediterranean Drought Fluctuation during the Last 500 Years Based on Tree-Ring Data," *Climate Dynamics* 31 (2008): 227-245. doi: 10.1007/s00382-007-0349-3.

48 The Index developed in 1965 by the US Meteorologist Wayne Palmer is based on a supply-and-demand model of soil moisture assessed from precipitation and temperature (cf. Wikipedia, "Palmer Drought Index," accessed December 23, 2014. <http://en.wikipedia.org/wiki/Palmer_Drought_Index>).

49 Nicault et al, "Medterranean drought," 7.

50 Jacobeit, Jucundus et al., "European Surface Pressure Patterns for Months with Outstanding Climatic Anomalies during the Sixteenth Century," *Climatic Change* 43.1 (1999): 201-221. doi:10.1023/A:1005594216878206.

FIGURE 5.2 *Spatial dimension of the extreme drought in 1540*

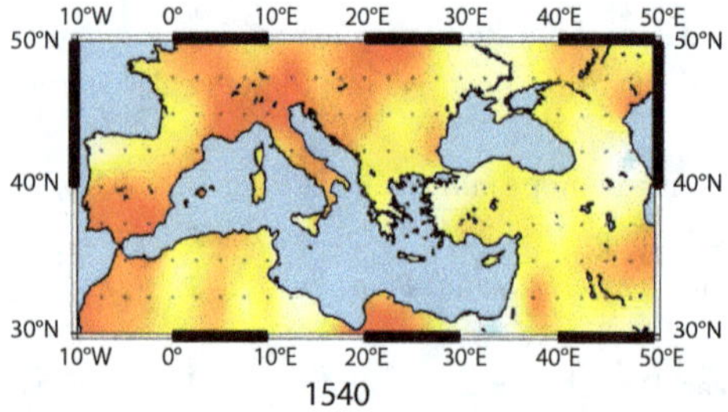

FIGURE 5.3 *Spatial dimension of the extreme drought in 1540 in the larger Mediterranean land area based on tree-ring evidence*

oriented ridge of high pressure over continental Europe connected to the Azores high widely blocking the passage of frontal zones into continental Europe.[51] According to this large-scale blocking situation, a multi-seasonal cold and wet spell might have prevailed in the region east of the anticyclone. Indeed, the famous chronicle of Novgorod (Western Russia) reports that

51 Jacobeit et al., "European Surface Pressure Patterns."

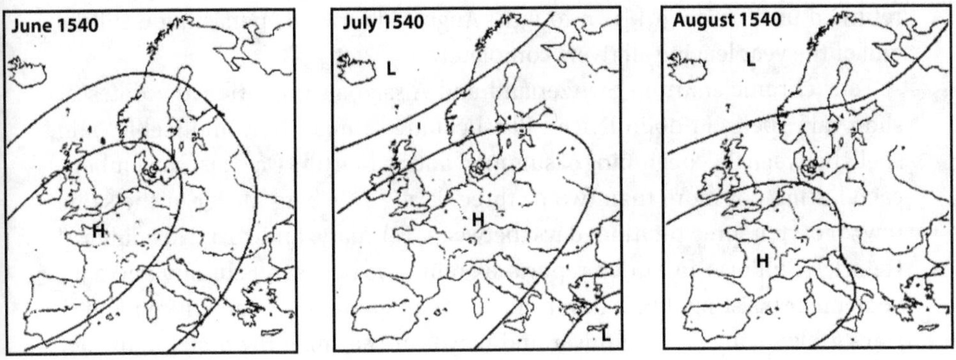

FIGURE 5.4 *Reconstructed monthly surface pressure patterns for 1540*

> Springtime was cold and during all summer there were floods, and the rye did not grow and was frozen in spring, and all the meadows at the banks of the rivers and lakes were flooded […]. In autumn there was a lot of rain, and the sun was not seen for two weeks until the eve before Filippov [15th November].[52]

Most chroniclers reported the date of the main rain-spells and the quasi-rainless period in between as a proxy for drought severity. For example, no rain fell in England "from June [about 10th July] to eight days after Michaelmas" [18th October].[53] In Louny (Bohemia), it only rained on 8th August within the period between 26th May and 13th October.[54] People in Northern Italy were worst affected, because it did also not rain between October 1539 and early April 1540.[55] In Varese Liguria (La Spezia, Italy) not a drop fell between mid-April and 30th June [10th July], when it rained for two days. Then the terrible drought

52 *Novgorodskija i pskovskija lětopisi* [The Chronicles of Novgorod and Pskov]. *Polnoe sobranie russkich letopisej* [Complete Edition of the Russian Chronicles], vol. 4, trans. Ursula Bieber (Saint Peterburg: v tipografii Ėduarda Praca, 1848, reprinted Düsseldorf: Brückner 1973), 303.

53 "This yeare was a hott sommer and drie, so that no raine fell from June till eight daies after Michaelmas" (19th September n. st.). Wriothesley, Charles, *A Chronicle of England during the Reigns of the Tudors, from A.D. 1485 to 1559*, ed. William Douglas Hamilton (Westminster: J.B. Nichols and Sons, 1875), 45.

54 Glaser et al., "Seasonal Temperature," 193.

55 "Chronicon Pontremulense ab Anno MDXXVI usque ad annum MDXLIII a Joanne Maria De Ferrariis–vulgo Ser Marione," in Sforza, Giolio, ed., *Memorie e documenti per servire alla Storia di Pontremoli* (Lucca: Giusti, 1887), 38.

returned until 1st to 3rd [11th to 13th] August, afterwards persisting until the end of the year leaving all rivers completely dry.[56]

Four chroniclers from Switzerland and Alsace also mention the dates of short rain spells. Fridolin Ryff in Basel counted three "short and feeble" rain spells between the "beginning of summer" and St. Martin's Day [21st November], each lasting not more than two or three days.[57] Adelberg Meyer in the same town got ten precipitation days between February and 21st November.[58] Heinrich Bullinger in Zürich reports six rain-days between February and 29th September insisting that it never rained for an entire day or an entire night.[59] Hans Stolz, being a vine-grower and mayor of the small town of Guebwiller (Alsace), was the most continuous chronicler in Western Central Europe in the second quarter of the sixteenth century. He wrote about 1540: "February was warm and dry throughout. The month of March except the first three days, when it rained, was consistently dry and sunny, but ice was found every morning."[60] Ulrich Meyer (Winterthur) reports persistent cold winds,[61] probably related to a strong Siberian anticyclone dominating the weather in Central Europe at that time. April and May were sunny throughout and completely rainless. At the end of May, the grass in the Swiss Plateau had dried out and the level of brooks and small rivers was already so low that many mills stopped working. Temperatures during that time were extreme, concluding from the rapid development of vegetation. Cherry trees flowered around 10th April in Ancy-sur-Moselle (France). One month later, cherries in the same region were

---

56 Relazione dell'origine et successi della terra di Varese [Varese Ligure] descritta dal R.P. Antonio Cesena l'anno 1558, (La Spezia: Academia lunigianese di Scienze Giovanni Cappellini, 1982), 36.

57 "Dasz esz von anfang desz sumersz [June 1st] bisz Martini (11th/21th Nov) nit über drümol regnet; so esze schon regnet, wert esz nit über zwen oder drü tag und necht lang und klein, senfft regen," Ryff, Fridolin, "Die Chronik des Fridolin Ryff 1514-1541, mit der Fortsetzung des Peter Ryff 1543-1585: Die Chroniken des Karthäuser Klosters in Klein-Basel 1401-1532," in *Basler Chroniken*, vol. 1, ed. Historische und Antiquarische Gesellschaft in Basel (Leipzig: S. Hirzel, 1872), 86.

58 "Die Aufzeichnungen Adelberg Meyers 374-1542," in *Basler Chroniken*, vol. 8, ed. Historische und Antiquarische Gesellschaft (Leipzig: S. Hirzel, 1902), 43.

59 Bullinger, Heinrich, "Diarium (Annales Vitae) der Jahre 1504-1574," in *Quellen zur Schweizer Reformationsgeschichte*, vol. 2, ed. Emil Egli (Basel: Basler Buch- und Antiquariatshandlung, 1904), 22.

60 Tschamser, Malachias, *Annales oder Jahrs-Geschichten der Baarfüsern oder Minderen-Brüdern S. Franc. ord. insgemein Conventualen genannt, zu Thann* (Colmar, 1864), 64.

61 "Die Aufzeichnungen Adelberg Meyers," 43.

already ripe.[62] The flowering of all grapes in Schaffhausen and in Biel-Bienne (Switzerland) had ended in the last days of May (prior to 10th June),[63] about five weeks earlier than usual. In early June, the heat was already unbearable, considering the fact that quarrymen in Besançon (France) got time off from hard physical work.[64] From 18th to 22nd June it was somewhat rainy.[65] Subsequently not a drop fell for 45 days until 6th August. Hans Stolz reports that "July was torrid and dreadful until the end".[66] People in Besançon (France), unable to stand the heat in the streets during the day, took refuge in cellars after 9 a.m.[67] The heat wave peaked in early August. On August 2nd, the town council of Ulm ordered the parsons to preach "about the hot and dry weather, begging God for rain".[68]

Trees came under drought stress as leaves withered and fell to the ground like in late autumn. Forest fires became rampant in many parts of the continent, such as in the Vosges mountains,[69] in the Black Forest,[70] in the Bohemian Forest, in Thuringia, in the Spessart mountains,[71] in Hungary and in Poland,[72] infernos that nobody could get under control. Reformer Martin Luther

62 Bouteiller, Ernest de, ed., *Journal de Jean le Coullon, 1537-1587* (Paris, Nancy: D. Dumoulin, 1881), 22.

63 Huber, Oswald, *Schaffhauser Chronik* 1521-1582, ed. Carl August Bächtold (Schaffhausen: Buchdruckerei H. Meier und Cie 1906), 95; Bähler, Arnold, ed., *Bendicht Rechbergers Bielerchronik 1524-1566* (Biel, 1902, reprinted 1989), 18.

64 Delsalle, Paul, *La Franche Comté au temps de Charles Quint* (Besançon: Presses Universitaires de Franche Comté, 2004), 58.

65 Baechtold, Jacob, *Hans Salat: Ein schweizerischer Chronist und Dichter aus der ersten Hälfte des XVI. Jahrhunderts: Sein Leben und seine Schriften* (Basel: Bahnmaier's Verlag, 1876), 56.

66 Stolz, Wolfram, *Die Hans Stolz'sche Gebweiler Chronik: Zeugenbericht über den Bauernkrieg am Oberrhein* (Freiburg i.Br.: Steinmann Druck und Verlag, 1979), 373.

67 Frossard, Anatoile, "Livre de raison de la famille de Froissard-Broissia de 1532 à 1701," in *Mémoires de la Société d'émulation du Jura* (published 1886), 46.

68 Pfister, Christian, "When Europe Was Burning: The Multi-Seasonal Megadrought of 1540 and the Arsonist Paranoia," in *Disasters, Risks and Cultures: A Comparative and Transcultural Survey of Historical Disaster Experiences between Asia and Europe*, ed. Gerrit Jasper Schenk (Heidelberg: Springer, 2016).

69 Tschamser, *Annales*, 65.

70 Spicker-Beck, Monika, Räuber, Mordbrenner, *Umschweifendes Gesind: Zur Kriminalität im 16. Jahrhundert* (Freiburg i.Br.: Rombach, 1995), 231.

71 *Chronik der Stadt Kitzingen*, Stadtarchiv Kitzingen, Manuscript, Msc. 314.

72 Glaser et al., "Seasonal Temperature," 193.

anticipated the far-reaching consequences of this calamity in terms of rising prices for timber.[73] School teacher Hans Salat writes:

> On Mary and Magdalene's day [31th July] I went to Solothurn. It was unbearably hot, everybody complained about water shortages. Forests were burning everywhere around. The sun and the moon, being reddish at their rising and setting, looked pale during the day, because the sky was dark of mist and smoke. Mount Pilatus [above Lucerne] could hardly been seen in the morning due to fog conditions like in autumn.[74]

Similarly, the inhabitants of Schneeberg (Saxony) complained about breathing in stinking smoke during the night emitted from forest fires.[75] The scientist Marcin Biem reported from Cracow (Poland) that the sun often retained a reddish colour throughout the day due to the presence of smoke in the air. Glaser and co-authors argue that this smoke was probably the effect of a cover of forest fire aerosols,[76] as was observed during the Russian drought in 2010. Likewise, town fires in Germany, according to a unique statistic of 8,200 events set up by Cornel Zwierlein, were more frequent in 1540 than in any other peace year since AD 1000.[77] This gives reason to believe that settlements became fire-prone at about the same rate as forests. No thunderstorm was observed in contrast to summer 2003.[78] On the other hand, the Lucerne botanist and

73 On 26th July/4th August Martin Luther wrote to his wife: "ist ym Düringer walt mehr denn tausent acker holtz abgebrand vnd brennet noch, dazü sind heüte zeitung, das der wald bey werda aüch angangen sey. Vnd an vil orten mehr, hilfft kein lesschen. das wil theür holtz machen." Luther, Martin, „Luther an seine Ehefrau" [Eisenach,] 16/26, Juli 1540, in *D. Martin Luthers Werke: Kritische Gesamtausgabe*, ed. Otto Clemen, Abt. Briefwechsel, vol. 9 (1540-1528), Februar 1542, Nr 3519 (Weimar: Hermann Böhlaus Nachfolger, 1941), 204-205.

74 "Um M. Magdalene [31. July n. st] gieng ich gen Soloturn […] [es war] unbillich heiss, clagt sich all welt fast um wasser, was am uf- und nidergang sunn und man bluot rot; schinendn ouch ganz bleich, dann der himel was tunkel von itel hitznebel. Es brunnend die weld an vil orten, […] an eim morgen was es uf der wyti nit anders von rouh und hitz, als im herbst mit nebel, dass man Pilatusberg kum sehen moht." Baechtold, *Hans Salat*, 56.

75 Melzer, Christian, *Bergkläufftige Beschreibung der Churfürstl. Sächß. freyen und im Meißnischen Ober-Erz-Gebirge löbl.* Bergk-Stadt Schneebergk (Schneeberg, 1684).

76 Glaser et al., "Seasonal Temperature," 193.

77 Zwierlein, Cornel, *Der gezähmte Prometheus: Feuer und Sicherheit zwischen Früher Neuzeit und Moderne* (Göttingen: Vandenhoeck & Ruprecht, 2011).

78 Concurrent observations by Ryff, "Die Chronik;" Stolz, *Die Hans Stolz'sche Gebweiler Chronik* and an anonymous chronicler from Lindau quoted in Burmeister, Karl Heinz, "'Der heiße Sommer' 1540 in der Bodenseeregion," in *Schriften des Vereins für Geschichte*

politician Renward Cysat reports eyewitness accounts that Alpine meadows were literally "irrigated" every morning by abundant dew. This effect was possibly generated by the intensive daytime evaporation of the then larger glaciated areas and the subsequent condensation during the night.[79]

Hans Salat reports that it rained in Lucerne for some days after 11th August so that meadows became green again.[80] Farther north the rain was more intensive and longer lasting causing floods in the Main, middle Rhine, Elbe and Danube River system.[81] Afterwards, rain spells were again sporadic for the rest of the year. Vine grower Hans Stolz noted a second bloom of fruit trees in early September 1540 in Alsace,[82] which reflects similar observations being made in Munich at the same time in 2003.[83] Based on observations of a record breaking second flowering of vines on 9th October and cherries reaching maturity for a second time in Lindau on the shore of Lake Constance[84] it is concluded that September and October were warmer than in 2003. The reports by several chroniclers agree that weather in the remaining part of the year was sunny and warm "like in April" without any frost and snow covering the ground until Christmas (Julian Style), i.e. 4th January 1541. At that time, several adults demonstratively swam across the Rhine at Schaffhausen (Switzerland)[85] to capture the attention of chroniclers. They were eager to include such physical evidence into their narratives to demonstrate how extraordinarily warm the water still was at the end of this longest known bathing period in European history.[86] Taking into account the preceding record-breaking spring-summer temperature anomaly and the outstandingly warm conditions until December 1540, we assume that water temperatures might have been at about 15°C at the end of the year which still is somewhat below comfortable water temperatures

des Bodensees und seiner Umgebung, vol. 126, ed. Jürgen Klöckler (Ostfildern: Jan Thorbecke, 2008), 61.

79 Cysat, Renward, "Stationes Annorum: Witterung, Missjahre, Teuerung," in *Collectanea Chronica und denkwürdige Sachen pro Chronica Lucernensi et Helvetiae*, Abt. 1, vol. 1, part 1, ed. Josef Schmid (Luzern: Diebold Schilling Verlag, 1969), 935.

80 Baechtold, *Hans Salat*, 61.

81 Glaser, Rüdiger et al., "The Variability of European Floods since AD 1500," *Climatic Change* 101 (2010): 235-256.

82 Stolz, *Die Hans Stolz'sche Gebweiler Chronik*, 374.

83 Wikipedia, "Hitzewelle 2003," last modified January 21, 2014, accessed April 6, 2017. <http://de.wikipedia.org/wiki/Hitzewelle_2003>.

84 Burmeister, "'Der heiße Sommer' 1540," 62.

85 Huber, *1521-1582: Schaffhauser Chronik*, 95.

86 Wetter and Pfister, "An Underestimated Record Breaking Event," 51.

for swimming.[87] Maximum water temperatures of the Rhine measured at this time of the year within the period 1978 to 2011 were about 11°C in December 2006 and about 9°C in January 2007.[88] Updated European averaged autumn and winter air temperature time-series indicate that temperatures for autumn 2006 and winter 2007 were likely the highest since 1540.[89]

Because reported dates of rainfall are not entirely consistent among the four key chroniclers, all rain-days mentioned at least by one chronicler were aggregated to get believable though perhaps overrated seasonal values. To be sure, values in spring, summer and autumn are even below the absolute minima of the instrumental period (since 1864). The annual estimate for 1540 including ten precipitation days for the missing January observation is forty percent below the lowest instrumental measurement since 1864. Reliable data on precipitation frequency from East Central Europe were obtained from the only weather diary so far known. It was kept by Marcin Biem, who was a theologian and president of the renowned Cracow University.[90]

Considering the cumulative deficit of precipitation days from the 1901-2000 average (figure 5.5[91]), the 1540 drought was significantly more persistent and extreme than the recent events in 2011 and 2003.

The amount of precipitation depending on the observed rain-days was estimated through a hierarchy of statistical models.[92] From figure 5.6[93] it is concluded that the precipitation in Western Europe remained significantly below one hundred-year minimum levels in 1540 throughout spring, summer and autumn. The reconstructed annual precipitation of 236 millimetres (24 percent) is somewhat below of the 'guesstimate' attempted by the author in 1984.[94] No event of similar severity is known within the instrumental period. In Poland, the drought likewise persisted over three seasons, but it was somewhat less severe, as precipitation was possibly above the one hundred-year minimum (including upper uncertainty amounts). The year 1540 is the warmest

87 Loc cit.

88 Data from the Swiss Federal Office for the Environment (FOEN).

89 Luterbacher, Jürg et al., "The Exceptional European Warmth of Autumn 2006 and Winter 2007: Historical Context, the Underlying Dynamics and its Phenological Impacts," *Geophysical Research Letters* 34.12 (2007): 1-6. doi:10.1029/2007GL029951.

90 Wetter, and Pfister, "An Underestimated Record Breaking Event."

91 Source: Wetter et al. "The Year-Long Unprecedented European Heat and Drought of 1540," 355.

92 Ibid., 354.

93 Temperature reconstruction based on the homogenised Swiss GHD series covering the period from 1444–2011.

94 Pfister, *Das Klima der Schweiz*, 138, guesstimated "a third at the most".

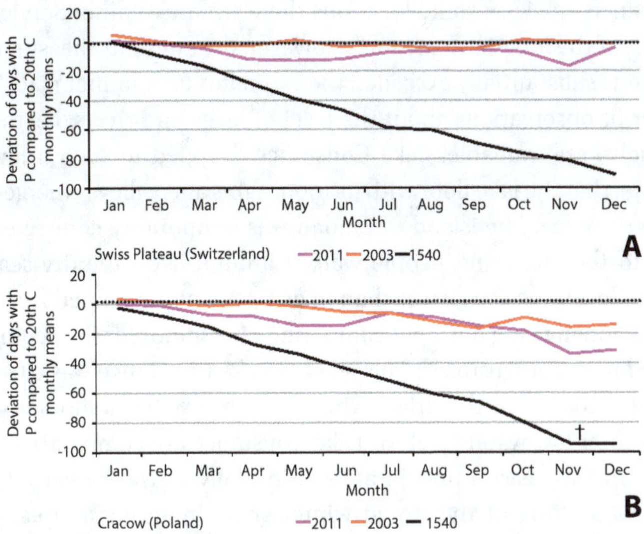

FIGURE 5.5 *Cumulative deviations of the number of precipitation days in 1540 in Northern Switzerland and in Cracow (Poland) in comparison with twentieth century values*

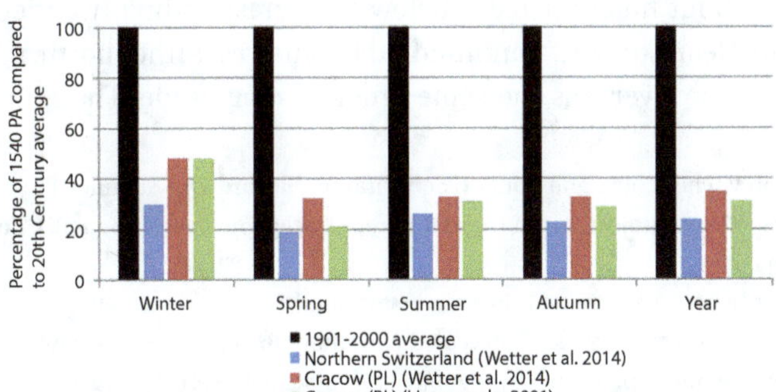

FIGURE 5.6 *Estimated 1540 seasonal precipitation in Northern Switzerland and in Cracow (percent) compared to the twentieth century*

since 1500 in the Central European temperature series[95] and the driest in the Czech Republic.[96] The 2015 Central European summer precipitation was the

95 Dobrovolný et al, "Monthly, Seasonal and Annual Temperature Reconstructions for Central Europe Derived from Documentary Evidence and Instrumental Records since AD 1500." *Climatic Change* 101.1-2 (2010): 69-107. doi:10.1007/s10584-009-9724-x.

96 Dobrovolný et al, "Precipitation Reconstruction for the Czech Lands, AD 1501-2010."

lowest on record since 1901.[97] In some locations the German Weather Service only measured half of the long-term mean.[98] However, the amounts measured for two Swiss stations substantially exceeded the estimates for summer 1540.[99]

A great number of observations about the level of lakes and rivers, i.e. the hydrological drought, are available. Lake Constance dropped to such a low level in August, 1540, that the lake floor with its mountains and valleys emerged over considerable distance. The island of Lindau was temporarily connected by a land bridge to the coast and people walked around it on the dry sea-ground.[100] The lake level was thus lower than in 2003. A woman named Anna Schmid was lucky enough to dig out a pot containing nine hundred silver coins from the time of Emperor Augustus.[101] This level of Lake Constance was close to the lowest ones recorded in the driest winters since 1550.[102] It should be pointed out that winter low water levels of Lake Constance are comparatively common, because in this season precipitation, especially in the higher altitudes, falls as snow. Melting of the stored winter snow in spring and early summer usually fills the lakes at the foothills of the Alps, so that the extreme late summer low water level in 1540, also considering the summer precipitation maximum in the Alps, was indeed a record-breaking event. The discharge deficit of the river Rhine in Basel, still fed by melt-water from the then more glaciated Alps, was about ninety percent below its corresponding twentieth century mean value. Near Cologne, four hundred kilometres farther north, the relative discharge of the river was the same order of magnitude. The Rhine

97 Orth, René, Jakob Zscheischler, and Sonia I. Seneviratne, "Record Dry Summer in 2015 Challenges Precipitation Projections in Central Europe," *Scientific Reports* 6 (2016). doi: 10.1038/srep28334.

98 Deutscher Wetterdienst – Wetter und Klima aus erster Hand, "Deutschlandwetter im Jahr 2015," last modified December 30, 2015, accessed November 8, 2016. <https://www.dwd.de/DE/presse/pressemitteilungen/DE/2015/20151230_deutschlandwetter_jahr2015_news.html.

99 Bundesamt für Meteologie und Klimatologie Meteoschweiz, "Datenportal für Lehre und Forschung," Summer (JJA) 2015: Zürich 242 millimetre, Basel 191 millimetre, accessed May 13, 2017. <http://www.meteoschweiz.admin.ch/home/service-und-publikationen/beratung-und-service/datenportal-fuer-lehre-und-forschung.html>. Summer (JJA) 1540: 91 millimetre (estimated) (Wetter et al. 2014, Supplementaries, Table S4).

100 "Der See war im Sommer so klein, daß man rings um die Stadt gehen und im Wasser Berg und Thal sehen konnte." *Schnell'sche Chronik, Manuscript*. Stadtarchiv Lindau, Lit 29, 99.

101 Burmeister, "'Der heiße Sommer' 1540," 77.

102 Pfister, Christian, Rolf Weingartner, and Jürg Luterbacher, "Hydrological Winter Droughts over the Last 450 Years in the Upper Rhine Basin: A Methodological Approach," *Hydrological Sciences-Journal-des Sciences Hydrologiques* 51.5 (2006): 966-985. doi:10.1623/hysj.51.5.966.

downstream of Basel, the Seine near Paris and the Elbe near Meissen could be waded through in some places. The water of the river Meuse, being the second largest tributary of the Rhine, was green and unsuitable for drinking, due to a very slow flow and high water temperatures promoting the growth of green algae,[103] as is known from Mediterranean rivers in summer.[104] Around Liège, Belgium, cattle were seen wading in the nearly stagnant water.[105] Likewise, the Odra River "was green" according to a Czech source.[106] The flow of the river Thames in London was so feeble that the salt water penetrated further upstream than London Bridge. Navigation on the river Po in Northern Italy ceased completely. Smaller watercourses such as the river Doubs in France were completely dry.[107] In July, the water shortage became acute,[108] considering that fountains ran dry that so far had never failed. This shortage points to a substantial drop in ground-water tables. In Canton Lucerne, desperate people were digging for water 1.5 metres deep in the bed of a small river without finding a single drop.[109] People in Parma, Italy, had to cart water from river Po over a distance of twenty kilometres, because their wells were all dry.[110] Those in Avignon, Southern France had to get their water from the river Rhone.[111] All

---

103 "Aussi en fut la Moselle si courte, Que l'eaue à la Ville estoit très courte, Elle changea & devint si verde Qu'on en pouvoit en œuvre metre." The level of river Moselle was so low near the town [of Metz, Lorraine] that the water became green and could not be used any more (Calmet, Augustin, "Chronique de la noble cité de Metz: Preuves servant à l'histoire de Lorraine," in *Histoire de Lorraine*, vol. 3, ed. Augustin Calmet, François Jean-Baptiste Noël and Auguste Prost, col. Cccxxviij (Paris: Éditions du Palais Royal, 1973). (Interpretation provided by Prof. Dr. Alfred Johny Wüest, EAWAG, CH 6047 Kastanienbaum, Switzerland.)

104 Power, Mary E. et al., "Algal Mats and Insect Emergence in Rivers under Mediterranean Climates: Towards Photogrammetric Surveillance," *Freshwater Biology* 54.10 (2009): 2101-2115. doi:10.1111/j.1365-2427.2008.02163.x.

105 Buisman, Jan, and Aryan F.V. van Engelen, *Duizend jaar weer, wind en water in de lage landen*, vol. 3 (*1450-1575*) (Franeker: Van Wjnen, 1998), 466.

106 Brázdil, Rudolf et al., Climate of the Sixteenth Century in the Czech Lands, History of Weather and Climate in the Czech Lands (Brno: Masaryk University, 2013).

107 Delsalle, Paul, *La Franche-Comté*, 58.

108 At the end of July "giengend brunnen und wasser ab, so bi mentschen tenken nie abgangen warend". Baechtold, *Hans Salat*, 56.

109 Baechtold, *Hans Salat*, 56.

110 Sforza, G., ed., *Memorie e documenti per servire alla Storia di Pontremoli* (Lucca: Giusti, 1887), 59.

111 Mémorial d'Aix en Provence quoted in: Pichard, George, and Emeline Roucaute, "Sécheresses du Bas Rhone et invasions de sauterelles," accessed August 25, 2014. <http://histrhone.cerege.fr/squelettes/pdf/Chronologie_Secheresses_Bas_Rhone.pdf>.

this could be taken to imply that most people living some distance from lakes or large rivers were left high and dry. Local conflicts over water were probably frequent, but the respective local sources have so far hardly been examined under this context.

In conclusion, Western and Central Europe suffered from a dust-dry spring in 1540, which was then followed by a torrid summer and an almost rainless autumn, making 1540 to the worst drought year of the last five centuries. On the basis of these results, it becomes plausible that temperatures in 1540 might have been higher than in 2003, as will be subsequently demonstrated.

In a previous study, Wetter and Pfister investigated the severity of the heat wave going along with the megadrought. At first, the authors attempted an in-depth view of the Dijon vine harvest series from the archives. Firstly, it turned out that the outstanding heat wave in August 2003 could not have mattered for the Dijon grape harvest date of 15th August. According to the renowned oenologist John Gladstones, "temperatures of the first two or three growing season months [...] can usually predict quite closely the dates of [...] maturity to follow [...]. The later phenological intervals [...] tend to be constant from year to year".[112] Gladstone's observation agrees with the statistical analyses that August temperatures are not significant for the harvest date. Secondly, two historians, Thomas Labbé and Damien Gaveau, discovered through a critical reinterpretation of the Dijon series in the municipal archives that the correct harvest date for 1540 was 3rd September, instead of 4th October given by Isabel Chuine and co-authors.[113] Thirdly and most importantly, the beginning of grape harvest in 1540 was postponed in many places beyond grape maturity. Vine-growers in Schaffhausen (Switzerland) were "long waiting for rain to begin the harvest", as chronicler Oswald Huber relates. Nevertheless, he writes, "they finally tackled the work because the plants withered."[114] Vine growers at the shores of Lake Constance and in the Upper Alsace interrupted the vintage after picking the juicy grapes, because the remaining ones had already become raisins. The vintage was only resumed after a two-day rain spell on St. Michael's Day [8th October].[115] By waiting for rain, vine growers hoped to get more juice into the berries, as their income, regardless of prices, depended on the quantity of liq-

112 Gladstones, John, *Wine, Terroir and Climate Change* (Kent Town: Wakefield Press, 2011), 7.

113 Labbé, Thomas, and Fabien Gaveau, "Les dates de bans de vendange à Dijon: Etablissement critique et révision archivistique d'une série ancienne," *Revue Historique* 657.1 (2011): 19-51. doi:10.3917/rhis.111.0019; Chuine, Isabelle et al., *Burgundy Grape Harvest Dates and Spring–Summer Temperature Reconstruction*, published by IGBP PAGES/World Data Center for Paleoclimatology, 2005, 2 Data Contribution Series #2005-007. Boulder Co. 2005.

114 Huber, 1521-1582: Schaffhauser Chronik, 95.

115 Burmeister, "Bodenseeregion," 64; Stolz, *Die Hans Stolz'sche Gebweiler Chronik*, 379.

uid they got from the wine press. The situation in France was hardly different.[116] Pierre de Teysseulh being a capitular of the church of Limoges (Central France) notes that "the grapes were like roasted and the leaves of the vines had fallen to the ground like after a severe frost."[117] The unintentional sweet-late vintage yielded a sherry-like wine, which made people rapidly drunk.[118]

Wetter and Pfister set up a long series of grape harvest dates for Switzerland (1444-2011) to re-assess temperatures in France and Switzerland in 1540. It turned out that grape harvest dates are not available for 1540 in Switzerland, probably because the current practice of the harvest ban could not be maintained under the extraordinary situation in that year. The likely date of full grape maturity in the Swiss Mittelland was thus estimated from grape phenological evidence to be between 12th and 24th August. The reconstruction of mean temperatures from April to July yielded an estimate of 4.7°C and 6.8°C (+/−0.5°C) above the 1901-2000 regional mean depending on the assumed value for full grape maturity.[119]

Even the lower of these estimates is significantly higher than the anomaly of +2.86°C measured for April to July temperatures in 2003 (figure 5.7[120]). In interpreting this result, it needs to be conceded that the two dates for full grape maturity underlying the temperature assessment are statistically shaky. Using climate model output based on the relationship between the number of dry days and temperatures, René Orth and co-authors recently re-examined the question whether the summer in 1540 was warmer than in 2003. The results suggest a high probability that maximum temperatures in 1540 exceeded those in 2003, whereas for mean temperatures, such a probability is rather low.[121] These results concur with the fact that August temperatures could not be statistically estimated. It is also consistent with the fact that widespread forest and settlement fires were observed towards the end of the rainless July 1540. In 2003 forest fires mainly ravaged in the Mediterranean basin, particularly in Portugal, where temperatures of more than 40°C were measured during the

116 Wetter et al., "The Year-Long Unprecedented European Heat and Drought of 1540," 359.

117 Leroux, Alfred, and Auguste Bosvieux, ed., "Extraits du Journal de Me Pierre de Teysseulh : chanoine de l'église de Limoges, 1533-1568," in *Chartes, Chroniques et Mémoriaux pour Servir à l'Histoire de la Marche et du Limousin*, ed. Alfred Leroux and Auguste Bosvieux (Tulle: Crauffon, 1886), 259.

118 Kessler, Josua, *Chronologie Santgallischer Begebenheiten vom Jahr 1540 bis Ende des Jahres 1645, Handschriften Nr. 74*, Manuscript (St. Gallen: Stiftsarchiv St. Gallen, n. d.).

119 Wetter, and Pfister, "An Underestimated Record Breaking Event," 41.

120 A questionable comparison of the outstandingly hot years 2003 and 1540.

121 Orth, René et al., "Did European Temperatures in 1540 Exceed Present-Day Records?," *Environmental Research Letters* 11.11 (2016). doi: 10.1088/1748-9326/11/11/114021.

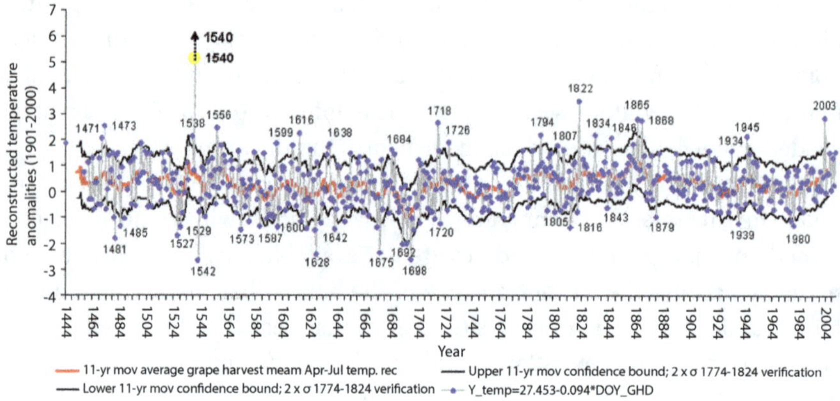

FIGURE 5.7 *Temperature reconstruction based on the homogenised Swiss GHD series covering the period from 1444-2004*

peak of the heat wave.[122] Similarly high temperatures might have been reached at the peak of the 1540 heatwave in Central Europe.

On the other hand, the conclusion that the period from April to July 1540 was warmer than in 2003 seems plausible considering the physical reasons for severe droughts and the outstanding environmental context of the event. Spring droughts are an important precondition for the triggering feedbacks of soil dessication and temperature underlying extreme heat waves. Undoubtedly, the spring drought in 1540 was longer lasting and more intense than in 2003 and it was then followed by the summer drought, which further intensified the feedback between soil dessication and temperature. Eye observations about extreme soil dessication and soil cracking confirm the hypothesis of a record-breaking soil moisture deficit. Some cracks were so wide that people could put their feet into them.[123] In contrast to 2003, the drought in 1540 persisted until the end of the year. Annual temperatures in 1540 for Germany, Switzerland and the Czech Republic estimated from documentary "Pfister indices" were 2.3°C (±0.4°) higher than in 1901-2000, which is the highest value since 1500.[124]

Several authors showed that summer heat waves in Central Europe are associated with a precipitation deficit in winter and spring in Southern Europe, as droughts in the South cause unusual low cloud cover which is then transported

122 Trigo et al., "Atmospheric Conditions."

123 Goldschmid, Johann Jakob, *Erzellung seltsammer Natur-Geschichten* [...] *so sich bey uns in der Statt Wintertur zugetragen haben* (*bis 1543*), vol. 3, Ms Fol. 3, Stadtbibliothek Winterthur.

124 Dobrovolný, Petr et al., "Monthly, Seasonal and Annual Temperature Reconstructions."

northward, enhancing existing anti-cyclonic conditions.[125] This cause-and-effect relationship can also be demonstrated for the 1540 megadrought. Chroniclers in Northern Italy report that winter 1540 was "like in July"[126] with no rain falling from 10th [20th] November 1539 to 6th [16th] April 1540, leaving wells dessicated.[127]

In conclusion, the record-breaking heat and drought of 1540 was an analog case to the 2003 event, but it was more intense, spatially more extended and longer lasting. This result was challenged by Büntgen and co-authors (2015) based on a large sample of tree-ring width measurements. The authors argued that Wetter and co-authors (2013, 2014) "probably overstated the intensity and duration of the 1540 drought event."[128] On the other hand, it needs to be noted that an earlier tree-ring study led by the same author trees demonstrated that a very small tring ring width emerges in 1541, which was interpreted as the lagged result of the severe drought in 1540.[129] This interpretation is supported by the tree-ring study of Nicault and co-authors who concluded that trees in the larger Mediterranean land area including the Alps suffered from a severe drought in 1540 (Figure 3).[130]

4 Vulnerabilities of Pre-modern and Present-day to Severe Drought

Climate impacts are joint products of the interaction of climate and society. Social vulnerability exists within human systems independently of external hazards. It determines whether a society is more or less vulnerable to climatic shocks. Under these premises, similar climatic impacts will have different

125 Stefanon, Marc, Fabio D'Andrea, and Philippe Drobinski, "Heatwave Classification over Europe and the Mediterranean Region," *Environmental Research Letters* 7 (2012): 1-9. doi:10.1088/1748-9326/7/1/014023.5.

126 Burigozzo, Giovanni Maria, "Cronaca di Milano dall'anno 1500 sino al 1544," in *Archivio Storico Italiano,* vol. 3, ed. G.P. Viesseux (Florence: Leo S. Olschki, 1842), 539.

127 Fossati, Francesco, ed., *Opere scelte di Benedetto Giovio* (Como: Fostinelli, 1887, reprint 1982), 36.

128 Büntgen, Ulf et al., "Limited Tree-Ring Evidence for a 1540 European 'Megadrought'. Commentary to Wetter et al. (2014)," *Climatic Change Letters* (2015).

129 Büntgen, Ulf, et al., "Combined Dendro-Documentary Evidence of Central European Hydroclimatic Springtime Extremes over the Last Millennium," *Quaternary Science Reviews* 30 (2011). doi:10.1016/j.quascirev.2011.10.010; Pfister, Christian, Wetter Oliver, et al., "Tree-Rings and People—Different Views on the 1540 Megadrought. Reply to Büntgen et al. (2015)," *Climatic Change Letters* (2015).

130 Nicault et al., "Mediterranean Drought Fluctuation," 7

socio-economic consequences under different sets of social conditions. Many different concepts of vulnerability are known in the literature. This analysis uses the generally applicable vulnerability scheme worked out by Hans-Martin Füssel, distinguishing between four groups of related vulnerability factors:

1. A focus of analysis such as an human-environment system, a geographical region or an economic sector;
2. an attribute of concern of the analysed system such as income or health;
3. a hazard which threatens attributes of the system; and
4. a temporal reference.

In a first step, the analysis explores the vulnerability of pre-modern societies to the megadrought. In the second step the vulnerability of key branches of present-day economies is investigated from the example of 2003. In the last step, the physical boundary conditions assessed from the analysis of the 1540 megadrought are used to consider the vulnerability of present-day societies for an extreme event of this magnitude.

Subsequently, the impacts of the megadrought on contemporaries are described. Water scarcity was the worst effect. Where wells and fountains were dry, people had to get water over considerable distances, usually at night.[131] For example, the residents of the small village of Goldiwil situated five hundred metres above the town of Thun (Canton Bern) had to get water from Lake Thun on donkeyback.[132] Well water was priced in some regions. In Lake Constance, upstream of the Rhine valley, it was more expensive than wine.[133] High mortality is reported in some chronicles, which, considering the heat and lack of water, might mostly have been caused by dysentery.[134] "Many people fell ill," we read in a chronicle from Gent (Belgium), "[...]and there was red melisoen [illness] which in particular affected people's defecation very much,

131 Cysat, "Stationes Annorum," 905.

132 Züricher, Gertrud, ed., Karl Friedrich Ludwig Lohners Chronik der Stadt Thun (Bern: Haupt, 1935), 74.

133 Walser, Gabriel, *Neue Appenzeller Chronick oder Geschichte des Landes Appenzell der Innern und Äussern Rhoden* (St. Gallen: Selbstverlag, 1740), 484.

134 "In Folge der grossen Hize des vorigen Jahres entstand in diesem eine höchst mörderische Krankheit, welche der grosse Tod genannt wurde und nach unverbürgten Nachrichten an 3.000 Personen in unserer Stadt, sowie zwei Fünftheile der Bewohner unserer Landschaft wegraffte." Cf. Im-Thurn, *Chronik der Stadt Schaffhausen*, 68. "In y latter ende of this moneth, was vniversally through the realme greate death, by reason of newe hote agues and Flixes, and some Pestilence." Cf. Holinshed, Raphaell, *The Description and Histories of England* (London: Henry Denham, 1587), 841.

because many people died from mid-August until December which was caused by the great vehement heat."[135]

Three branches of the economy particularly suffered from the drought: agriculture, energy production and transportation. Watering of cattle and small livestock became critical. Animals were driven to drink over long distances up to and beyond ten kilometres.[136] Hay harvests completely failed and prices skyrocketed.[137] Cattle all over Europe died of thirst and hunger, as chroniclers from England, Alsace (France), Cologne (Germany) and Northern Italy reported.[138] In many cases, animals may have died from heat stroke. Many more were probably slaughtered. Hence, prices for milk and dairy products surged.[139] Spring grains, fruit and legumes failed completely. The harvest of winter grain, which was the major food crop, was satisfactory in Central Europe, because yields and related water needs were much below present-day values, but this crop failed in Northern and Central Italy, because the drought there already began in autumn 1539 leading to severe famine and widespread (e) migration.[140]

---

135 Anonymous, MS 2543, University Library Ghent, kindly provided by Adriaan de Kraker, Utrecht University. Likewise: Stolz, *Die Hans Stolz'sche Gebweiler Chronik*, 373, mentions "many people [in Alsace] dying from an epidemic".

136 "Das Baursvolck muoßt an etlichen enden weit her [...] mit dem Viehe zuotrencken/ fahren." Cf. Wurstisen, *Basler Chronik*, 79. "In divers partes of this realme the people caried their cattle six or seven miles to watter them." Cf. Wriothesley, *A Chronicle*, 123.

137 "Und ward das heuw thur, und das veche fast wolfeill." Anonymus, ed., "[...] Die Anonyme Chronik bei Schnitt, sammt Fortsetzung: 1495-1541 [...]," in *Basler Chroniken*, vol. 6, ed. Historische und Antiquarische Gesellschaft in Basel (Leipzig: S. Hirzel, 1902), 220.

138 For England: "Muche cattell died for lacke of water." In Grafton, Richard, *Grafton's Chronicle, or history of England* [...] *from the year 1189 to 1558, inclusive, Volume 2* (London: J. Johnsen, 1809), 475. For Guebwiller (Alsace): "Und starb vill vieh an etlichen orten wasserhalb." Stolz, *Die Hans Stolz'sche Gebweiler Chronik*, 376. For Cologne: "Vil beisten storben". Höhlbaum, Konstantin, ed., *Das Buch Weinsberg: Kölner Denkwürdigkeiten aus dem 16. Jahrhundert. Erster Band* (Leipzig: Dürr, 1886), 45. For Modena (Northern Italy): "Dominica a di 25 Questo dì [...] è stato et è uno excessivo caldo [...] fu uno secho antiquamente tanto grande che li poci [pozzi] se secorno per tutto el modoneso e moriva bestie assai de fame e de sete" (Sunday 25th July/4th August was an unbearable heat and a drought of such extent that the wells got dry in the entire region and animals died from hunger and thirst), in Borghi, Carlo, ed., *Tomasino de Bianchi detto de Lancellotti: Cronica Modenese, 1538-1540. Monumenti di Storia Patria delle province modenesi*, vol. 7 (Parma: Fiaccadori, 1868), 56.

139 Ryff, "Die Chronik," 86.

140 Many chronicles from Northern and Central Italy mention a terrible famine (for example, Borghi, *Tomasino de Bianchi detto de Lancellotti.*

Likewise, the transportation sector was affected. Waterways were important for transportation prior to the creation of railway networks. For example, preindustrial river transport before the early 19th century accounted for ninety percent of goods transported on the Middle Rhine.[141] Highly priced goods made up the lion's share of the upstream cargo and bulky goods dominated the downstream cargo.[142] Erich Weber demonstrated in his illustrative dissertation that river navigation on the Middle Rhine was impossible during "extreme low flows"[143] that were, however, above those described in 1540. The situation of river navigation was not much better on the High Rhine. The small transport ships between Lake Constance and Schaffhausen carried less than half their usual cargo,[144] whereas goods hauled downstream from Schaffhausen to the Basel Fair were entirely transported by land.[145] Navigation on the river Po and its tributaries in Northern Italy was discontinued in the drought until late November, at which time water levels were usually high due to abundant rains in the Apennine Mountains.[146]

The main energy source collapsed with the ebbing of water. For preindustrial societies, the worst effect of this situation was a standstill of mills except those within big rivers. Mills pulled by horse driven capstans were set up as a substitute in some places. In others, people resorted to the use of hand-mills. Surging prices for flour and bread and the lack of water resulted in life-threatening circumstances, especially for the poor.[147] Trades such as hammer-mills became idle leaving the workers unemployed and dependant on begging.[148]

---

141 Weber, Erich M., *Untiefen, Flut und Flauten: Der Güterverkehr auf dem Rhein zwischen 1750 und 1850: Die Modernisierung der vorindustriellen Rheinschifffahrt aus einer wirtschafts-, sozial- und umweltgeschichtlichen Perspektive betrachtet* (Bern: Selbstverlag, 2005), accessed November 20, 2013. <http://www.zb.unibe.ch/download/eldiss/05weber_e.pdf>.

142 Ibid., 429.

143 Ibid., 457.

144 "Der Rhein war so klein und dünn, dass die schiff nicht halb geladen mochten herab kommen". Huber, *1521-1582: Schaffhauser Chronik*, 96.

145 "Uff diese Baszler mesz diesz fierzigisten jorsz alle gutter über lant und mit wegen musten obenherab gfurt werden." Cf. Ryff, "Die Chronik," 86.

146 "Il fiume Po è in secca proprio nel periodo autunnale [che generalmente è caratterizzato da piene, legate alla piovosità degli Appennini e della Pianura Padana]." [River Po was dry in autumn, whereas usually it reaches high flood levels due to rainfall in the Appenin mountains and in the Po valley], in Borghi, *Tommasino de Bianchi detto de Lancellotti*, 58.

147 Glaser et al., "Seasonal Temperature," 192.

148 "Die Flössen/Mühlen und Hammer=Hütten [near Meissen, Germany] stunden müssig/ die Hammer=Arbeiter liessen betteln." Lehman, Christian, [...] *Historischer Schauplatz*

Impacts on the natural environment are only casually mentioned. Oswald Huber reports that graylings (*Thymalus thymallus L.*) in the Rhine, unable to find zones with cool water, were caught by hand belly up in large numbers,[149] as was also reported for 2003.[150] Considering the drying up of most brooks and smaller rivers over large parts of continental Europe together with the extremely low level and warming up of major rivers, the multi-seasonal heat and drought in 1540 may have caused the greatest fish mortality in the second half of the last millennium. The congruent descriptions about stinking smoke made by Salat in Lucerne and by Biem in Cracow—two cities separated by a horizontal distance of nine hundred kilometres—bear witness to the fact that the fire disasters in 1540 affected a far larger area than any other known natural hazard such as a severe flood or a violent winter storm. Hence, large parts of Europe might well have had a peacetime maximum of town and village fires in 1540 in an analogy of Zwierlein's findings for Germany.[151] These calamities may have acted as a major break on economic development. According to Eric L. Jones, "net capital formation was held down in the preindustrial world, not merely by lower incomes and lower savings propensities, but by a weaker capacity to control and recover from natural and social calamity", including urban and rural fires.[152] The longer-term economic impact of the drought in 1540 was, in addition to burnt-down settlements, multiplied by large-scale destructions of forest and substantial losses of cattle, thereby reducing the availability of draught-power and manure needed for grain production over subsequent decades.

The long-lasting heat and drought promoted a widespread epidemic phenomenon called the "Big Death", which may have mainly consisted of dysentery and other intestinal diseases linked to contaminated water.[153] Following the extreme drought of 1719, a dysentery epidemic in France took 450,000 lives,

---

derer natürlichen Merckwürdigkeiten in dem Meißnischen Ober=Erzgebirge (Meissen: Lankisch 1699).

149 "Der Häwmonat war so häiss, dass die Jfer und Escher im Rhein ans land schwummen, kalt wasser zusuochen, und ehe sie wider recht ins wasser kommen mochten, fielen sie für grosser hitz an den ruggen, dass die fischer die in grosser menge mit den händen fiengen, waren faisst und guot." Huber, *1521-1582: Schaffhauser Chronik*, 96.

150 BUWAL, Auswirkungen des Hitzesommers 2003 auf die Gewässer, 137-139.

151 Zwierlein, *Der gezähmte Prometheus*, 104.

152 Jones, Eric L., *The European Miracle* (Cambridge: Cambridge University Press, 2003), 39.

153 Christian Pfister, "When Europe was Burning. The Multi-Season Mega-Drought of 1540 and the Arsonist Paranoia," in *Disasters, Risks and Cultures. A Comparative and Transcultural Survey of Historical Disaster Experiences between Asia and Europe*, ed. Gerrit J. Schenk, (Springer, in press.).

which corresponds to two percent of the French population at that time.[154] Assuming an equally excessive mortality in 1540, the "Big Death" may have killed about one million Europeans assuming a population of forty million people were affected by the drought.[155]

Climatic extremes and nature-induced disasters are known hotspots of cultural history. Thus, the environmental 'emergency situation' of 1540 entailed an emergency within society. The widespread forest and settlement conflagrations involved a loss of control over fire outside people's experience and beyond their imagination. Conspiracy theory offered plausible and convincing explanations. Settlement fires were frequently attributed to arsonist plots allegedly inspired by political opponents and put into effect by paid vagrants and beggars. In people's imagination, these groups living outside of parish structures formed a criminal counter-society hidden from the authorities and the public.[156] In the German Reich, the Schmalkaldian Alliance of Protestants blamed the League of Catholic Princes for setting towns aflame using paid "Mordbrenner" (murder arsonists). Organised arson was regarded as the most threatening political crime of the early modern period, as attacks were not directed toward individuals, but aimed at spreading insecurity and chaos within communities and territories, similar to terrorist attacks today.[157] Trials against alleged "terrorists" allowed the application of torture to learn the names of other members of the sect.[158] Gender stereotypes played an important role in the construction of the arsonist conspiracies. Even though there were of course itinerants of both sexes, the members of the imaginary arsonist gangs were all supposed to be male. Women were non-existent in the conspiracy. However, there was a competing picture of gender-specific crime, that of the witch.[159] Indeed, the disastrous "Witches Hammer," published in 1486 by Heinrich Kramer, propagated the idea that witches were in fact an occult sect enabled by their pact with the devil to commit horrible crimes against a fictitious collective in the form of crop destruction via hailstorms, untimely frost,

---

154 Lachiver, Marcel, Les années de misère: La famine au temps du Grand Roi, 1680-1720 (Paris: Fayard, 1991), 414-417.

155 Pfister, "When Europe was Burning"; de Vries, Jan, "Population," in *Handbook of European History 1400-1600. Late Middle Ages, Renaissance and Reformation*, vol. 1, ed. Thomas A. Brady, Heiko A. Oberman, and James D. Tracy (Leiden: Brill, 1995), 13.

156 Dillinger Johannes, "Organized Arson as a Political Crime: The Construction of a 'Terrorist' Menace in the Early Modern Period," *Crime, History & Societies* 10.2 (2006): 102-119.

157 Dillinger, "Organized Arson," 111.

158 Rummel, Walter, and Rita Voltmer, *Hexen und Hexenverfolgung in der Frühen Neuzeit* (Darmstadt: Wissenschaftliche Buchgesellschaft, 2008), 238-240.

159 Dillinger, "Organized Arson," 110.

and cold rains. Thus, there seems to have been a dichotomy between male arsonist conspiracy, blamed for "warm" disasters in the human realm, and female magic generating "cold" disasters in the natural world that needs to be further investigated.

In general, however, premodern societies were remarkably resilient considering the magnitude of the impact: epidemics, cattle mortality, water shortages, frequent settlement and forest fires and rising bread prices due to stalling mills. In contrast to Italy, famines could be avoided in Central Europe.

Agriculture, transportation and energy were also particularly vulnerable to heat and drought in 2003, leaving aside the death of about 70,000 mostly elderly people in Europe during the peak of the heat wave.[160] The EU arable sector showed an overall slump in production of more than ten percent[161] causing an approximate loss of US$12.3 billion.[162] Cattle breeding is not mentioned among the vulnerable branches of agriculture, apart from Germany, where the loss of feed on meadows was estimated at €1 billion. The French cattle industry suffered from a sixty percent drop of green fodder.[163] Inland navigation was restricted in analogy to 1540 due to the low level of major rivers. Ships on the river Rhine could only use twenty to thirty percent of their loading capacity. German railways were not able to fully compensate for the insufficient capacity of river transport, which led to delays and rising cost of transport.[164]

Energy is the Achilles heel of modern economies. A reliable electricity supply is the essential resource basis for modern life.[165] Thermal (fossil and nuclear) power plants are the most vulnerable energy producers, because they depend on an abundant water supply for cooling purposes. Electricity generation through fossil and nuclear thermal power plants requires about 100 l/kWh of freshwater for cooling purposes, process, and other supply water. Thermal power stations in Germany are using a staggering annual amount of 25 billion m^3 freshwater representing 64 percent of the average available surface water.[166]

160 Poumadère, Marc et al., "The 2003 Heat Wave in France: Dangerous Climate Change Here and Now," *Risk Analysis* 25.6 (2005): 1483-1494. doi:10.1111/j.1539-6924.2005.00694.x.

161 García-Herrera et al., "A Review," 292f.

162 SwissRe (ed.), "Natur- und Man-made-Katastrophen im Jahr 2003: Zahlreiche Todesopfer, vergleichsweise moderate Versicherungsschäden," *Sigma* 1.1 (2004): 10.

163 Food and Agriculture Organization of the United Nations, "Climate Change and Food Security: Risks and Responses," accessed November 10, 2016. <http://www.fao.org/3/a-i5188e.pdf>.

164 Bundesanstalt für Gewässerkunde, *Niedrigwasserperiode*, 175.

165 Ward, David M., "The Effect of Weather on Grid Systems and the Reliability of Electricity Supply," *Climatic Change* 121.1 (2013): 103-113. doi:10.1007/s10584-013-0916-z.

166 Bundesanstalt für Gewässerkunde, *Niedrigwasserperiode*, 184.

Extreme temperature-related impacts, like unusually long periods of hot, dry weather, lead to a shortfall in water supply and high-river temperatures. Rising water temperatures first lead to a higher withdrawal of water in order to meet legal and environmental thresholds for the temperature of discharged water without large reductions in efficiency. If the higher amount of cooling water is still not sufficient, the energy conversion consequently needs to be reduced in order to meet environmental regulations. There are numerous examples in 2003, from the United States in 2002, Switzerland, Germany, and France, as well as in 2006 from France, Spain, and Germany, where high ambient water temperatures resulted in reduced power output at several thermal power plants.[167] Allowances were given to thermal power plants most severely affected by the heat wave in the federal state of Baden-Württemberg (Germany) to discharge cooling water into the rivers, which was beyond the limits of the environmental regulations.[168] Europe's main electricity exporter, "Electricité de France", running 58 nuclear power plants, had to cut its electricity exports more than half[169] when water temperatures exceeded 23°C the last day of the August heat wave. Most wind turbines stood still because it was almost windless. Water power stations couldn't compensate for the shortfalls because their capacity was equally reduced due to low water levels. Power exported from the Swiss storage power stations declined by twenty percent.[170] Power prices skyrocketed as a consequence, although power cuts and blackouts could be avoided. In conclusion, due to the enormous cooling water need of thermal (fossil and nuclear) power stations, energy production turned out to be particularly sensitive to long periods of hot and dry weather.

Though the recurrence probability of a 1540-like megadrought might be small, even in the course of intensified Global Warming, some considerations about the possible consequences of such a worst-case event are attempted. In comparison to most scenarios, which have mainly speculative character, the climatic and hydrological boundary conditions underlying the case of 1540 are evidence-based. It was demonstrated that the heat and drought in 1540 was more persistent than in 2003, lasting about eleven months. It was also more extensive, comprising Europe from the Atlantic to Poland and from Tuscany to the North Sea. Finally, the heat wave was also more intense, considering the

---

167 Sieber, Jeannette, "Impacts of, and Adaptation Options to, Extreme Weather Events and Climate Change Concerning Thermal Power Plants," *Climatic Change* 121.1 (2013): 55-66. doi:10.1007/s10584-013-0915-0.

168 [168] Bundesanstalt für Gewässerkunde, *Niedrigwasserperiode*, 188.

169 Poumadère et al., "The 2003 Heat Wave," 1492.

170 Bundesanstalt für Gewässerkunde, *Niedrigwasserperiode*, 190.

drought-related extreme soil desiccation and its interactions with temperature, and also the continent-wide spread of forest fires.

The major impacts of a recurrent 1540-like worst-case event would be related to water scarcity and its multiple ripple effects across other interlinked systems. Inland navigation might have might have to be cancelled completely, considering discharge deficits of about ninety percent in large rivers. Railways would even less be able to compensate for the insufficient transport capacity of river navigation than in 2003. Besides severe crop failures, the megadrought would heavily affect cattle breeding and hence milk and meat production. There were about 65 million head of cattle in the EU in 2012. At a temperature of 28°C, which may be realistic for average conditions in summer 1540, a high-production cow consumes a daily amount of 110 to 130 litres of water. Even without any milk production, the animal still consumes 78 litres a day, while the need of feeder cattle is between 23 and 66 litres and that of calves somewhat less.[171] Assuming a moderate daily water need of fifty litres per animal, the forty million cattle living within the 1540 drought perimeter would want about two billion m^3 of water per day. There is still the question of how to feed forty million cattle from completely withered meadows. Probably, milk production would collapse and a considerable amount of cattle would have to be slaughtered. Forest fires might hardly be extinguished for lack of water and air traffic might be severely disturbed by widespread forest-fire aerosols.

Energy production would bear the brunt of the disaster. It is an open question how thermal power stations in Europe, consuming at present the lion's share of the average available surface water for cooling, might maintain their production capacity during a 1540-like megadrought. Besides the staggering quantitative deficit of about ninety percent, the water might be considerably warmer than in 2003. With regard to the situation in the federal state of Baden-Württemberg, which was most affected by the 2003 heat wave, scenarios of the consequences of further restrictions in cooling water availability envisaged a complete power failure (blackout) leading to massive disturbances in the integrated European network.[172] To what extent solar energy, in contrast to wind power, might compensate for the drastic reduction in thermal power and hydropower generation needs to be investigated. In any case, photovoltaic (PV) modules are not vulnerable to extreme drought. However, their output drops by about 0.5 percent for every 1°C increase in temperature. This means that

171 "Wasser – das wichtigste Futtermittel für Kühe," published by Tiergesundheit und mehr, last modified October 9, 2013, accessed April 6, 2017. <http://www.tiergesundheitundmehr.de/wasser-fuer-kuehe.pdfx>.

172 Bundesanstalt für Gewässerkunde, *Niedrigwasserperiode*, 188.

high ambient air temperatures in situations with high direct solar irradiation can have a significant impact on the maximum possible power output.[173] In the case of extraordinary supply situations, such as in summer 2003, the needed reserve power might be transmitted over greater distances. However, the transmission grid is only limitedly suited for substantial energy transmissions over large distances and over longer time periods.[174]

Whereas regional short-term power blackouts are experienced frequently around the world, societies are not ready to face large-scale, long-lasting power blackouts as might be experienced during a 1540-like megadrought. The 2003 blackout in the United States (which was, however, not climate-related), illustrates

> that after 3 to 6 hours without power most fuel stations and the refineries had to close down, leaving the public without fuel for cars or backup generators as the pumps did not operate. Most critical systems such as hospitals, water and sewage systems have backup power generation in place. However, these typically have only enough fuel for several hours to a maximum of a few days. [...] Governments have typically, however, implemented emergency fuel storages to keep most critical facilities alive for several weeks up to a month. After one month with no electrical power, water, transportation, emergency services, critical manufacturing, and chemical sectors can face widespread outages within the affected region.[175]

Water may still be available for drinking and a minimum of personal hygiene, albeit hardly for the full use of water systems. In combination with power blackouts, extreme water shortages would lead

> to many cascading effects. Hospitals, schools, nursing homes, restaurants, and office buildings all rely on water to operate. Water is used for drinking, sanitation, and heating and cooling systems in those facilities. Many

173 Patt, Anthony, Stefan Pfenninger, and Johan Lilliestam, "Vulnerability of Solar Energy Infrastructure and Output to Climate Change," *Climatic Change* 121.1 (2010): 93-102. doi:10.1007/s10584-013-0887-0.

174 Bundesanstalt für Gewässerkunde, *Niedrigwasserperiode*, 188.

175 Bruch, Michael et al., *Power Blackout Risks: Risk Management Options*, CRO Forum Position Paper, published 2011, accessed August 26, 2014. <https://www.allianz.com/v_1339677769000/media/responsibility/documents/position_paper_power_blackout_risks.pdf>.

> manufacturing operations either use water as an ingredient in their processes or rely on wastewater systems to remove and process their manufacturing waste. Access to safe water is necessary for providing mass care services and preventing the spread of disease.[176]

Under conditions of water shortages and power blackouts, excess mortality might take on staggering dimensions.

The cost of energy blackouts could be enormous. Researchers at the Hamburg Institute of Global Economics concluded that a one-hour blackout in mid-afternoon Berlin would cost the city €23 million. In the unlikely event of a one-hour blackout across Germany at peak time the cost would be €593 million.[177] For Switzerland, an analysis of the electricity industry estimates a minimum daily 2.4 billion Swiss francs' GDP loss and a 1.3 billion loss of daily earnings. In addition, longer term physical, financial, physiological, and psychological consequences also have to be considered. [178] In 2015 the Swiss Federal Office of Civil Protection estimated the cost of a longer term thirty percent power deficit due to an extended winter drought at one hundred billion CHF without, however, considering the 2014 study about the outstanding summer drought in 1540.[179] All in all, the vulnerability of modern societies in the face of a 1540-like megadrought is many times higher than that of pre-modern societies.

5 Conclusions

Nobody is able to imagine the magnitude and severity of low probability, high-impact events that are expected more frequently in the future as the result of continuing Global Warming, is the expert's tenor in an influential study edited

176 Ibid.

177 Peter Terium, head of energy provider RWE, warns of blackouts in Europe: "Energy Boss Warns of Blackouts in Europe," published by The Local, accessed November 7, 2013. <http://www.thelocal.de/20131029/energy-boss-warns-of-blackouts-in-europe>.

178 Verband Schweizerischer Elektrizitätsunternehmen, "Elektrischer Blackout: Basiswissen Stand November 2015," accessed April 11, 2017. <https://www.strom.ch/fileadmin/user_upload/Dokumente_Bilder_neu/010_Downloads/Basiswissen-Dokumente/29__Blackout.pdf>.

179 Bundesamt für Bevölkerungsschutz, "Katastrophen und Notlagen Schweiz: Technischer Risikobericht 2015," accessed April 11, 2017. <https://www.newsd.admin.ch/newsd/message/attachments/40201.pdf>.

by the World Bank.[180] This is the first paper to provide coherent evidence on what Europe might expect in the case of a worst-case high impact event. Nature has made an appropriate experiment in 1540 which hundreds of chroniclers across the continent recorded in words as a memory for future generations. However, because the individual texts were so far not properly interpreted and integrated into a coherent picture, their message went unheard.

The drought in 2003 was not extreme according to the lowest twentieth century values measured in France, Germany and Switzerland, in contrast to 1540. Seasonal precipitation for this year was estimated from the number of precipitation days obtained from Swiss and Alsatian chronicles as well as for a weather diary from Cracow (Poland). Rain-days in spring, summer and autumn were considerably below those of the 20th century average—in Alsace and Switzerland even below the absolute minima of the instrumental period. The amount of precipitation estimated from the observed rain-days remained significantly below one hundered-year minimum levels in 1540 throughout spring, summer and autumn. No event comprising a persistent extreme drought is documented within the instrumental period.

Likewise, the year 2003 does not rank among the smallest annual values in many very long German discharge series. Discharge deficits of more than fivty percent were not registered for large rivers. Those of the river Rhine were −37 percent, while in 1540 they were about ninety percent both in Basel and Cologne. The levels of other large rivers—Seine, Elbe, Thames—were also below their 2003 minima, as can be estimated from coherent qualitative descriptions. In conclusion, Western and Central Europe suffered from a dust-dry spring in 1540, which was then followed by a torrid summer and an almost rainless autumn making 1540 the worst drought year of the last five centuries. In the discussion about the significance of soil-dessication feedbacks for extreme heat waves there is consensus that spring droughts are an important precondition for the triggering of such events.

We demonstrated that the study on grape harvest dates in Burgundy[181] overestimated spring-summer temperatures in 2003 in comparison with 1540. Likewise, it was shown that the tree-ring study on Lötschental (Switzerland) showing average summer temperatures in 1540[182] in contrast to the maximum value in 2003 is inconsistent with later research.

180 Field, Christopher B. et al., ed. Managing the Risks of Extreme Events and Disasters to Advance Climate Change Adaptation: Special Report of the Intergovernmental Panel on Climate Change (Cambridge: Cambridge University Press, 2012).

181 Chuine et al., "Historical Phenology."

182 Büntgen et al, "Summer Temperature."

Premodern societies were remarkably resilient to heat and drought, even in the face of the Megadrought. They mainly suffered from water scarcity, which affected cattle breeding, milling and transportation. Forest and settlement fires together with widespread cattle mortality were probably the worst consequences. The winter grain harvest was satisfactory, because the yields and water needs of the crop were much below present-day values.

The vulnerability of present-day societies to a 1540-like drought was assessed by looking at the known impacts of the hot 2003 summer. Agriculture, transportation and energy were the main sectors of the economy impacted by the heat wave. Energy production turned out to be particularly sensitive due to the enormous need of cooling water for thermal (fossil and nuclear) power stations. The major impacts of a 1540-like worst-case event on present-day societies would be related to water scarcity and its multiple ripple effects across other interlinked systems. In any case, the drought would heavily affect cattle breeding and milk production. Whether thermal power stations, consuming the lion's share of the average available surface water for cooling purposes, might maintain their production capacity is open to debate. The power cuts in Baden-Württemberg experienced in August 2003 suggest that large-scale, long-lasting, disruptive power blackouts might not be excluded, involving cascading effects on critical infrastructure such as wastewater systems, gasoline availability and medical support, not including the staggering cost from lost production. Resource conflicts for water both between economic branches within and between countries could hardly be avoided.

From these considerations, it becomes obvious that "Europe's largest nature-induced disaster"[183] is far from being merely a topic of academic discussion about the remote past. Rather, it meets the criteria of the implausible black swan being a low-probability, high-impact event of an appalling magnitude. In the present, it would likely jeopardise critical national infrastructures and human security in Europe. The reasons why climatologists and risk analyists so far have not given due consideration to this hazard is rooted in several interrelated factors:

1. The quality of the raw data and the methodology were not checked in the case of the aforementioned *Nature* paper about the Dijon grape harvest dates. Perhaps this is because the review of man-made evidence in the

183 Bojanowski, Axel, "Hitze-Jahr 1540: Wetterdaten enthüllen Europas größte Naturkatastrophe," published by Spiegel Online Wissenschaft, July 2, 2014, accessed October 14, 2014. <http://www.spiegel.de/wissenschaft/natur/hitze-und-duerre-1540-katastrophe-in-europa-im-mittelalter-a-978654.html>.

archives and their interpretation "should be seen as a delicate task requiring a lot of endurance and accurateness" as historians Thomas Labbé and Damien Gaveau, who were critically looking at the Dijon series, have emphasised.[184]

2. In the face of the record-breaking heat and drought, the grapes' response was outside the normal range of biological and probability laws. Considering the comparative study of documentary and tree-ring data, trees responded to the 1540 megadrought with very small rings in 1541.[185] On the other hand, the record breaking tree-ring for 2003 in the series from Lötschental (Switzerland)[186] may well be related to both extremely high temperatures and a substantial availability of moisture,[187] whereas earlier hot summers in Switzerland were dry.[188]
3. With regard to the pre-instrumental low probability of high impact events such as 1540, models still leave decision makers with little practicable information on the scales over which they have jurisdiction[189] and they are rarely included in risk analyses. Yet "the characterization of such events and the related climate system functioning is particularly relevant in the context of global warming and the corresponding increase of extreme heat wave magnitude and occurrence frequency."[190]
4. Climate change studies have so far put a focus on temperature reconstruction at the expense of precipitation, although infrequent but catastrophic droughts—often connected with heat waves—may lead to far more devastating impacts than from heat waves alone.[191] Precipitation is the most important source of uncertainty derived from the Global

184 Labbé, and Gaveau, "Les Dates de Bans de Vendange à Dijon."

185 Büntgen Ulf et al., "Combined Dendro-Documentary Evidence," 3947-3959.

186 Büntgen et al, "Summer Temperature."

187 Wiliam Jolly and coauthors found a divergent vegetation growth response during the 2003 event in the Swiss Alps characterized by high elevation growth enhancement and low elevation growth suppression in response to the extreme summer temperatures and low-altitude drought conditions. Jolly, William M.et al. "Divergent Vegetation Growth Responses to the 2003 Heat Wave in the Swiss Alps," *Geophysical Research Letters* 32 (2005): L18409, doi:10.1029/2005GL023252.

188 Bundesamt für Umwelt, *Hitzesommer*.

189 Oreskes, Naomi, David A. Stainforth, and Leonard A. Smith, "Adaptation to Global Warming: Do Climate Models Tell Us What We Need to Know?," *Philosophy of Science* 77.5 (2010): 1023.

190 Orth et al., "European Temperatures in 1540"

191 Ibid., 1014.

Circulation Model predictions, and this uncertainty is further magnified by runoff-related issues.[192]

5. The great potential of documentary evidence for detailed studies on extreme events and their human impacts is not yet known, simply overlooked or not fully recognised. This epistemological blind spot has deep roots in the culture of the sciences, in which solely quantitative evidence and statistical methods lead to valid conclusions.

What occurred five hundred years ago may happen again in the near future. Last but not least, we have already witnessed record-breaking seasons in the new millennium, of a kind which shaped the eleven-month long heat and drought in 1540, albeit in different years and in a somewhat less extreme form. Spring 2011 was warm and extremely dry,[193] while summer 2003 and autumn 2006 were by far the hottest within the period of instrumental measurement.[194] In order to put such estimates on a broader empirical basis, further black swans need to be investigated in the period prior to 1540.

---

192 Mukheibir, Pierre, "Potential Consequences of Projected Climate Change Impacts on Hydroelectricity Generation," *Climatic Change* 121.1 (2013): 67-78. doi: 10.1007/s10584-013-0890-5.

193 Quesada, Benjamin, Robert Vautard, Pascal Yiou, Martin Hirschi, and Sonia I. Seneviratne, "Asymmetric European Summer Heat Predictability from Wet and Dry Southern Winters and Springs," *Nature Climatic Change* 2 (2012): 736-741. doi:10.1038/nclimate1536.

194 Luterbacher et al, "European Warmth."

PART 3

Memoirs

CHAPTER 6

The Birth of Climate History

Emmanuel Le Roy Ladurie

I would like to posit that climate history—its "historicised history", to borrow Fernand Braudel's term, albeit in a derivative sense; alternatively, climate history collaboratively written by historians (differing slightly from that written by climatologists)—originated as a more general branch of ecological knowledge, in short, as a discipline, as modest as it may be, in 1955, in December of that year, in fact.

Properly speaking, it is a matter here of *historiography*, since, for the followers of Clio, factual climate history began well before 1000 AD, following a prehistory in the Neolithic period and the first Metal Ages. This prehistory is, however, beyond the scope of this article, which is necessarily limited to a particular time period for obvious conceptual and thematic reasons. I will thus restrict my comments in the following to the second millennium and the beginning of the third.

In terms, then, of climate *historiography*, its beginning can be traced back to December 1955, or, to be more precise, to October–December 1955, the date the important French journal *La Météorologie* published an article by Marcel Garnier about the role of phenology in the study of climate variations. Specifically, Garnier described the way in which grape-harvest dates can function as indicators of the temperatures from April to September. This, in turn, allows for a study of climate conditions well before the first temperature measurements, which have only been conducted continuously and without interruption since 1659 in England and 1676 in France. Along similar lines, *Sciences et Avenir*, a popular science magazine, published an article during that same period, 1955, that was one of the first to introduce the study of tree rings to the French public. These rings can provide information on certain aspects of the climate of each year of the current period, but also of older periods going back well before the 17th century. It thus seemed to me—I was 26 years old at the time—that given these two sources of information, a climate history of the period prior to the appearance of thermometers (which did not come into use in France until the personal reign of Louis XIV and were not used systematically until much later) was conceivable.

In order to elucidate this point, allow me to provide some biographical and personal background—since, indeed, I was the first historian in France, at

 | DOI 10.1163/9789004356825_008

least in the second half of the 20th century, to be interested in climate history from this angle. This interest was to become manifest in 1967 with the publication by Flammarion of my book *Histoire du climat depuis l'an mil* (translated as: *Times of Feast, Times of Famine: A History of Climate Since the Year 1000*). In reality, however, I had been very active in this field, in complete solitude, since 1956, as a result of my impassioned reading of, and fascination with, the two articles just mentioned, particularly the one by Marcel Garnier. My vocation as a climate historian, nevertheless, is part of a much broader personal context. Throughout a long career, I have in fact worked on other topics related to the ecological, but also to the rural, urban, social, village and political past in France and Western Europe in general (see for example my multinational studies of the Platter family in the 16th century).

But there is also a personal reason for my inclination towards certain aspects of the climate from the time of my adolescence onward: my father and mother were farmers in Normandy, in the Calvados region, on a 120-hectare mixed farm where they bred animals and grew cereal crops on granite soil of low fertility.

The drying of the crops on the fields, after the harvest, was essential for a good yield of grain. Today, with combine harvesters, the process takes at most a few days—and a few nights—of work. The noise of the motors at night wakes up a few Parisian tourists, who complain, but the farmer does not care, and rightly so—what matters to him is to be able to gather in his harvest. In my parents' time, in the Thirties, Forties, even Fifties, the process was entirely different: the wheat was cut and bundled into sheaves with the help of a reaper binder, the improved successor of the first McCormick reaper (the reaper binder would ultimately be replaced by today's combine harvester).

Next the sheaves were arranged in small groups in the shape of what was called a *treiziaux*, a dozen sheaves standing upright with an additional sheaf placed horizontally on top, the thirteenth (*treizième*) sheaf from which the name derives. This sheaf was probably meant to provide some basic protection from the rain, or it might have been the portion of the harvest that, before the Revolution, was given to the clergy as a tithe.

In the case of extended good weather with sunny blue skies, as in the famous Bruegel painting *The Harvesters*, the drying and bringing in of the grain and straw was an easy matter. But if July and August were very rainy, it could spell catastrophe—and this was frequently the case in the wet climate of my native Normandy. The downpours damaged the crops, which would not dry; the grain began to sprout and rot. There were attempts to dry it artificially under unfavourable conditions and it was a disaster.

The experience, growing up, of such unhappy events served me well in understanding the famines of the Middle Ages and the food shortages of later periods. By the 1930s, food shortages were a thing of the past, as it had become easy to import wheat from North America. But the 'model', from a historical perspective, remained valid. Consider, for instance, the great European famine of 1315, which was caused by rains of this kind in the summer half-year (April to September) and even in the preceding seasons.

A too rainy winter, autumn or even spring could be dangerous for both the planting in the autumn as well as for the harvest. In the northern half of France, excessive rains were the greatest enemy of the wheat producers, but there was also the risk, albeit more rare, of heat damage scorching the grain in June, or the Great Frost of 1709 which killed the young shoots of wheat which had barely broken through the soil.

These reflections coupled with memories of the summers of my youth did not, however, play a significant role in my beginnings as a historian (I would return to them later), when, in 1954 or thereabouts, I was looking for a thesis topic. In the early Fifties, I was still steeped in a certain Marxism, of which I have preserved some rudiments of social analysis, no more, having rejected in the mid-to-late Fifties the eschatological contents of this doctrine, which posited the Last Judgement as social, communist, etc. revolution.

Still under the strong impression left by accounts of the 1929 crisis (which would long remain a pet theme of certain secondary school history teachers), I considered embarking on a study of a great economic depression analogous—albeit, of course, less severe—to that of 1873, even if the latter crisis paled in comparison to the unavoidable standard of 1929, not to mention those of 1973 and 2007. But I was living in Montpellier at the time, where I was teaching secondary school, and the local and regional Languedoc archives would not have provided me with much information about the great shifts in 19th-century conditions, industrial or otherwise.

I would have to look elsewhere, then. My friend Raymond Dugrand, one of the great geographers at the University of Montpellier geographers, advised me to consider working on the rural history of Occitania under the Ancien Régime, a topic that had received very little attention until then. This is how I came to complete a lengthy doctoral thesis on the peasants of Languedoc. I defended it at the Sorbonne in 1966.

In the process of the archival research I began conducting for my thesis in 1955, I was inevitably drawn to the question of the ecology of this peasant world—more specifically, to the question of the climate's influence on harvests. It was at that time that the experiences I had growing up on a farm came back to me.

My teacher Fernand Braudel had already come across this problem in his work on the Mediterranean. He had observed the effect of the climate on the economy of the 16th and 17th centuries and, quite remarkably, drew attention well before any other historian (none of whom were particularly troubled by this fact) to the almost spectacular advancement of the alpine glaciers around the late 16th century. His sources were an excellent Italian geographer of the 1930s, U. Monterin, and a *Frankfurter Zeitung* article he came across in the prisoner of war camp where he was held on German territory after France's defeat in 1940.

This re-glaciation, which indicates a slight cooling of the climate, figures among the countless causes Braudel identified behind what would later be called the economic crisis of the 17th century (which may in fact have begun as early as the late 16th century). It is worth mentioning here that the existence of the 17th-century crisis has not yet been sufficiently substantiated. Pierre Chaunu has given a detailed account of the extraordinary achievements of this same century, which are certainly well known in the fields of letters, arts and sciences (the world written in the language of mathematics…).

Suffice it to say that, crisis or no crisis, the 17th century did see a plateauing of the French population and perhaps also of its neighbouring countries: around 1340, the population in the area of today's metropolitan France was around twenty million; following the calamities of the conflicts referred to as the Hundred Years' War, plagues, etc., it numbered only roughly a dozen million around 1450; by 1560 it was back at its old level of twenty million, where, apart from some minor fluctuations, it stayed until about 1715.

Is this stasis more or less the effect of the *crisis* of a long century or is it simply a process of adjustment, balancing, and stabilisation? Given that such a process would have been due to all sorts of demographic slowdowns in which the climate may or may not have played a minor role, the debate could go on forever.

At the time, however, I was interested in the—perhaps badly articulated—idea of a 17th-century crisis, and I wondered whether there had also been, as Fernand Braudel suggested, a kind of climate crisis, or climatic aggression against crops, humanity, etc. A badly posed question, one might say. Be that as it may, Christian Pfister investigated this very issue in a series of highly interesting articles on the 1570-1630 crisis, which, indeed, he connects to the onset in the Germanic world, or even in Western Europe as a whole, of the gruelling period of the Little Ice Age.

Based on what I knew at the time, then, I was interested in climate history, a subject that continues to interest me today, even if the foundations upon which I now engage with the topic have evolved since my youth in the mid-to-late fifties. From this period onward, climate history was part of my thinking. As a rule, however, my colleagues—the young ones and the not so young ones—avoided the subject: it was considered too dangerous, and susceptible to being torn apart by the at times rather cruel criticism of the academic establishment and of what was more generally (and unfairly) referred to as "the Sorbonne"—an entity that was all the more formidable for being poorly defined. This criticism came mostly from young careerist academics who vigorously rejected any suggestion that a simple fluctuation in the weather could have played a de facto role, no matter how modest, among the countless (non-ecological) causes of historical phenomena in a particular modern or contemporary period.

This negative attitude was particularly surprising in the case of the Marxists, of whom there were many among the young historians at the time. Given this Marxian generation's interest in materialist historical concepts, one would have expected a greater concern for one of the basic 'matters' they could not simply ignore: the climate, the weather, whose influence on harvests and thus on all of rural history was considerable.

Friedrich Engels, Marx's inseparable companion, was in fact interested in what he called, in an interesting turn of phrase, the "dialectics of nature". But who recalled this among these militants in the making, a peculiar brand of realists who were blind to one of the most basic aspects of nature in the past? "Nature is a temple in which living pillars sometimes give voice to confused words…" if we are to believe Baudelaire.

But one of the pillars was missing, and the conception of a global and complex history of events and concepts ran the risk of being affected by this lacuna. I remember a terse comment made by Pierre Jeannin, one of the standard bearers of the new economic history of the 1950s: "I am wary of explanations based on the climate." And yet it was not necessarily a matter of explaining the past—in terms of the weather or anything else—but rather of an analysis of a context, in this case an ecological one. Suffice it to say that nature was considered a fixed unit exempt from even minimal fluctuations and that there were more important things to talk about. It is true that the climate has been more or less stable for several thousands of years, but it has also been subject to fluctuations, whether they are moderate and measured on a centennial scale, relatively severe on a multi-decennial scale, or tragic on an annual scale (e.g., the Great Frost of 1709). I refused to resign myself to the notion of absolute fixity and continued to ask questions that for the time being went unanswered.

It was then at the very end of 1955 and in early 1956 that I became aware of the two crucial articles mentioned above, which had just been published. One of them was pure research, the other of a popular nature, albeit very innovative in view of the rigid and archaic traditionalism dominating the field of ecological history in France at the time.

The first article, by Marcel Garnier, concerned the use of phenology in determining the history of weather fluctuations. Phenology is the study of the dates when plants flower, fructify and ripen; more precisely, of the dates of harvest, which are essential, albeit necessarily imperfect, indicators of average spring and summer temperatures (the period from April to September). Garnier, a professional meteorologist, even used phenological dates—specifically grape-harvest dates—to correct certain errors in the thermometric curve for Île-de-France in the 18th and 19th centuries. An audacious extrapolation, perhaps, but it demonstrates the ways in which the chronology of grape-harvest dates could be valuable to climatologists.

Garnier's article made a profound impression on me. As a young researcher, I felt like a door had suddenly opened, shedding new light on what until then had seemed the unknowable past of the weather fluctuations experienced—but not quantified—by our predecessors, the historians or ordinary peasants of the economic Ancien Régime.

As for the article on tree rings, it concerned mostly the arid west of the United States (Arizona, New Mexico, etc.). Simply put, the dryer the year in this semi-arid climate, the thinner the tree ring, and vice versa: in rainy years, the rings were thicker. Along with the attendant calculations, this provided a kind of natural rain gauge, which was admittedly somewhat crude, but all the more informative about past centuries, or even millennia if the samples were taken from sequoias, whose lifespan, as everyone knows, is very long.

I was fascinated by this second article. I got a hold of a kind of drill that allowed me to remove a sample cross-section from a tree's core to its bark. The only thing I managed to do with it was hurt my hand, however, so I decided to limit my research to studying scientific publications, which soon came to include those treating European—and in particular French—forests. Much later, I did personally examine the beams of the Collège de France with the aim of determining the date of construction of this establishment under Henry IV and Louis XV, with results, incidentally, which were precise almost to the year. But that is another story.

I turned almost immediately (in 1956) to the grape-harvest dates that are recorded in so many wine-making regions in France, from Île-de-France to Languedoc. I was living in the latter region at the time and soon had collected a very large amount of data from the local and village archives. The principle of this research was simple (even if its realisation was not always easy): the milder the spring and the hotter the summer of the year in question, the earlier the harvest. Conversely, if the spring was cold and/or the summer was cool during what is called the summer half-year (which, again, extends from April to September), the harvest was delayed. Thus the 1816 grape harvest was postponed due to a long summer half-year that was very chilly, caused in part by the planetary shield created by aerosols from the maximum explosion of the Tambora volcano in Indonesia in April 1815; on the other hand, the 1846 harvest was very early due to an exceptionally hot and dry summer half-year.

Marcel Garnier, the author of that aforementioned very memorable article, was an excellent meteorologist. He published some remarkable pluviometric series for the 18th, 19th and 20th centuries. He did not believe in global warming, which, it is true, was hardly noticeable at the time, even if the rise in temperatures over the century had already left its mark on French winters during the long time span since 1893. Nor did he believe that the glaciers were retreating. I assured him on this point that there were photos to prove it, but he remained sceptical.

Be that as it may, Garnier's work on grape-harvest dates remains indispensable, even if other researchers such as Angot in the late 19th century and Duchaussoy in the 20th paved the way for him. Thanks to Garnier, beginning in 1956-1957, several historians (myself the first, it must be said) became interested in grape-harvest dates as climate indicators, notably Jean Nicolas for the Savoy region, Bartholomé Bennassar for Valladolid and the American David Herlihy for Tuscany.

In 1956, I made my decision. I would become a historian of the climate, among other specialisations that would follow me throughout my career (rural, urban and multi-regional history, etc.). If I try to reconstruct my decision-making process as a novice historian interested in the climate, I come up with the following explanation: When I was in my last year at the École normale supérieure, Professors Meuvret and Palmade lectured often on the great economic crisis of the late 16th/17th century—whether interpreted rightly or wrongly is of no consequence here. Braudel, on the other hand, singled out the glacier crisis and extreme climate conditions of the late 16th century. Was the latter the cause of the former? Did the climate wreck the economy? The question, it might seem, is badly posed, but the powerful historical motivation was there under the guise of a wide-ranging historical ambition. In the following years and decades, I had no choice but to allow myself to be guided by my intuition. Note the common denominators that characterise this first construct of a kind of climate history and its ensuing iterations: a 1929-style economic *crisis* in the early versions of a young man's thesis prospectus after passing his *agrégation* in 1953/54. The violent determinism of an extreme crisis of the environment and of glaciers generated the long economic *crisis* of the 17th century. A true or false claim? No matter, I am describing here the itinerary of a still young researcher. Prolonged studies by the same, now established, scholar of subsistence *crises* of the old type, came out during the ensuing forty years. Finally, at the outset of the third millennium, a prodigiously expanding human *crisis* arose due to a dangerous warming of the earth's atmosphere and an increase in greenhouse gases. Today there are many scientists and historians around the world devoting themselves to an investigation of climate fluctuations and global warming. I am no longer as solitary as I once was. All the pioneers have to do is join forces with this new, ever-growing group.

In any case, in 1956 I set out on this as yet barely cleared path with the modest project of a "phenologico-viticultural" study. I was followed years or decades later by my colleagues Jean Nicolas and a few others. A first paper, of a historical and confidential nature, was published in 1957 under my name in the proceedings of the Historical Federation of Languedoc and Roussillon, a learned society for regional scholarship, whose work at the time remained in the discreet safekeeping of its venerable members. On the other hand, irrespective of their merit, the earlier publications about grape harvests, such as those by Marcel Garnier, were only concerned with the thermometric—in short, physical—correlations of these harvest dates.

The essential source for the young French researcher was thus the major article by Alfred Angot published in the 1880s. The piece includes a considerable collection of annual grape-harvest dates, beginning in the final third of the 14th century and continuing almost without interruption until the Third Republic. Angot compiled the 19th-century dates by sending circulars and letters to the mayors and town councils of a large number of villages and wine-making towns in France as well as in western Germany and western Switzerland.

As part of the research for my dissertation on the peasants of Languedoc (mentioned above), I also compiled a relatively large number of unpublished grape-harvest dates from wine-growing localities in southern and central France. I then began comparing these dates with the annual thermometric data in France and England from the late 17th century to the present day. Marcel Garnier was a huge help to me in establishing these parallels between temperature and viticulture. I also began building contacts with top-level English climatologists, such as Hubert Lamb, Gordon Manley and the lesser known but very respectable Derek J. Schove.

Among professional climatologists, Lamb is the true founder of a dynamic and fluctuating climatology of the last millennium. I met him in Aspen, Colorado, at a conference on the climate of the 11th and 16th centuries. Incredibly enough, the American organisers of the conference, all of them meteorologists, asked me to publish the graphs created during this seminar held in the western United States.

In 1962, the journal *Annales: Economies, sociétés, civilisations*, founded by Lucien Febvre and Marc Bloch, and continued by Fernand Braudel, agreed to publish this collaborative 'American' work. A British university awarded me an honorary doctorate based on my first articles and early research on climate history as well as on various subjects related to rural history, which, for me, was indissolubly linked to an ecological orientation in my research that encompassed peasant history and the environment. The latter, over the course of the centuries, unites the joint destinies of the biosphere and the atmosphere.

I also knew Gordon Manley, author of a celebrated month-by-month, year-by-year temperature series of central England. With its multiple additions made by younger researchers following Manley's death, this series is now the oldest thermal archive based on mercury in existence, beginning in 1659 and extending to the present day. It will nevertheless be equalled, even slightly surpassed historically, by the French, or, to be precise, Île-de-France series being published and gradually perfected by Daniel Rousseau, climatologist for Météo France, the French national meteorological service. At present Rousseau's series dates back to 1676 and also extends to the present day.

Finally, I had the chance to personally get to know Derek J. Shove, a very knowledgeable amateur, headmaster of a secondary school in Beckenham, Kent, and diligent discoverer of unpublished archives on climate history. Shove is the author of an important publication on the Maunder Minimum, a period from 1645 to 1715 in which sunspots all but disappeared. The climate experts across the Channel thus bring together the hard science associated with their profession and the historical knowledge so highly valued by our colleagues at Oxford, Cambridge, and the Sorbonne.

Today it is English-speaking climatologists, from the UK and the US, who have gone furthest in developing our physical knowledge about climatic change and the warming of the air that surrounds us—notably such eminent researchers as Phil Jones and Michael Mann. At times, however, some of them seem to have lost touch with the discipline of history in its traditional forms.

From a historian's perspective, and particularly for me in 1956, the use of grape-harvest dates, introduced by several Swiss, and in particular, by Alfred Angot in 1883 and continued by Marcel Garnier until 1955, constituted a stunning breakthrough: it allowed for a study of climate phenomena going back to well before 1659, the start date of the long English temperature series. Using grape-harvest dates offered the possibility of exploring weather data going all the way back to the late 14th century. Today, Valérie Daux has been able to study late Medieval and Renaissance climate conditions very closely by combining the archival science of phenology and the physics of oxygen-18.

Besides, there was the question of the glaciers. In my reading of German, Anglo-American and Scandinavian sources (the latter in English or German translation), I came across mention of the retreat of Alpine glaciers and the different phases of moderate re-glaciation (17th century) and glacial retreat (20th century), the latter in all likelihood indicating a certain degree of warming. The observation of this retreat gave rise, from the 1970s onwards, to the famous notion of 'global warming', which has been much debated in our day and age.

I decided to not content myself with second-hand sources, journalistic or otherwise, as seductive as they may be. Alone or accompanied by my wife, I explored the terminal moraines of the Savoie glaciers of Chamonix one after the other; then, in Switzerland, those of Grindelwald, Aletsch and the source of the Rhone River; in Italy and Austria, the Brenva, Ruitor and Vernagt glaciers. My wife Madeleine, more athletic than I was, took me along on glacier excursions, somewhat daring ones in my view, but which would have made a seasoned mountain climber smile. Nonetheless, these trips involved several occasions of reckless risk, taken on far too lightly for my taste by the lady guiding me. These walks, which, in the end, were done for sport, opened up a whole new world for a historian from Normandy who until then had climbed no higher than the summit of Mont Pinçon (362 meters elevation), the highest point of the department of Calvados.

All the while, I remained an archive historian and a scholar of old paperwork. In the early years of the Fifth Republic, departmental archives were still sent by post, unlike today for reasons of loss prevention. Barely legible documents from the old municipal collections of Chamonix were sent to me care of the Montpellier archives, the city where I was then living. The research conducted by Mougin, an alpine engineer for the French Water and Forest Administration in the early 20th century, provided me with a starting point, and I was thus able, after laboriously deciphering these unpublished papers, to pinpoint the geography and chronology of the glacial thrust in Chamonix between 1570 and 1590, as well as throughout the 17th century and beyond.

This work by a young researcher would later be cited with high praise, I am proud to say, by the most eminent Mer de Glace specialists: Grove in England and Zumbühl and Nussbaumer in German-speaking Switzerland. My research from the years between 1950 and 1960 on the late 16th and first half of the 17th centuries, published in *Times of Feast, Times of Famine*, is still pertinent. What I wrote on the subsequent phases of the large alpine glaciers, on the other hand, was rendered obsolete on several points by the rigorous studies conducted later on by my Swiss colleagues. I also explored Bonanay and Le Châtelard, hamlets crushed or knocked over by the Mer de Glace around 1600, which allowed me to define more precisely the Little Ice Age as referring to the period of significant alpine glacier expansion between 1570 and 1854.

I also applied myself to the study of a history of seasonal events, in particular of 16th century winters. I was able to detect a certain cooling of the alpine climate on the order of several tenths of a degree or more, distinguishing the cooler period between 1560 and 1600 from the slightly milder period between 1500 and 1560.

This research with its grape-harvest dates, thermometric and glaciological series, and series of events found its way into articles I began publishing in 1960 in *Annales ESC*, *Revue Historique*, *Revue de Géographie Alpine* etc. This 'articulated' ensemble was crowned, on might say, by two works: *The Peasants of Languedoc*, the principal thesis I defended in 1966, one of the first chapters of which concerned climate history. Around the same time, I also sat a viva on a secondary thesis on climate history, covering in detail all of my work in that field to date. Thanks to Fernand Braudel, this secondary thesis was expanded into an actual book, *Times of Feast, Times of Famine*, published in French by Flammarion in 1967. In this book, I argued for defining the Little Ice Age as extending from the second half of the 16th century to the middle of the 19th. I also suggested that there had been a certain degree of warming of European winters in the 20th century—despite the slight, momentary drop in annual global temperatures beginning in 1950.

I revised this book for an English translation published a few years later, followed by new additions and corrections for the French paperback edition, also published by Flammarion in two very handy volumes in 1978. The book was not badly received. A private radio station reported on it. A contributor to a weekly newspaper was also interested in my work, but told me that he did not have the time to read the book on climatology I had just published. He asked me to write a paper summarising it, which I did.

Pierre Chaunu wrote an exhaustive, convincing review review in a historical journal. This excellent historian had at the time broken with Fernand Braudel, although relations between the two were still amicable. The *Annales* School would go on to split into two different factions—not scientifically, but politically: on one side there was the Chaunu group (with Bardet, Poussou, Denis Crouzet and a few others), politically more conservative but historically audacious in the best sense of the word; on the other, a group formerly surrounding Braudel, whose participants as well as their direct successors to this day continue to edit the still excellent journal *Annales ESC* (*Economies Sociétés Civilisations*). The name has changed but the spirit remains.

In general, my book on the climate was a critical success, but for the best minds—Pascal Acot for instance—it was a bombshell, particularly because of the audacious use of glacial data and, even more so, of grape-harvest dates. To repeat: the latter made it possible to go back in time well before the first available thermometric data, which in England dates back to 1659 and in France to 1676, as can be seen in the work of Gordon Manley and Daniel Rousseau, respectively.

Henceforth, it was possible to determine, with a certain degree of precision, the weather fluctuations of the years 1600 to 1650 and even of the 16th century—indeed, even of the 15th. It was Valérie Daux, mentioned above, who skilfully worked out the historical possibilities as they presented themselves in the 1960s, or even beginning in 1957 if we include my first article on the subject—modestly buried, it is true, in the pages of a regionalist journal about the Languedoc region, which was scholarly but almost entirely unknown.

1967 saw me continuing with my research, still with the encouragement of Fernand Braudel and with the collaboration of my excellent friends from the École des Hautes Etudes, sixth section, such as Jean-Pierre Peter and, in particular, Jean-Paul Desaive. It was during this period, furthermore, that I began my—also totally new—work in physical anthropology and, in particular, physical anthropometry concerning the height of French conscripts in the 19th and 20th centuries, which I showed to be increasing at a more or less rapid rate.

In terms of the climate, my friend Jean Meyer introduced me to the archives of the French Academy of Medicine, dating back to the reign of Louis XVI and the beginnings of the Revolution. A great scholar of that period, Félix Vicq d'Azir, had become inspired by theories of aerism, in vogue in the 18th century under Louis XV and his successor—but out of date, or as good as out of date, today. These theories posited that the atmosphere—in other words, the climate–weather complex—was responsible for numerous diseases. With the support of the monarchy, Vicq d'Azir initiated a large national, even international, survey. Volunteer researchers were armed with forms and sent far and wide to record daily temperatures as well as observations of epidemics spreading in their regions. J.P. Desaive, mentioned above, and I, along with other colleagues, began examining these old records, still safely preserved for the 21st century at the Academy of Medicine.

My particular interest in these files concerned the thermometric series recorded for many towns in France and abroad. As part of our collective project, these series allowed me to draw attention to the hot summer half-years (April to September) from 1778 to 1781 and beyond. These unusually warm summers provided an easy explanation for that period's wine overproduction crisis and simultaneous collapse in prices, the production of wine having been significantly stimulated by these heat waves and hours of sunshine, *ipso facto* favourable to the grape yield.

This finding changed the import of Ernest Labrousse's great Sorbonne thesis on the French economic crisis on the eve of the Revolution. Rather than a nation-wide economic crisis leading up to 1787, there had been an excess of wine for several years around 1780, an actual crisis only for the wine-growers. Thus modified, Labrousse's work takes on a more limited, albeit more solid significance than at the time of its publication. Next I conducted more research on grape-harvest dates, particularly in 16th-century Île-de-France, with the help of the remarkable historian Micheline Baulant. Among other areas of specialisation, I had become known by then as a historian of the climate—a very subordinate kind of history to be sure, but useful nevertheless as a modest help to scientists studying the meteorology of past centuries.

Thanks to this moderately recognised distinction, in 1976 I was invited to take part in a television programme about the great drought of that year. I held back in the debate, yielding to the recognised specialists in fields directly concerned with the subject. The discussion remained courteous, unlike the gladiator fights that erupt around questions of global warming and CO_2 today.

I have already mentioned the contributions of other historians who were kind enough to put their confidence in my research: indeed, grape-harvest dates came to figure in the great regional theses of Bartholomé Benassar, Jean Nicolas and David Herlihy. But these were men of my generation, not young scholars. Quantitative history was losing its prestige, as a result of the offensive led by the generation of 1968, who were enamoured with cultural history and hostile to the technocracy of the number, whether it be thermometric or pluviometric. I sensibly toed the line, all the while discreetly persevering in my environmental, that is climate, research.

Having been burned by the new tendencies in my field, I kept my studies on the history of rain and fair weather to myself, for the most part. I buried myself in work on a village monograph—another confidential genre if there ever was one. Spending my days working as a teacher, I devoted my evenings to studying a tiny village no one had ever heard of, Montaillou in the department of Ariège. For reasons I still have not understood, this miniscule monograph was an enormous success in France and in many other countries.

In terms of the climate, one of my students, Pierre Alexandre, produced an excellent work on medieval climate history. In terms of seasonal history, the silence—albeit active—I forced upon myself gave a well-deserved chance to teams outside of France. In the final quarter of the 20th century and beyond, my friend Christian Pfister dominated the field of study.

A host of German-speaking Swiss researchers surrounding Pfister took up his work, quickly demonstrating their originality. The glaciologist and mountaineer Hanspeter Holzhauser was part of this group; his scientific work shed light on the history of the Aletsch glacier from 1500 BCE to the present day. Jürg Luterbacher drew attention to the weather fluctuations of the late 17th century and subsequent time periods. He was also interested in the three harsh winters of 1940, 1941 and 1942, whose causes he traced all the way to the El Niño oscillations across the American continent. In Czechoslovakia and then the Czech Republic, Rudolf Brázdil made a very noteworthy appearance after the Iron Curtain was lifted, showing himself to be one of our best historical climatologists. Moreover, the team of Swiss 'glaciologist-historians', whom I have had repeated occasion to cite, produced remarkable work on the glacier at Grindelwald and on the Mer de Glace. By combining visual and text-based information drawn from old maps, prints and engravings, paintings, photographs and other data collected on site with accounts from English and Continental travellers and tourists, these fossil and new ice archivists created a chart tracing the evolution of the Grindelwald glacier from the 16th century to the present day, in almost yearly increments. Their diagram clearly shows the hot and mild periods of the second third of the 16th century, the Little Ice Age in the Alps from 1570 to 1850, the successive retreats of the glacier between 1860 and 1880 and the on-going, now irreversible retreat beginning in 1930.

Heinz Zumbühl, a great authority in the field, initiated this vast survey; he was subsequently joined by Samuel U. Nussbaumer and the two continued to work on it into the 21st century. The same authors recently completed an analogous study in English on the Mer de Glace with its three maximum LIA extents of 1644, 1821 and 1852. After this, the 'Mer' also suffered a catastrophic regression between 1853 and around 1880 and, once again, from 1930 until the present day. A French glaciologist from the Savoie region, Christian Vincent, joined the two Swiss to produce an even more sophisticated book on the development of the Chamonix glaciers, among which the 'Mer' continues to be the most beautiful ornament, despite its ongoing retreat. French-speaking Switzerland joined into this polyphony without borders with the publication of a lovely little book entitled *La Suisse se réchauffe* [Switzerland is getting warmer] by Martine Rebetez.

Christian Pfister in any case remains the uncontested leader in the field of historical climatology of Switzerland and the German-speaking world in general. The authority of the Bern (and Zurich) school remains uncontested on the old continent, even if Pfister saw fit to introduce one of the editors of this volume as the "founding father of historical climatology". Pfister's predominance since the 1970s is furthermore confirmed by his involvement in the prestigious American journal *Climatic Change*. His English-language articles have appeared in a large number of specialised periodicals both in the United Kingdom as well as in the United States such as *Science*, *Nature*, etc. His books written in German are not well known internationally, for reasons of linguistic incompetence in relation to this language, which is widespread both in France and abroad. This work, however, is immensely important, with its review of climatically remarkable years going back more than five centuries and his precise study of the links between historical weather catastrophes and health demographics in Europe. A small book by Pfister on the Swiss *Patrioten* of the Enlightenment is equally important for an understanding of agrarian and climatic conditions in the stretch of the country between Basel and Geneva in the second half of the 18th century. Camuffo and Bonardi in Italy, Barriendos in Spain and Franz Mauelshagen in Germany along with many others have also distinguished themselves in the field of historical climate research. In Scandinavia, Moberg produced a remarkable work based on dendrological and sedimentary data. He stresses the contrast between the mildest periods of the Little Climatic Optimum (LCO) around the early 11th century and the cold periods of the Little Ice Age in its most frosty phase around the late 16th century. According to Moberg, the difference amounts to more than half a degree in average temperatures, on the plus side for the extreme LCO and on the minus side for the extreme LIA.

These inevitably incomplete assessments were accompanied by important new publications appearing in England and in the US around the same time, beginning in the mid-to-late 1970s. British climatologists such as Lamb, Manley and Schove were humanists proficient in continental languages, particularly French and German. Despite being highly sophisticated scholars, they were interested in English and European history.

The English and American climatologists of our period—among the most distinguished are Phil Jones and Michael Mann—are scholars of the highest order and internationalists. They are among the most responsible leaders of the Intergovernmental Panel on Climate Change (IPCC). They keep track, step by step, of a) the regrettable progress of global warming, as well as of b) the increase in CO_2 and greenhouse gases in the earth's atmosphere; with b), as far as we know, being one of the most important causes of a). With their globalist approach, those in charge at the IPCC have brought a planetary universalism to the discussion, to no one's regret—but one still cannot help being nostalgic: for the gains of the previous generation, those of Lamb, of Manley, who loved British localism and the European context, climatic and otherwise. The IPCC organisers, including Jones and Mann, understand this, and the panel's next report will place more of an emphasis than in the past not only on global but also on regional conditions, including unusual behaviour and meteorological phenomena in Western Europe and other parts of the world.

What role does France play in all of this? For a long time, North American researchers were literally contemptuous of the most important French climatology periodical, *La Météorologie*. One of the reasons for this disdain was that American researchers in particular were completely ignorant of the French language and were thus unable to familiarise themselves with the articles published in the journal—which, in fact, were hardly insignificant.

But it is also true that our great national meteorology magazine remained resolutely fixist for many years. Until 1980 and even beyond, the editors rejected the notion of centennial warming, with the sole exception of a very brief note from Germany on the Hamburg region, which was, admittedly, accompanied by a long 'warmist' and prescient article by the German scholar Hans Flohn, translated into French here. But *La Météorologie* resolutely changed course around the turn of the third millennium. It now welcomes articles by Jean-Marc Moisselin, Daniel Rousseau, Yves Tourre and many others who situate themselves—not without bringing their own perspectives into play—in line with the IPCC and share the view that global, as well as national, temperatures are rising, in particular due to the greenhouse effect.

This radical turn among French meteorologists is also evidenced by the appearance of several new research centres and think tanks at the beginning of the 21st century. At Aix-Marseille University, Georges Pichard authored a monumental thesis on the history of the southern French environment, including its climate. A large laboratory in Provence under the direction of Édouard Bard, professor at the Collège de France, is studying the respective influences of the greenhouse effect and solar radiation on the global warming of the planet, without in any way succumbing to the melancholy of climate scepticism.

In Toulouse, under the aegis of Météo France, Daniel Rousseau has published a thermometric series for the Île-de-France region with annual and monthly data from 1676 to the present day, mentioned above. I have also referred to the work done by the Grenoble group of glaciologists together with the Swiss on the Mer de Glace, which is complemented by publications on Antarctica, the Arctic and Greenland by Claude Lorius, Jean Jouzel and their students. Jean Jouzel also serves as the vice-chair of one of the IPCC working groups.

At Gif-sur-Yvette, the Climate and Environment Sciences Laboratory (LSCE) is staunchly committed to historical climatology and, quite simply, to the history of the climate in the form that it has been practiced since 1956 and then 1967 by one of the authors of the present volume. But the sophistication of the

LSCE and its 'Pfisterite' analyses clearly extends beyond the early stages of the discipline in the 1960s.

The enormous collection of grape-harvest dates was first taken up again by Isabelle Chuine at Montpellier in the form of an influential article published in *Nature* in 2004. Next, Valérie Daux picked up where Chuine left off, assembling all the grape-harvest dates Alfred Angot had collected from hundreds of French, Swiss and West-German wine-making localities (see above). She was able to supplement this collection with numerous series of grape-harvest dates provided to her by sources from current or former French wine-growing areas as well as from the Collège de France. This information was used to create various average-temperature series for France, Switzerland and Germany, which are based on even larger datasets, as each series includes figures from multiple wine-making localities in each area. The most remarkable and most well-established series is the Burgundy series, extending from 1370 to the present day.

Finally, Daux supplemented the grape-harvest data with information derived from measuring the oxygen-18 isotope contained in tree rings from the harvest years in question.

•••

The time has perhaps come to enlarge our perspective by looking at what could be called the French historical-climatology landscape (on a global scale, it would be a matter of painting a large picture indeed, too ambitious for the scope of this volume).

As far as France and its neighbouring countries are concerned, what is remarkable is the reception this young discipline has enjoyed among literary theorists and art historians. Thanks to Professors Sermain and Berchtold, as well as Anouchka Vasak, several symposia about the climate and weather phenomena in literature (Shakespearean tempests), painting (Ruysdael), music and history have taken place in Paris at the Singer-Polignac Foundation and in Poitiers. Quite recently, researchers at the Sorbonne have revived 18th-century analyses of the harsh winters of the Middle Ages and the early modern period, composed by this or that German Enlightenment scholar. These chronological series concerning a long-ago past have lost none of their value or freshness. The aesthetics of thunderstorms, the paintings of Turner, but also a real attention to the history of the climate as such come together at these conferences, which are multidisciplinary in the best sense of the word.

Geographers are also at the forefront of this discussion. Some of them subscribe to the 'warmist and CO_2' analyses put forward by the IPCC, such as

Professor Vinet at Montpellier. Others are, to a greater or lesser degree, climate sceptics. The bottom line, however, is the work both groups are doing on our meteorological history. Thus Martine Tabeaud, professor at Panthéon-Sorbonne University's Institute of Geography, has used the Paris Observatory's archives to conduct excellent research on the variations in the French capital's cloud cover over the past several centuries. Tabeaud also has some excellent students, such as Alexis Metzger, who has conducted a very close study of the important multi-volume work published in Dutch by J. Buisman and A. Van Engelen on the weather of the Netherlands and its neighbouring regions, tracked year for year from the Middle Ages until the present day. Other researchers are, as it were, free radicals, and all the more efficient for it. Eric Levavasseur, a pharmaceutical sales representative by day, has produced excellent studies of the Parisian climate from the reign of Louis XV to the Revolution. He is currently doing work on the 18th-century Languedoc thermometric series, compiled by Badon, also conserved in the Montpellier archives.

Among professional historians, the situation is a little different. Most of them are not interested in the history of the climate, despite the fact that this history is important for anyone considering the question of subsistence, which is completely dependent on the vicissitudes of farming weather and is at the heart of political agitation. This disinterest sometimes verges on ignorance: thus a distinguished figure, holder of a doctorate and a professorship, was heard rejoicing over the regression of the glaciers, as this would liberate additional quantities of water that could benefit industry and humanity. One hardly dares inform this person that this water will ultimately flow into the ocean and thus be lost for humankind. Climate history is sometimes viewed by historians as an opportunistic discipline, taking advantage of the cachet it has with scientists: the latter may grant financial resources to climate historians who furnish them with old weather data from the archives.

We would be well served, however, to raise the level of the debate. Among our most distinguished climate historians today is Emmanuel Garnier, author of an important book on the subject and also mentor to a new crop of students in the field. With a background in forest history, he recently published a book on weather disturbances and has written numerous articles in the speciality that concerns us here. Garnier participates in his own right in ad hoc scientific committees and English-language symposia; he publishes in British and American journals, and, after applying only once, was promptly admitted into the Institut Universitaire de France.

Still, this budding discipline is very far away from the prodigious accomplishments of the history of religions in France, which, through the impetus of Jean Delumeau, boasts a large number of doctoral students. It is of little importance whether these young researchers are anti-clerical or sympathetic to Catholicism—the point is the valuable work produced in this way. In an entirely different context, a simple association like the Fondation Napoléon brings together hundreds of people belonging to the imperial or the traditional aristocracy, to the intelligentsia, to the circle of academic historians, and to the most diverse range of professions. A symposium on the history of the climate drawing a few dozen people pales next to the Napoleonic crowds assembled in such a judicious way.

I have mentioned Emmanuel Garnier and Valérie Daux several times. I should also mention Pascal Yiou at the Climate and Environment Sciences Laboratory (LSCE), a specialist in atmospheric physics and very well known in English-speaking and Japanese scientific circles. He has been more keenly aware than anyone of the need to involve historians in the research conducted in his laboratory. He is one of the most eminent researchers in the field of historical climatology in France, in the best—that is, the strictly scientific—sense of the term.

PART 4

Present

CHAPTER 7

EU Climate Leadership? Europe's Role in Global Climate Negotiations

Clara Brandi

Abstract

This chapter analyses European climate leadership at the global climate negotiations of Copenhagen and the summits after Copenhagen, culminating in the successful adoption of the Paris Agreement in 2015. An analysis of the Union's conduct on climate issues indicates that by setting an example, directional leadership has been the most important leadership mode of the European Union (EU). The chapter underscores that the EU is an important player shaping global climate policy, but shows that the Union's leadership has come under pressure. Both the failure of the Copenhagen climate summit in 2009 and global and European economic and political crises have weakened EU climate leadership in recent years. At the climate summits following Copenhagen, however, the EU managed to partially revive some of its leadership position.

The chapter starts by presenting the analytical framework for the analysis of EU climate leadership by identifying three key conditions for the EU to be a successful leader in multilateral negotiations: preference cohesion (speaking with one voice), credibility, and opportunity. The chapter then illustrates that EU climate leadership came under duress in Copenhagen but has recovered since then. At the same time, EU leadership is still troubled by a number of challenges, which have to be addressed in order to safeguard and reinforce the European leadership role. For the future of climate policies and the implementation of the Paris Agreement, the EU's role is all-important—especially in light of the position of US President Donald Trump on climate change. Looking ahead, in order to strengthen its leadership, the EU should play a key role in fostering the implementation of the Paris Agreement and in promoting it. Moreover, the EU should integrate pertinent measures to address climate change with the Agenda 2030 for Sustainable Development and the Sustainable Development Goals (SDGs) it enshrines.

 | DOI 10.1163/9789004356825_009

1 Introduction

Leadership is often regarded as necessary for tackling transnational challenges, such as climate change, and for establishing global governance arrangements.[1] For more than a decade, the European Union (EU) has been characterised as a leader in international climate policy-making.[2] Yet the 15th United Nations Framework Convention on Climate Change (UNFCCC) Conference of the Parties (COP) meeting in Copenhagen in December 2009 ended disappointingly,[3] especially for the EU.[4] The failure at Copenhagen called into question the EU as a leader on climate issues and raised concerns about EU climate leadership. Why did EU climate leadership come under pressure given that the Union is often seen as the most ambitious actor in the fight against climate change?

Against this background, this chapter analyses EU climate leadership at the climate talks in Copenhagen and at the summits after COP 15 on the road to the Paris Climate Change Conference in 2015. The chapter underscores that the EU has become an important player shaping global climate policy, but shows that the Union's potential to shape global climate policy has been challenged. Both

1 Young, Oran R., "Political Leadership and Regime Formation: On the Development of Institutions in International Society," *International Organization* 45.3 (1991): 281-308; Sjöstedt, Gunnar, "Leadership in Multilateral Negotiations: Crisis or Transition?" in *International Negotiation: Actors, Structure/Process & Values*, ed. Peter Berton, Hiroshi Kimura, and William I. Zartman (Hampshire: Macmillan, 1999), 223-253.

2 For a recent review of the scholarly literature on the EU's role in the context of climate policies, see also Groen, Lisanne, "European Foreign Policy on the Environment and Climate Change," in *The SAGE Handbook of European Foreign Policy*, ed. Knud E. Jorgensen et al. (London: SAGE, 2015), 884-897. See also Bretherton, Charlotte, and John Vogler, *The European Union as a Global Actor* (London: Routledge, 1999); Groenleer, Martijn L.P., and Louise G. Van Schaik, "United We Stand? The European Union's International Actorness in the Cases of the International Criminal Court and the Kyoto Protocol," *Journal of Common Market Studies* 45.5 (2007): 969-998; Oberthür, Sebastian, "The Role of the EU in Global Environmental and Climate Governance," in *The European Union and Global Governance*, ed. Mario Telò (London: Routledge, 2009), 192-209; Rayner, Tim, and Andrew Jordan, "The European Union: The Polycentric Climate Policy Leader?" *Wiley Interdisciplinary Reviews* 4.2 (2013): 75-90. In this article, the European Union (EU) is defined as the combination of both the EU institutions that represent the EU at the climate negotiations (the European Commission, the Council Presidency) and the EU Member States.

3 Dimitrov, Radoslav S., "Inside Copenhagen: The State of Climate Governance," *Global Environmental Politics* 10.2 (2010): 18-24.

4 Metz, Bert, "The Failure of Copenhagen: What Now for the EU?" *European Voice*, January 22, 2010, accessed April 20, 2011. <http://www.europeanvoice.com/article/2010/01/the-failure-of-copenhagen-what-now-for-the-eu-/66967.aspx>.

the failure of the Copenhagen climate summit in 2009 and recent global and European economic and political crises have weakened EU climate leadership in recent years. At the climate summits following Copenhagen, however, the EU succeeded in partially re-strengthening its leadership.

The revival of EU climate leadership was especially evident during the Durban climate talks in 2011 and also during the Paris negotiations in 2015. In 2011, the EU contributed to safeguarding the UN process of climate negotiations, which would have unravelled in the absence of an agreement at Durban.[5] Although the final agreement of the Durban negotiations, the Durban Platform, did not generate many immediate specific outcomes, it is one success on the way towards a global rules-based climate agreement. Numerous commentators regarded the Durban Platform as a triumph for European climate leadership.[6] This is remarkable given that European climate leadership had increasingly been questioned.

The EU also played a key role in the negotiations during COP 21 in Paris in 2015—negotiations that were celebrated as a diplomatic success for the EU.[7] After the failure in Copenhagen to secure a legally binding agreement, it was felt that negotiations in Paris could not afford to fail. But there were concerns that the outcome of COP 21 would be excessively watered down and not ambitious enough to address the challenges surrounding climate change.[8] In the end, the outcome of the Paris negotiations exceeded expectations, and the adoption of the Paris Agreement by the Parties to the UNFCCC in Paris in December 2015 was lauded as a historic achievement and as a revival of successful multilateral cooperation to address climate change. If the Paris Agreement is implemented consistently, it will truly mark the beginning of the end of the fossil fuel era, ushering in a radical transformation in the global economy and society.

The EU not only played an important role during the negotiations to secure the Paris Agreement but also for its ratification. In September 2016, EU

5 Harvey, Fiona, "Doha Climate Talks: EU Weakened over New Emissions Targets." *The Guardian*, November 23, 2012.

6 Keating, Dave, "Shaky Leadership on the World Stage", *European Voice*, May 1, 2013, accessed November 25, 2013. <http://www.europeanvoice.com/article/imported/shaky-leadership-on-the-world-stage/77126.aspx>.

7 For example, see Dröge, Susanne, and Oliver Geden, "After the Paris Agreement New Challenges for the EU's Leadership in Climate Policy," *SWP-Comments* (Berlin: Stiftung Wissenschaft und Politik, 2016).

8 International Institute for Sustainable Development (IISD), "Summary of the Paris Climate Change Conference," *Earth Negotiations Bulletin COP21* (2015), accessed October 18, 2016. <http://www.iisd.ca/vol12/enb12663e.html>.

ministers approved the ratification of the Paris Agreement by the EU, thereby ensuring that the Paris Agreement was ready to enter into force in November 2016.

At the same time, there have been concerns since the Copenhagen negotiations that the EU might be losing its ability and willingness to be a climate leader.[9] It remains to be seen whether the EU can reassert itself as an important player in climate policy après Paris.

This chapter starts by presenting the analytical framework for the analysis of EU climate leadership by identifying three conditions that are necessary for the EU to be a successful leader in multilateral negotiations: preference cohesion (speaking with one voice),[10] credibility[11] and opportunity.[12] With these conditions in mind, the chapter then illustrates why EU climate leadership came under duress in Copenhagen. First, despite the facade of unity suggested by the EU's bold climate policies that were agreed upon before COP 15, the Union had no unified voice at Copenhagen. Second, EU leadership in Copenhagen was undermined by credibility challenges. Third, the constella-

---

9 Brugger, Silvia, "European Climate Leadership under Siege—the 2030 Climate and Energy Debate", published by Heinrich-Böll-Stiftung, October 7, 2013, accessed January 7, 2014. <http://www.boell.de/en/2013/10/07/european-climate-leadership-under-siege-2030-climate-and-energy-debate>; Bals, Christoph, et al., *The End of EU Climate Leadership* (Bonn: Germanwatch, 2013); Groen, Lisanne, Arne Niemann, and Sebastian Oberthür, "The EU's Role in International Climate Change Policy-Making: A Global Leader in Decline?" in *Global Power Europe. Vol. 2. Policies, Actions and Influence of the EU's External Relations*, ed. Astrid Boening, Jan-Fredrik Kremer, and Aukje Van Loon, (Heidelberg: Springer, 2013), 37-54.

10 Parker, Charles F., and Christer Karlsson, "Climate Change and the European Union's Leadership Moment: An Inconvenient Truth?" *Journal of Common Market Studies* 48.4 (2010): 923-943; Falkner, Robert, "The Political Economy of 'Normative Power' Europe: EU Environmental Leadership in International Biotechnology Regulation," *Journal of European Public Policy* 14.4 (2007): 507-526; Groenleer, and van Schaik, "United We Stand?"; Andresen, Steinar, and Agrawala Shardul, "Leaders, Pushers and Laggards in the Making of the Climate Regime," *Global Environmental Change* 12.1 (2002): 41-51.

11 Young, "Political Leadership and Regime Formation," 281-308; Underdal, Arild, "Leadership Theory: Rediscovering the Arts of Management," in *International Multilateral Negotiation: Approaches to the Management of Complexity*, ed. William I. Zartman (San Francisco: Jossey-Bass, 1994), 178-197; Elgström, Ole, "Outsiders' Perceptions of the EU in International Trade Negotiations," *Journal of Common Market Studies* 45.4 (2007): 949-967.

12 Vogler, John, and Hannes Stephan, "The European Union in Global Environmental Governance: Leadership in the Making?," *International Environmental Agreements: Politics, Law and Economics* 7.4 (2007): 389-413; Falkner, "The Political Economy of 'Normative Power' Europe," 507-526; Bretherton and Vogler, *The European Union as a Global Actor.*

tion of actors at COP 15 undercut promising opportunities for the EU to act as a global leader on climate change.

The next section moves on to show how EU climate leadership has evolved, why it partially recovered after Copenhagen and how it evolved on the road to Paris. While Europe can only maintain a credible leadership role if it spurs ambitious climate actions at home, the EU's aspiration to be a leading actor on climate issues is still troubled by a number of domestic policy challenges that are undermining EU leadership. At the same time, while the cohesion of preferences among EU member states had hardly changed after Copenhagen, the division among EU member states did not restrain the EU from adopting a coherent negotiating stance. Moreover, the opportunity structure for EU leadership changed in the Union's favour. Additionally, the EU focused on alliance-building, above all with the Alliance of Small Island States (AOSIS) and least developed countries (LDCs), which proved crucial for reaching agreement on the Durban platform, and in the context of the EU-led "high ambition coalition", which was essential for the successful adoption of the Paris Agreement. The final section offers concluding comments on EU climate leadership.

2 EU Climate Leadership

Leadership can be defined as "an asymmetrical relationship of influence in which one actor guides or directs the behaviour of others towards a certain goal".[13]In order to understand the role of leadership in international cooperation, it is helpful to distinguish between three different types of leadership.[14] Directional leadership focuses on making the first move and demonstrating a commitment to act: by setting an example, the prospective leader offers a model that others may want to follow and also eliminates insecurity about whether the leader is in fact committed to what is at stake.[15] Structural leadership is based on the prospective leader using power resources that generate

13 Underdal, "Leadership Theory," 178-197. This section and the following builds on Brandi, Clara, "EU Climate Leadership in Copenhagen and Beyond," in *The European Union and Global Development: An 'Enlightened Superpower' in the Making?*, ed. Stefan Gänzle, Sven Grimm, and Davina Makhan (Houndmills: Palgrave Macmillan, 2012), 175-189.

14 Parker, and Karlsson, "Climate Change and the European Union's Leadership Moment," 923-943.

15 Underdal, "Leadership Theory," 178-197. See also Schunz, Simon. *Beyond Leadership by Example: Towards a Flexible European Union Foreign Climate Policy*, SWP Working Paper FG8, 2011/1 (Berlin: Stiftung Wissenschaft und Politik, 2011).

new incentives and change the costs and benefits that are associated with different possibilities for action in that specific issue area.[16] Idea-based leadership refers to prospective leaders naming and framing, thereby supporting solutions to the collective problems at issue.

An analysis of the Union's conduct on climate issues indicates that, while it has relied on all three types of leadership,[17] thus far directional leadership has been its most important mode.[18] First, the EU has deployed its structural weight in several ways over the past years,[19] for example in the context of the Copenhagen negotiations, by offering inducements to developing countries in the form of funding actions to mitigate and adapt to climate change in exchange for them entering into a binding climate agreement.[20] Second, the EU has deployed idea-based leadership[21] by embracing the 2°C goal.[22] Regarding directional leadership, before the Copenhagen negotiations, the EU passed ambitious internal climate legislation, namely its December 2008 energy and climate package that insisted on mitigation and renewable energy

16 Young, "Political Leadership and Regime Foundation," 281-308; Underdal, "Leadership Theory," 178-197.

17 Parker, Charles et al., "Fragmented Climate Change Leadership and the Ambiguous COP-15 Outcome: From Hopenhagen to Nopenhagen," (paper presented at the 3rd Uppsala Forum Workshop on the Politics, Policy, and Ethics of Global Climate Change, University of Uppsala, September 28, 2010).

18 Oberthür, Sebastian, and Claire Roche Kelly, "EU Leadership in International Climate Policy: Achievements and Challenges," *The International Spectator* 43.3 (2008): 35-50; Elgström, Ole, "EU Leadership in an Emerging New World Order" (paper presented at the 5th Pan-European Conference on EU Politics, Porto, June 24-26, 2010); Parker and Karlsson, "Climate Change and the European Union's Leadership Moment," 923-943. For a review of the driving factors behind the EU's activism in global climate politics since the mid-1990s in terms of norms versus interests and an assessment of the normative orientation that has dominated the EU's 'leading by example' strategy in global negotiations, see van Schaik, Louise, and Simon Schunz, "Explaining EU Activism and Impact in Global Climate Politics: Is the Union a Norm- or Interest-Driven Actor?," *Journal of Common Market Studies* 50.1 (2012): 169-186.

19 Parker, and Karlsson, "Climate Change and the European Union's Leadership Moment," 928-929.

20 Council of the European Union, *Energy and Climate Change—Elements of the Final Compromise, 17215/08* (Brussels: Council of the European Union, December 12, 2008), 6-7.

21 Parker, and Karlsson, "Climate Change and the European Union's Leadership Moment," 931.

22 European Commission, *Limiting Global Climate Change to 2 Degrees Celsius: The Way Ahead for 2020 and Beyond. Communication to the European Parliament, the Council, the European Economic and Social Committee and the Committee of the Regions, COM 2* (Luxembourg: Office for Official Publications of the European Communities, 2007).

commitments. Shortly thereafter, the EU released its ambitious goals for Copenhagen, calling for developed countries to slash their collective emissions by a minimum of thirty percent of 1990 levels by 2020 and for the large emerging economic powers, China, India and Brazil, to limit growth of their collective emissions by 15 to 30 percent below business-as-usual levels by 2020.[23]

In order to examine EU performance as a leading actor in multilateral negotiations, a number of conditions for success are found in the literature, which "are taken to be those conditions the EU is seen to have met when it has been judged either a leader or, more implicitly, externally effective in making a difference in negotiations".[24] Conditions for successful EU performance in multilateral negotiations can be summarised in three categories: preference cohesion (speaking with one voice), opportunity and credibility.[25]

The first condition for success is preference cohesion or speaking with one voice.[26] Speaking with one voice requires not only the competence and capability to act during negotiations, but also the cohesion of preferences amongst the EU member states.[27] Regarding climate change, competence is shared between the EU and its member states. This means that both are entitled to represent their areas of internal competence in international negotiations. The agreements that emerge from negotiations in areas of shared competences are signed and ratified by the Commission and all member states. Where competences are shared, the European Council can empower the Commission to negotiate international agreements. Yet due to the importance of and the public's concern about climate change, the Council has decided against that, so

---

23 European Commission, "Copenhagen Conference Must Produce Global, Ambitious and Comprehensive Agreement to Avert Dangerous Climate Change," IP/09/1867, published December 2, 2009, accessed June 15, 2011. <http://europa.eu/rapid/pressReleasesAction.do?reference=IP/09/1867>.

24 Dee, Megan, "Evaluating European Union Leadership in Multilateral Negotiations: A Framework for Analysis" (paper presented at the European Union Studies Association (EUSA) 12th Biennial International Conference, Boston, March 3-5, 2011), 3.

25 For a detailed analysis of the shortcoming regarding the use of the concept of EU leadership in international negotiations and suggestions for how to address them, see Dee, "Evaluating European Union Leadership in Multilateral Negotiations".

26 Ibid.; Parker and Karlsson, "Climate Change and the European Union's Leadership Moment," 926; Falkner, "The Political Economy of 'Normative Power' Europe," 507-526; Groenleer and van Schaik, "United We Stand?" 988-989; Damro, Chad, "EU–UN Environmental Relations: Shared Competence and Effective Multilateralism," in *The European Union at the United Nations*, ed. Katie V. Laatikainen and Karen Smith (Hampshire: Palgrave Macmillan, 2006), 181; Andresen and Shardul, "Leaders, Pushers and Laggards," 48.

27 Dee, "Evaluating European Union Leadership in Multilateral Negotiations," 11; Groenleer and van Schaik, "United We Stand?" 988-989.

that the Presidency, rather than the Commission, represents and negotiates on behalf of the EU on the basis of common agreement by the member states prior to the international negotiations. Agreeing on a common position that is supported by all member states, thereby speaking with one voice, requires significant internal negotiations, and is necessary for EU leadership in multilateral negotiations.

Second, earlier research indicates that credibility is a key factor of, or even requirement for, effective leadership: if an actor is regarded as credible, its potential to be a leader increases while an actor that is not perceived as credible is unlikely to be regarded as playing a leadership role.[28] The EU's focus on the directional form of leadership—leading by example—requires more credibility: directional leadership is highly susceptible to being jeopardised by a lack of credibility, for example because of deficient policy coherence, which could be due to unsuccessful implementation.[29] For the Union to successfully play a directional leadership role, the EU's external ambitions as a climate leader have to be underscored by internal policies that show that the Union keeps its promises.

A third condition for EU success is opportunity, which refers to a favourable distribution of power.[30] For example, a frequently cited reason for the EU's successful leadership before Copenhagen was the US's abdication from the Kyoto protocol, which eliminated competition for leadership[31] and left a power vacuum.[32] The EU was able to step forward and assume the leadership role.

While the EU aspires to be a climate leader, demand for EU leadership ambition has recently decreased. This is crucial since leadership is always "a

---

28 Young, "Political Leadership and Regime Foundation," 281-308; Underdal, "Leadership Theory," 178-197; Elgström, "Outsiders' Perceptions of the EU," 949-967.

29 Parker and Karlsson, "Climate Change and the European Union's Leadership Moment," 941.

30 Dee, "Evaluating European Union Leadership in Multilateral Negotiations," 10; Vogler and Stephan, "The European Union in Global Environmental Governance," 410; Falkner, "The Political Economy of 'Normative Power'," 512; Bretherton and Vogler, *The European Union as a Global Actor*, 106; Hovi, Jon, Tora Skodvin, and Steinar Andresen, "The Persistence of the Kyoto Protocol: Why Other Annex I Countries Move on Without the United States," *Global Environmental Politics* 3.4 (2003): 14.

31 Vogler and Stephan, "The European Union in Global Environmental Governance," 410.

32 Falkner, "The Political Economy of 'Normative Power' Europe," 512; Bretherton and Vogler, *The European Union as a Global Actor*, 106; Hovi, Jon, Tora Skodvin, and Steinar Andresen, "The Persistence of the Kyoto Protocol: Why Other Annex I Countries Move on Without the United States," *Global Environmental Politics* 3.4 (2003): 14.

relationship between leaders and followers".[33] Recent empirical data collected during COP 14 in Poznań and COP 15 in Copenhagen indicates that the demand for EU leadership has diminished.[34] In Copenhagen, the EU was recognised as having a leadership role by just 46 percent of the COP 15 respondents. It had lost the position of the most commonly recognised leader that it had at Poznań, where 62 percent of the COP 14 respondents regarded the EU as a climate leader.[35] At COP 15, the US, with its then recently elected President Obama, surpassed the EU as the most recognised leading actor.[36] Yet the EU still views itself as a climate leader. This self-conception is based on the EU portraying itself as being both coherent and credible.[37]

Below, this chapter assesses the degree to which the EU's self-conception is justified regarding its coherence and credibility. In order to assess the EU's potential for leadership on climate change, the next four sections analyse the extent to which ambition, preference cohesion (speaking with one voice), credibility and opportunity are present in the context of the Union's approach to combating climate change.

3 EU Climate Leadership in Copenhagen

The failure at Copenhagen called into question the EU as a leader on climate issues and raised concerns about EU leadership. Why did EU climate leadership come under pressure?

3.1 *Preference Cohesion in Copenhagen*

The first condition for success is preference cohesion (speaking with one voice). [38] In the case of the Kyoto negotiations in 1997, a major reason for successful EU leadership was that the Union entered the Kyoto negotiations with a common and unified position, above all regarding burden sharing and emission reduction targets.[39] Yet in the run-up to Copenhagen and during COP 15, the situation turned out to be different. At first sight, the EU seemed to be able

33 Underdal, "Leadership Theory," 181.

34 Parker et al., "Fragmented Climate Change Leadership"; Karlsson, Christer et al., "Looking for Leaders: Perceptions of Climate Change Leadership among Climate Change Negotiation Participants," *Global Environmental Politics* 11.1 (2011): 89-107.

35 Parker et al., "Fragmented Climate Change Leadership," 11.

36 Ibid, 12.

37 Elgström, "EU Leadership".

38 See also Brandi, Clara, "EU Climate Leadership in Copenhagen and Beyond," 179-181.

39 Andresen and Shardul, "Leaders, Pushers and Laggards," 48.

to negotiate as a united actor. The EU member states and the European Commission agreed that the Union's goal was to be a leading actor in the COP 15 negotiations and that an ambitious agreement had to be reached.[40] Prior to Copenhagen, the EU embraced a unilateral binding target to reduce emissions by twenty percent from 1990 levels in 2020 and appealed to others to boost their ambitions for COP 15.[41] Moreover, governments of the member states managed to put into writing an EU negotiating mandate for the Copenhagen negotiations that contained the major fundamental objectives of the EU for each issue on the COP 15 negotiation agenda.[42]

However, whereas the common objective of an ambitious external climate policy somewhat bolstered the EU's seemingly united stance, there were underlying disagreements between EU member states rooted in diverging national interests, which kept the EU from attaining a high degree of preference cohesion on a number of issues.[43] Important examples are the emission reduction target and climate finance, in which the wording of the EU negotiating mandate for Copenhagen obscured deeper disagreements. It implied that the negotiating mandate included no common EU position on these points, which in turn meant that EU negotiators were not in a position to advance deals with other negotiating parties regarding these issues.[44]

To illustrate, one key issue of disagreement among EU member states concerned the question of whether the EU should commit itself to an emission reduction target of thirty percent compared to 1990 levels by 2020.[45] Although

40 "European Commission, "Copenhagen Conference Must Produce Global, Ambitious and Comprehensive Agreement."

41 Barroso, José Manuel Durão, "Statement of President Barroso at the UN High Level Segment COP 15 Copenhagen, 16 December 2009," speech/09/583, accessed June 15, 2011. <http://europa.eu/rapid/pressReleasesAction.do?reference= SPEECH/09/583>.

42 Council of the European Union, *Environment Council of Ministers Conclusions, 14790/09.* (Brussels: Council of the European Union, October 21, 2009).

43 Groen, Lisanne, and Arne Niemann, "The European Union at the Copenhagen Climate Negotiations: A Case of Contested EU Actorness and Effectiveness," *International Relations* 27.3 (2013).

44 Ibid.; Groen, Lisanne, and Arne Niemann, "Challenges in EU External Climate Change Policy-making in the Early Post-Lisbon Era: The UNFCCC Copenhagen Negotiations," in *EU External Relations Law and Policy in the Post-Lisbon Era*, ed. Paul James Cardwell (The Hague: T.M.C. Asser Press, 2012).

45 Groen and Niemann, "The European Union at the Copenhagen Climate Negotiations," 312. See also Groen, Lisanne, Arne Niemann, and Sebastian Oberthür, "The EU as a Global Leader? The Copenhagen and Cancun UN Climate Change Negotiations," *Journal of Contemporary European Research* 8.2 (2012) and Groen, Niemann and Oberthür, "The EU's Role in International Climate Change Policy-Making."

the target of thirty percent, which was made conditional on comparable emission reduction commitments by other countries, had already been set in 2007[46] and was re-stated in the negotiating mandate for Copenhagen,[47] this goal continued to be disputed by a number of EU member states, among them Italy and Poland.[48] Moreover, the negotiating mandate did not identify the specific conditions under which the EU would commit to the thirty percent reduction target. Due to the EU member states' lack of consensus on this issue prior to COP 15, the precise EU reduction target had to be agreed on during the course of the negotiations in Copenhagen, which considerably impaired the EU's ability to act at all, let alone to act as a global leader.[49]

There were also differences of opinion among EU member states regarding climate finance and their financial contributions for adaptation and mitigation measures in developing counties.[50] In the end, the EU negotiation mandate did not mention any specific financial contribution.

These two examples, EU reduction targets and contributions to climate finance, are representative of a considerable number of issues in the EU negotiating mandate before COP 15 for which no sufficient agreement was reached in time. This partially paralysed EU negotiators and undermined EU leadership.

The EU's strategy, as set out in the Union's negotiating mandate, was to persuade the other major players to embrace the same ambitious targets as the EU. But the huge gap between the positions of the EU and that of the US and China made this impracticable.[51] Moreover, the EU negotiators were not allowed to alter the mandate before the EU member states had unanimously agreed upon the relevant alterations, which undermined the EU's ability to adjust its position during negotiations and was a crucial factor in the EU's

46 European Council, *Presidency Conclusions 8-9 March 2007, 7224/1/07* (Brussels: European Council, May 2, 2007), 12.

47 Council of the European Union, *Environment Council of Ministers Conclusions*, 5.

48 Charter, David, and Rory Watson. "EU Climate Change Push in Disarray as Italy joins Iron Curtain Revolt," *The Times*, October 17, 2008.

49 Groen and Niemann, "The European Union at the Copenhagen Climate Negotiations."

50 Groen and Niemann, "Challenges"; Groen and Niemann, "The European Union at the Copenhagen Climate Negotiations," 313.

51 Rapp, Tobias, Christian Schwägerl, and Gerald Traufetter, "The Copenhagen Protocol: How China and India Sabotaged the UN Climate Summit," *Spiegel Online International*, May 5, 2010, accessed October 21, 2013. <http://www. spiegel.de/international/world/the-copenhagen-protocol-how-china-and-india-sabotaged-the-un-climate-summit-a-692861 .html>.

leadership being pressured in Copenhagen, especially during the negotiations' frantic countdown.[52]

When the final negotiations moved to the level of heads of state and government, even the large number of day-to-day coordinating meetings did not manage to create enough cohesion within the Union to prevent the EU from disintegrating at the final stage of the negotiations when the leaders of France, the UK and Germany took over.[53] During her confirmation hearings as new Commissioner for Climate Action, Connie Hedegaard claimed that during "the last hours in Copenhagen, China, India, the US, Russia, Japan each spoke with one voice while Europe spoke with many different voices".[54] Behind the facade of unity suggested by the EU's seemingly cohesive stance on climate change is a less rosy image: the lack of cohesion, especially preference cohesion, played an important role in explaining the moderate degree of leadership during COP 15.

3.2 *Credibility in Copenhagen*

The second condition for EU success is credibility.[55] With the first mandatory carbon market, growing investment in energy policies and its comprehensive 2020 Energy and Climate Package, the EU was by far the most active player on emission reductions for a number of years. Yet, the EU's aspiration to be a leading actor on climate issues is beset with various 'inconvenient truths' that endanger the EU's continued commitment and ability, and thus credibility, to be a leading actor in the fight against climate change.

Firstly, the EU's commitments prior to the COP 15 negotiations lacked credibility.[56] Due to its ambitious reduction targets, the Union was conventionally viewed as having influence by leading by example. However, as underlined by observers of the process, in Copenhagen the EU proposed "a reduction target that conflicted with its claim to leadership".[57] For example, because of the

52 Groen and Niemann, "Challenges"; Groen and Niemann,"The European Union at the Copenhagen Climate Negotiations"; International Institute for Sustainable Development (IISD), "Summary of the Copenhagen Climate Change Conference," *Earth Negotiations Bulletin* 12.459 (2009), accessed April 19, 2011. <http://www.iisd.ca/vol12/enb12459e.html>.

53 Groen and Niemann, "The European Union at the Copenhagen Climate Negotiations".

54 Curtin, Joseph, *The Copenhagen Conference: How Should the EU Respond?* (Dublin: Institute of International and European Affairs, 2010), 11.

55 See also Brandi, Clara, "EU Climate Leadership in Copenhagen and Beyond," 181-183.

56 Kilian, Bertil, and Ole Elgström, "Still a Green Leader? The European Union's Role in International Climate Negotiations," *Cooperation and Conflict* 45.3 (2010): 255-273.

57 Spencer, Thomas, Kristian Tangen, and Anna Korppoo, *The EU and the Global Climate Regime. Getting Back in the Game, Briefing Paper* 55 (Helsinki: The Finnish Institute of International Affairs, 2010).

recession that followed the global financial crisis, the EU's twenty percent target became considerably easier to reach than had been assumed when proposed by the Commission in 2007. Insofar as the EU's reduction targets were perceived as modest rather than ambitious, the Union was not able to credibly assume a leadership role based on its traditional directional approach in terms of setting an example for others.

Moreover, although the EU aspires to be a global leader, the Union's internal policies before COP 15 were far from the forefront of the combat against climate change. Although the January 2008 Climate Change Package[58] and the negotiated agreement at the December 2008 summit[59] made the case that the new 20-20-20 targets would reduce the Union's overall emissions to at least twenty percent below 1990 levels by 2020, experts indicated that this package was very unlikely to have these desired effects.[60] In the end, the climate policy performance of the EU and its member states raised doubts about the "Green Europe myth".[61]

3.3 *Opportunity in Copenhagen*

The third requirement for EU success is opportunity, and above all, a favourable distribution of power. [62] In Copenhagen, the low degree of successful EU leadership can be explained partly by a lack of opportunity for EU leadership because of the constellation of actors at COP 15. Challenging external factors included the EU's relatively restricted structural power compared with the US and China and other emerging economies, these more powerful players' much less ambitious negotiating stances, and the less developed countries' relative lack of support.

First of all, in terms of overall emissions, the EU is a comparatively less important power than the US, China and other emerging economies—the

58 European Commission, Communication from the Commission to the European Parliament, the Council, the European Economic and Social Committee and the Committee of the Regions—*20-20 by 2020—Europe's Climate Change Opportunity*, COM 19 (Brussels: European Commission, January 23, 2008).

59 European Commission, *Brussels European Council, Presidency Conclusions, 17271/08* (Brussels: European Commission, December 11-12, 2008).

60 Helm, Dieter, "EU Climate-Change Policy—A Critique," in *The Economics and Politics of Climate Change*, ed. Dieter Helm, and Cameron Hepburn (Oxford: Oxford University Press, 2009), 222-244.

61 Lenschow, Andrea, and Carina Sprungk, "The Myth of a Green Europe," *Journal of Common Market Studies* 48.1 (2010): 133-154.

62 See also Brandi, Clara, "EU Climate Leadership in Copenhagen and Beyond," 183-185.

players that really matter in any meaningful climate agreement.[63] This situation undermines the Union's potential for structural leadership and hampers its capacity to act as a global leader on climate issues.[64]

In addition, the actor constellation at COP 15 was very challenging to the EU and its leadership goals because the US and the BASIC countries (Brazil, South Africa, India and China) had much more modest negotiating offers.[65] In the absence of domestic legislation, for example, President Obama and the US delegation had little to offer. Moreover, just before the start of the COP 15 negotiations, following China's lead, the BASIC countries jointly agreed to leave the negotiations if the developed countries attempted to convince them to set more ambitious targets. Overall, the EU's negotiating strategy did not adequately allow for the negotiating realities of the US and the BASIC countries' comparatively meagre commitment offers. In addition, as mentioned above, because of deficient preference cohesion and the unanimity rule, EU member states were unable to agree on major revisions of the EU negotiating mandate that could have allowed them to adjust their strategy in light of the changing circumstances and to act more strategically towards the US and the BASIC countries in order to try to convince them to take more ambitious stances.[66]

Last but not least, the developing countries gave no warm welcome to the EU proposals during COP 15.[67] The EU asked developing countries to reduce the growth of their emissions by 15 to 30 percent below business-as-usual levels by 2020. Countries like China and India, however, stressed that they would not sign any climate agreement that would undermine their economic growth. The EU was also attacked by developing countries for setting conditions, for example, demanding that comprehensive national strategies be in place before climate finance for adaptation is provided.

All in all, a number of external factors challenged the EU's leadership position in Copenhagen. The Union was not adequately prepared to address the actor constellation that prevented it from achieving its goal of reaching an ambitious agreement. Deficient preference cohesion and the Union's unanimity rule subverted EU leadership, above all with regards to its ability to convince

63 Curtin, *The Copenhagen Conference*, 7; Egenhofer, Christian, and Anton Georgiev, *Why the Transatlantic Alliance Climate Change Partnership Matters More Than Ever* (Brussels: Centre for European Policy Studies (CEPS), 2010).

64 Elgström, "EU Leadership," 14.

65 Groen and Niemann, "The European Union at the Copenhagen Climate Negotiations," 317-318.

66 Ibid., 318.

67 Parker and Karlsson, "Climate Change and the European Union's Leadership Moment," 940.

the US and the BASIC countries to adopt more ambitious negotiating positions. In the end, these powerful players marginalised the EU in the final negotiations. German Chancellor Merkel, former French President Sarkozy and former Spanish Prime Minister Zapatero had to sit through the plenary with all the other negotiators until after the heads of state and government leaders of the BASIC countries, Wen, Sing, Lula and Zuma, had finished—before they were permitted to chip in, during the very final stages of the negotiation. For the most part, the US and the BASIC countries had already concluded the Copenhagen Accord, and the EU had no alternative but to approve the disappointing outcome, which did not include any of the items that were dear to the EU, in particular specific commitments to reducing emissions.[68]

4 EU Climate Leadership from Copenhagen to Paris and après Paris

At the climate summits following Copenhagen, the EU succeeded in partially reviving its leadership. The Cancún COP 16 negotiations, held in December 2010, one year after the Copenhagen negotiations, were able to take away a large part of the distrust between developing and developed countries, culminating in an agreement which formalised many elements of the Copenhagen Agreement by incorporating them in a UNFCCC decision.[69] Agreement was reached, among others, on administrative UNFCCC mechanisms on adaptation, technology transfer and Reducing Emissions from Deforestation and Forest Degradation (REDD+). Moreover, the parties decided to establish a Green Climate Fund. Of the US$30 billion committed by donor nations for the 2010-2012 period, the EU offered—albeit merely as a political aspiration—almost a third of the total global commitment. The resulting Cancún Agreement recognises for the first time that overall cuts in emissions of at least 25 to 45 percent based on 1990 levels are required to reach the 2°C goal. Yet given the wish to reach agreement—any agreement—the final text included numerous

68 Curtin, *The Copenhagen Conference*; Rapp, Schwägerl, and Traufetter, "The Copenhagen Protocol".

69 Metz, Bert, "International Climate Change Agreements," in *Controlling Climate Change*, ed. Bert Metz (Cambridge: Cambridge University Press, 2011), 318-350; Barroso, José Manuel Durão, and Connie Hedegaard, "Statements by the President of the European Commission, José Manuel Barroso, and Connie Hedegaard, European Commissioner for Climate Action on the Cancùn Agreement on Climate Change," MEMO/10/673, published December 11, 2010, accessed May 28, 2013. <http://europa.eu/rapid/pressReleasesAction.do?reference=MEMO/10/673&format=HTML&aged=0&language=EN&guiLanguage=fr>.

loopholes and ambiguities, disregarded major issues about the document's final legal form and postponed setting emission cuts for all countries.

Overall, during COP 16 in Cancún, the EU reasserted itself as a key player in climate negotiations. On the basis of more moderate objectives, the EU played a more influential role at the Cancún negotiations than at Copenhagen, being more involved in the decision-making process and having a firmer grip on the outcomes.[70]

The outcome of COP 17 in Durban, called the "Durban Platform", put the world on track to be able to reach a global agreement to reduce emissions—albeit not until 2015 and not taking effect until 2020.[71] This timeline was agreed in part due to the "hard-nosed leadership of the EU".[72] At the same time, it was an open question as to whether there would in fact be sufficient consensus on a global agreement in 2015 and whether the future global agreement will be a protocol, another legal instrument or an agreed outcome with legal force. Another outcome of Durban was that it marked the formation of a progressive alliance, the so-called Cartagena Dialogue for Progressive Action, held together by the recently "much derided factor: European climate leadership", which stood together to generate the impetus for an ambitious agreement at Durban.[73]

The EU congratulated itself on the outcome of COP 17 climate negotiations in Durban,[74] and many commentators concur that EU climate leadership had recovered,[75] suggesting the Union to be "back in form, and leading the game in the global fight to combat climate change".[76] However, it has to be also kept in

70 Fuhr, Lili et al., *COP 17 in Durban: A Largely Empty Package* (Berlin: Heinrich-Böll-Stiftung, 2011); "Summary of the Copenhagen Climate Change Conference"; Oberthür, Sebastian, "Global Climate Governance after Cancun: Options for EU Leadership," *The International Spectator* 46.1 (2011): 5-13.

71 "Summary of the Copenhagen Climate Change Conference"; Malla, Katak, *The EU and Climate Treaty Negotiations after the Durban Conference, European Policy Analysis* March 2012 (Stockholm: Swedish Institute for European Policy Studies, 2012); Rajamani, Lavanya, "The Durban Platform for Enhanced Action and the Future for Climate," *International and Comparative Law Quarterly* 61.2 (2012): 501-518.

72 Keating, "Shaky Leadership."

73 Mabey, Nick, "Understanding Europe's Unexpected Durban Success," published January 2012, accessed October 21, 2013. <http://esharp.eu/big-debates/the-green-economy/16-understanding-europe-s-unexpected-durban-success/>.

74 Hedegaard, "New Phase in Global Climate Policy".

75 Keating, "Shaky Leadership"; Mabey, "Understanding Europe's Unexpected Durban Success."

76 Dupont, Claire, "Leading or Muddling Through? The EU and Durban," *Ideas on Europe: European Council and Foreign Affairs* (2012), accessed April 6, 2017. <http://www.ies.be/node/1297>.

mind that Durban was merely the start of a process and that substantial effort would be required in order to reach an effective global climate agreement in 2015. In effect, the outcome of the 17th COP and the role of the EU in achieving this outcome can only be regarded as an accomplishment if compared to the utter disappointment of the 15th COP in Copenhagen. Still, whereas EU climate leadership was strongly undermined in Copenhagen, the negotiations and the result of Durban are a sign of the recovery of EU's role and can be viewed as indicating "the resurgence of a leader that learned to lower its expectations and work incrementally towards its desired results".[77]

At the 18th COP in Doha, as part of the so-called Doha Climate Gateway, the Parties agreed that the second commitment period of the Kyoto Protocol could begin as scheduled in January 2013, that the Bali Action Plan negotiation track (the Ad Hoc Working Group on Long-term Co-operative Action under the Convention, known as AWG-LCA, established in December 2007 under COP 13) was completed and that, in the future, everyone will negotiate under the new Durban Platform for Enhanced Action (ADP) track to try reaching a global agreement by 2015, which will come into effect in 2020.[78] Doha reaffirms that the future global agreement must seek to attain the 2°C goal and initiates a process to review countries' emissions targets in order to close the emissions gap between present pledges and the targets that are required to achieve that goal.[79] The Doha Climate Gateway also established a new mechanism to compensate developing countries that are particularly vulnerable for the loss and damage caused by climate change.[80]

After very intense discussions, the 19th COP in Warsaw gave rise to a timeline for the negotiations of a new, global climate agreement that was supposed to be concluded 2015 in Paris and come into effect in 2020.[81] In addition, the parties made progress regarding the implementation of former decisions, for

---

77 Dupont, Claire, "Leading or Muddling Through? The EU and Durban."

78 Fuhr, Lili, and Liane Schalatek, "The 'Doha Climate Gateway': Will the Camel Go Through the Eye of the Needle?" published by Heinrich-Böll-Stiftung, December 12, 2012, accessed October 21, 2013. <http://www.boell.de/de/node/276935>.

79 Jacobs, Michael, "The Doha Climate Talks Were a Start, But 2015 Will be the Moment of Truth," *The Guardian*, December 20, 2012.

80 Fuhr, and Schalatek, "The 'Doha Climate Gateway'"; International Institute for Sustainable Development (IISD), "Summary of the Doha Climate Change Conference," *Earth Negotiations Bulletin* COP18 (2012), accessed October 21, 2013. <http://www.iisd.ca/vol12/enb12567e.html>; Jacobs, "The Doha Climate Talks."

81 International Institute for Sustainable Development (IISD), "Summary of the Warsaw Climate Change Conference," *Earth Negotiations Bulletin* COP19 (2013), accessed November 27, 2013. <http://www.iisd.ca/vol12/enb12594e.html>.

instance on finance, REDD+ and loss and damage. The negotiations were characterised by disagreement and hostility, and by the rise of a new alliance among developing countries that managed to dilute central elements of the final Warsaw agreement. At the same time, the EU was able to reach most of its negotiation objectives.

Negotiations at COP 20 in Lima put the focus on outcomes needed to make progress towards an agreement in Paris at COP 21. After long negotiations, COP 20 was able to provide the basis for Paris by adopting the Lima Call for Climate Action, which started the negotiations towards a new climate agreement, including the process required for submission of intended nationally determined contributions (INDCs), later to be converted into nationally determined contributions (NDCs), and enhancing pre-2020 ambition.

In 2015, COP 21 in Paris was regarded as an essential window of opportunity to organise effective collective action at the international level to stabilise global warming and facilitate the transition towards a zero-carbon economy. At the end of the negotiations, the 196 parties managed to secure the Paris Agreement, which many hailed as an ambitious, fair and universal climate agreement: the Paris Agreement seeks to keep the global temperature rise this century well below 2°C above pre-industrial levels and to pursue efforts to limit the temperature increase to 1.5°C. Moreover, the Paris Agreement seeks to promote the ability of countries to deal with the impacts of climate change and envisages bringing about a more robust transparency framework as well as adequate financial flows, a new technology framework and an enhanced capacity-building framework that supports action by developing countries and the most vulnerable countries. A key component of the Paris Agreement is the five-year cycle, which demands that each NDC cycle is to be more ambitious than the previous one, with a regular global "stocktake" providing information on efforts on mitigation, adaptation and support.[82] The goal of the post-Paris COPs is to foster the implementation of the Paris Agreement—and COP 22 in Marrakech was the beginning of that phase, which will show whether the Paris Agreement does indeed manage to foster the end of the age of fossil fuels.

Some commentators argue that the EU has lost its leadership role on climate policies insofar as others have managed to catch up or even move faster

82 International Institute for Sustainable Development (IISD), "Summary of the Paris Climate Change Conference," *Earth Negotiations Bulletin COP21* (2015), accessed October 18, 2016. <http://www.iisd.ca/vol12/enb12663e.html>. For insights into the negotiation process, see also Dimitrov, Radoslav S. "The Paris Agreement on Climate Change: Behind Closed Doors." *Global Environmental Politics* 16.3 (2016): 1-11.

than the EU in terms of climate policies.[83] However, in the global climate negotiations after Copenhagen and in the run-up to the successful adoption of the Paris Agreement, the EU partly reasserted itself as an important player. The following sections will show that while the EU's leadership aspirations continue to be under duress due to credibility challenges, after Copenhagen, the EU revived its leadership by managing to agree on a more united position prior to the negotiations and by using the more favourable opportunity structure it encountered during COP 15 and during the subsequent negotiations on the road to the Paris Climate Change Conference to contribute building successful alliances for the fight against climate change.

4.1 *Preference Cohesion from Copenhagen to Paris*

After the climate talks in Copenhagen, the preferences of EU member states continued to diverge. Both the EU emission reduction target and climate finance remained controversial. However, the negotiations after Copenhagen were characterised by a key difference: despite the continuing divergence between EU member states, the EU succeeded in agreeing on one negotiating position.

Before Cancún, Durban, Doha, Warsaw, and Lima, it continued to be impossible to agree on increasing the European emission reduction target. The divergence of preferences endured. Moreover, EU member states did not move forward on the question of climate finance and financial contributions for adaptation and mitigation measures in developing countries. At the same time, at the climate talks after Copenhagen, the different preferences of EU member states did not prevent the EU from agreeing on a common negotiation stance, even though this stance was less ambitious than the Commission and the more progressive member states preferred.[84]

In 2014, the EU agreed on the 2030 climate and energy framework, aiming to remain in line with the longer-term perspective, as set out in the Roadmap for moving to a competitive low-carbon economy in 2050[85]—and announced its goal to reduce emissions to at least forty percent below 1990 levels by 2030.[86] In 2015, the European Environment Council reported this target as the EU's offi-

83 Bals et al., The End of EU Climate Leadership.

84 Groen, Niemann, and Oberthür, "The EU as a Global Leader?," 183-184. See also Groen, Niemann, and Oberthür, "The EU's Role in International Climate Change Policy-Making."

85 European Commission, "2050 Low-Carbon Economy," last modified March 24, 2017, accessed March 24, 2017. <http://ec.europa.eu/clima/policies/strategies/2050/index_en.htm>.

86 See European Council, Conclusions 23-24 October 2014. 2030 Climate and Energy Framework (Brussels: European Council, 2014).

cial INDC to the negotiations—the EU being the first major economy to communicate its INDC—and the Commission published a proposal for the design of the new climate agreement. Moreover, the EU member states agreed on a formal EU negotiating mandate for COP 21 in Paris.[87]

4.2 *Credibility from Copenhagen to Paris*

After Copenhagen, the EU's aspiration to be a leading actor on climate issues is still troubled by a number of climate policy challenges that undermine its credibility.

Firstly, one key problem on the Road to Paris was the lack of an ambitious EU reduction target. During the first rounds of negotiations after Copenhagen, the EU's reduction target for 2020 remained twenty percent.[88] At the same time, it became increasingly clear that the twenty percent target for the second Kyoto commitment period was actually outdated as it would be achieved without any additional actions.[89] Already in Doha, the EU entered the climate talks with a weakened bargaining position because it had already met its targets to cut emissions by twenty percent by 2020 eight years ahead of time and had no plans to put more ambitious cuts on the table.[90] The EU's twenty percent target was considered "inadequate" in terms of being compatible with the 2°C goal.[91]

In 2014, the EU updated its reduction target and officially put forward its INDC in 2015, which entails an economy-wide target of at least forty percent domestic emission reductions below 1990 levels by 2030. To safeguard the credibility of its leadership role, the EU has to translate its target—adopted in 2014 by the 28 heads of state and government—into binding legislation.[92] At the same time, the level of ambition of the EU target is not conducive to the EU's

---

87 Dröge, Susanne, and Oliver Geden, "The EU and the Paris Climate Agreement," in *SWP-Comments* (Berlin: Stiftung Wissenschaft und Politik, 2015).

88 According to a recent metastudy, developing country pledges in Cancún reflect more mitigation, on an absolute basis, than that of developed countries. See Kartha, Sivan and Peter Erickson, *Comparison of Annex 1 non-Annex 1 Pledges under the Cancun Agreements, Working Paper* 6, (Stockholm: Stockholm Environment Institute, 2011).

89 Verolme, Hans, *European Climate Leadership—Durban and Beyond*, Discussion Paper (Brussels: Heinrich-Böll-Stiftung, 2012).

90 Harvey, "Doha Climate Talks."

91 Vieweg, Marion et al., "Climate Shuffle—Climate Action Tracker Update," published by Climate Analytics, Ecofys, PIK, accessed December 22, 2013. <http://www.ecofys.com/files/files/ecofys-ca-pik-2013-climate-action-tracker-bonn-update.pdf>.

92 Dröge, Susanne, and Oliver Geden, "After the Paris Agreement New Challenges for the EU's Leadership in Climate Policy," in *SWP-Comments* (Berlin: Stiftung Wissenschaft und Politik, 2016).

credibility: the target of forty percent emission reductions by 2030 is based on a scenario of eighty percent decarbonisation by 2050, which puts the EU at the lower end of its long-term goal of 80 to 95 percent by mid-century.[93] If the EU does not increase the level of ambition of its emission reduction targets soon, this will undermine the spirit of the Paris Agreement: according to the Climate Action Tracker, the EU target can only be rated as "medium", as the overall level of emission reductions proposed in the European target is "not yet sufficient to fall within the range of approaches for fair and equitable emission reductions". In addition, according to the Climate Action Tracker, policies currently being implemented are projected to cut domestic emissions by 23 to 35 percent below 1990 levels, and hence, at least to date, do not "put the EU on a trajectory towards meeting either its 2030 nor its 2050 targets".[94] The EU's credibility is thus also undermined due to its current implementation gap. Secondly, a considerable problem for the EU's credibility as a climate leader is the near collapse of the European Emissions Trading Scheme (ETS), which has seen carbon prices fall sharply.[95] Large energy companies have taken windfall profits and many energy-intensive industries are, for the time being, granted a free ride in the auctioning system. Proposed measures to make the ETS more effective remain controversial, and the introduction of novel national targets for the sectors that are not covered by the ETS have spurred debate. But there is hope that the ongoing amendment process will contribute to strengthening the ETS, fostering the implementation of the Paris Agreement and underscoring the EU's ability and willingness to be a climate leader among EU member states.

To provide adequate leadership signals, Europe needs to set ambitious climate policy targets. Having increased the ambition of its emission reduction target, the EU has signalled to global partners that, despite the ETS's difficulties, the EU still has firm plans to reduce emissions. Moreover, the fact that the EU ratified the Paris Agreement in the fall of 2016 is fostering the credibility of the EU's leadership role in climate policy.

93 See also Brandi, Clara et al., *EU Climate Leadership: Five Building Blocks for Ambitious Action*, (Bonn: German Development Institute/Deutsches Institut für Entwicklungspolitik (DIE), 2015).

94 Vieweg, Marion et al., "Climate Shuffle—Climate Action Tracker Update," published by Climate Analytics, Ecofys, PIK, accessed October 17, 2016.

95 Verolme, *European Climate Leadership*.

4.3 *Opportunity from Copenhagen to Paris*

After Copenhagen, the opportunity structure for EU leadership changed in the Union's favour. In particular, two contextual factors helped the EU to revive its leadership role:[96]

Firstly, after the failure of Copenhagen, many parties viewed the next climate summit as a decisive moment for international climate change talks and most countries were keen to move forward and the EU was able to profit from this mood. For example, in Cancún, the US and the emerging economies acted more cooperatively than in Copenhagen—which was conducive to reaching agreement and in line with the EU's objective for Cancún climate talks. Secondly, after the failure in Copenhagen, expectations were substantially lower and the negotiations were less 'politicised' than the negotiations at Copenhagen, with far fewer heads of state and government attending and less media and civil society attention prior to and after the negotiations—and in this less politicised context, the EU worked actively towards its goals.

Moreover, after the exclusion of the EU from the inner circle of talks in Copenhagen, the EU became active in trying to recreate a progressive alliance with other countries.[97] For example, the EU became an active participant in the newly founded Cartagena Dialogue for Progressive Action with a commitment to multilateralism, greater ambition and a legally binding regime, which brings together developed and developing countries from all continents and includes representatives of the COP presidency as observers.[98]

In Durban, especially, the Cartagena Dialogue became a vocal and effective alliance and the EU emerged as a key player in the group. The Durban roadmap was largely accepted because the EU managed to build an alliance with the AOSIS group and LDCs and other members of the Cartagena alliance. The alliance made it harder for China and India, and other countries, to remain opposed to progress, and assured the EU a say in the 'inner circle' of the talks, from which it had been sidelined in Copenhagen.[99]

During COP 21 in 2015, the newly founded "high ambition coalition" that was formed in the run-up to Paris was highly successful in fostering the adoption of

96 For a more detailed discussion of the two factors outlined in the following paragraph, see Groen, Niemann, and Oberthür, "The EU as a Global Leader?," 184.

97 Ibid., 185; Verolme, *European Climate Leadership.*

98 Oberthür, "Global Climate Governance after Cancun," 10; International Institute for Sustainable Development (IISD), "Summary of the Cancun Climate Change Conference," *Earth Negotiations Bulletin* COP16 (2010), accessed October 2, 2013. <http://www.iisd.ca/vol12/enb12498e.html>; Verolme, *European Climate Leadership.*

99 van Schaik, Louise, *The EU and the Progressive Alliance Negotiating in Durban: Saving the Climate?* ODI Working Paper 354 (London: Overseas Development Institute, 2012).

the Paris Agreement—and the EU played a key role in contributing to this success.[100] Building on the progress made in Durban, the first meeting of the group took place in Berlin in May 2015 and was co-chaired by the EU Commissioner for Energy and Climate Action, Miguel Arias Cañete, and the Norwegian minister Vidar Helgesen. Over the course of the following months, the group continued to grow—above all due to the outreach and recruiting activities of EU Commissioner Cañete—and met several times discretely. During the COP 21 negotiations, key players such as the US and Brazil officially joined the coalition, which was also strongly driven by the Foreign Minister of the Marshall Islands, Tony De Brum. The high-ambition coalition, representing more than one hundred developed and developing countries, pushed for a legally binding and ambitious global deal on climate change. In the end, the coalition was celebrated for securing the Paris Agreement. In that context, the EU's attention has re-shifted towards coalition-building following the failure of the Copenhagen negotiations, which can be regarded as having opened the door to success for the Paris Agreement.[101]

5 Conclusion

This chapter has shown that the EU has the potential to be a leading actor in terms of combating climate change. Analysis of the EU's role regarding climate change reinforces the impression that the EU has become an international actor, but also indicates that its leadership aspirations have recently come under duress. Both the failure of the Copenhagen climate summit in 2009 and recent global and European economic and political crises have weakened EU climate leadership in recent years.

During the Copenhagen negotiations, the EU's performance as a leader was modest: the EU did not live up to its aim of being a leading actor during the COP 15 negotiations and to make considerable progress towards a comprehensive global agreement. The analysis in this chapter has shown that the extent of EU leadership at Copenhagen can be traced back to three requirements for

100 See European Commission, "Historic Climate Deal in Paris: Speech by Commissioner Miguel Arias Cañete at the Press Conference on the Results of COP21 Climate Conference in Paris." *SPEECH/15/6320*. Published December 14, 2015, accessed October 18, 2016. <http://europa.eu/rapid/press-release_SPEECH-15-6320_de.htm>. See also EU Climate Action, "How the EU Helped Build the Ambition Coalition," accessed October 18, 2016. <https://storify.com/EUClimateAction/how-the-eu-helped-build-the-coalition-ambition>.

101 Oberthür, Sebastian, "Where to Go from Paris? The European Union in Climate Geopolitic," *Global Affairs* 2.2 (2016): 119-130.

success—preference cohesion (speaking with one voice), credibility and opportunity—having been threatened at COP 15. Essential factors in the lack of leadership include restricted leadership opportunities due to an unsuitable actor constellation, deficient preference cohesion (speaking with one voice) and credibility challenges. The latter two factors are especially crucial due to the EU's reliance on the directional mode of leadership, which requires the EU to be unified and credible, and in the vanguard of the combat against climate change. Moreover, the challenging actor constellation made the EU's leadership objectives very difficult to attain, especially since the EU strategy did not sufficiently consider this constellation. In addition, the unanimity requirement kept EU member states from agreeing on major adjustments to the EU negotiating mandate that would have allowed for successfully addressing the external context and more flexibly interacting with the US and the BASIC countries to work towards a more ambitious outcome. Finally, many EU proposals were not welcomed by less developed countries. The EU was marginalised during the final negotiations, with decisions taken by the US and the BASIC countries.

At the climate summits following Copenhagen, the EU managed to revive some of its leadership position. The EU's leadership aspiration remains challenged by internal policy problems that undermine credibility. Yet, after Copenhagen, the EU succeeded in agreeing on a coherent negotiating stance. Moreover, the opportunity structure for EU leadership changed in the Union's favour. Importantly, in that more favourable context, the EU managed to build successful alliances, which turned out to be essential, for instance, for the agreement on the Durban platform and for the agreement on the Paris Agreement.

In the run-up to and during the COP 21 negotiations in Paris for a global climate agreement, Europe was looked upon to play a central role in cementing a new global climate deal. In the end, the EU's strong focus on coalition-building after the failed Copenhagen negotiations and the EU-led high-ambition coalition can be said to have paved the way for securing an ambitious Paris Agreement.

While it seems difficult to change the declining structural power of the EU, the Union is still in the position to restore its directional leadership. This could be done by continuing to strengthen the EU's internal climate policies. However, there are limits to the effectiveness of the EU leading by example, especially in light of the current external context. A promising way forward towards more successful climate leadership would be to continue seeking greater unity. To bolster its credibility, the EU would be wise to seek greater unity and keep its promises regarding climate actions. However, the EU's unity and credibility mean little if it cannot convince other international actors to support its position on climate change and pursue one common objective. Thus the EU should continue to focus on alliance-building.

For the future of climate policies and the implementation of the Paris Agreement, the EU's role is all-important—especially after the election of Donald Trump as US President. It must be an exemplar, a negotiator, a persuader, a leader for all reticent nations of the world to ensure that global emissions peak as soon as possible. Whether or not the EU can rise to this challenge in these crucial years remains to be seen. While the EU has partly reasserted itself as an important player in climate negotiations, in order to safeguard and reinforce its leadership role, the Union must address the challenges outlined above.

After the adoption of the Paris Agreement in 2015 and its ratification in 2016, the focus in the EU has shifted towards implementation—both during COP 22 in Marrakesh and beyond that[102]—and towards legislative activities that are relevant for addressing climate change, including revising the ETS Directive, revisiting rules on the accounting for the Land Use, Land-Use Change and Forestry (LULUCF) sectors and fostering renewable energy and energy efficiency.[103] Insofar as the EU is facing obstacles to increase its mitigation targets for 2030 and 2050 in the near future—above all due to ongoing controversies among the member states—Europe could still contribute to promoting the Paris Agreement, for example by topping up climate finance, by increasing cooperation with developing countries and by strengthening non-state and subnational climate actions.[104] In that context, EU climate policy should integrate pertinent climate measures with the Agenda 2030 for Sustainable Development and the Sustainable Development Goals (SDGs) it enshrines. The objective should be to gradually implement the Paris Agreement and to operationalise the UN Sustainable Development Goals in a climate-friendly manner as a means of promoting a radical structural transformation at the global level

102 Bauer, Steffen, and Clara Brandi, "Marrakesh Climate Conference: Time to Act!" *Diplomatisches Magazin* 10 (2016): 30-33.

103 Oberthür, Sebastian, "The Paris Agreement: Rebooting Climate Cooperation. Perspectives on EU Implementation of the Paris Outcome," *Carbon & Climate Law Review* 10.1 (2016): 34-45.

104 Chan, Sander, Clara Brandi, and Steffen Bauer, "Aligning Transnational Climate Action with International Climate Governance: The Road from Paris," *Review of European Community & International Environmental Law* (RECIEL) 25.2 (2016): 238-247, accessed March 29, 2017. <https://www.die-gdi.de/externe-publikationen/article/aligning-transnational-climate-action-with-international-climate-governance-the-road-from-paris/>.

in order to mark the end of the fossil fuel age and drive the decarbonisation of the world economy in the interest of sustainable global development.

CHAPTER 8

Energy Transition in Germany: Sonderweg or role model?

Claudia Kemfert

Abstract

The German energy transition is now in progress. In forty years, electricity generation which, for the most part, is currently based on fossil fuels such as coal and gas, will be almost entirely converted to renewable energy sources. Presently, the share of electricity produced from renewables is about 28 percent. Further, as part of the energy transition, a commitment has been made to phase out nuclear power early: the remaining nuclear reactors will be decommissioned by 2022. The energy transition is also focused on improving energy efficiency, both in the building energy sector and to achieve more sustainable mobility. The energy transition is designed to facilitate the development of a sustainable energy supply. There will be no blackouts, provided that sufficient funds are invested in improving energy efficiency, optimising the electricity grid management system, expanding the grid and storage capacity, and also in gas-fired reserve power plants during the transitional period. Only a slight increase in the price of electricity is anticipated since there are key factors exerting both a downward and an upward effect on prices. Although significant investment is required, this will, in turn, create added value and employment, however. Since Germany has sufficient plant, infrastructure, and power plant engineering as well as construction expertise, the German economy is in a better position than any other to profit from the energy transition, the boom in renewable energy, new power plants, improvements in energy efficiency, and sustainable urban development and mobility. The energy transition is expected to create hundreds of thousands of new jobs and thus undoubtedly brings more economic opportunities than risks.

1 Introduction

More than ever before, today's society is faced with major challenges. Fossil fuels such as oil, gas, and coal are finite and, when burned, emit greenhouse gases endangering the environment. Approximately three-quarters of greenhouse gases worldwide are produced by industrial regions such as the US,

 | DOI 10.1163/9789004356825_010

Europe, and Japan. Even now, the concentration of greenhouse gases in the atmosphere is causing global warming which, in future, will bring increasingly frequent and unpredictable extreme weather conditions in its wake.

In order to stem climate change, it is essential for modern developed economies to radically reduce climate gases and also replace fossil fuels with alternative energy sources. However, global demand for fossil fuels is continuing to increase dramatically. Fast-growing economies such as China, in particular, and also Russia and India have an enourmous hunger for energy. Coal consumption, in particular, is continuing to increase, resulting in inexorable growth in greenhouse gas emissions. At the same time, fossil fuel consumption is frequently subsidised, causing energy to be wasted.

2 Energy Transition

The energy transition aims to increase the share of renewable energy from almost twenty percent in 2011 to eighty percent in 2050. Presently, in Germany, electricity is primarily generated by coal (approximately almost 45 percent) and gas power stations (almost 14 percent)(see table 8.1[1]).

The nuclear phase-out envisages that all nuclear power stations currently still in operation will be gradually decommissioned by 2022. The plan is to increase the share of renewable energy from the current 20 percent to 35 percent by 2020, while, at the same time, significantly improving energy efficiency, particularly in buildings.

Germany's implementation of these targets is in line with EU directives. The EU Roadmap suggests that, by 2050, the EU should increase its share of renewable energy to eighty percent.[2] Further, the EU target is also to cut greenhouse gas emissions by between 80 and 95 percent in the same period. This is to be achieved primarily through the decarbonisation of electricity generation, i.e., a significant increase in the use of renewable energy combined with improved energy efficiency.

Germany has resolved to reduce greenhouse gas emissions by forty percent by 2020, compared to 1990 levels. The 25-percent drop in emissions achieved by 2010 already is evidence of success. The expansion of renewables and

1 Cf. AG Energiebilanzen, "Bruttostromerzeugung in Deutschland von 1990-2014," accessed July 14, 2015, accessed March 30, 2017 <http://www.ag-energiebilanzen.de/index.php?article_id=29&fileName=20150227_brd_stromerzeugung1990-2014.pdf>.

2 European Commission, "COM (2011) 885/2: EU Roadmap 2050," accessed July 14, 2015. <http://eur-lex.europa.eu/legal-content/EN/TXT/PDF/?uri=CELEX:52011DC0885&from=EN>.

improvements in energy efficiency, particularly in the transport and mobility sectors, are key to meeting the climate targets.

In 2010, the conservative–liberal coalition (CDU–FDP) coalition government under Chancellor Angela Merkel initially extended the lifespan of Germany's nuclear power plants by approximately eight years. However, in spring 2011, the nuclear disaster in Japan, which struck in the aftermath of an earthquake, caused Germany to rethink its nuclear policy. As part of a moratorium, also in spring 2011, eight nuclear power stations were immediately and irreversibly taken off the grid. This saw Germany return to the decision it originally made under a social democrat–green coalition in the year 2000, which, also as part of a nuclear consensus, committed to phasing out nuclear power by revising the Atomic Energy Act. As early as 2002, an amendment to the Act stipulated that nuclear power plants would only be permitted to produce limited volumes of electricity, which, assuming a normal service life would equally have brought about the gradual closure of nuclear power stations by 2022.

In public debates, the *Energiewende* (Energy System Transition) is often narrowly associated with a power sector transformation. In fact, the *Energiewende* aims at an almost carbon neutral economy. Thus, relying on the political priority of decarbonization, the concept aims at an overall change of the energy system to facilitate sustainable climate protection. Stimulation of economic growth based on new, low carbon technologies constitutes one of its key motivations and drivers. Decarbonization of non-energy industrial processes, buildings, agriculture, heat, and transport is also considered. Therefore, an accelerated upgrade of relevant infrastructures is adopted as a complementary goal.

The Energy Concept of the German Federal government of 2010 and the subsequent decisions set specific de-carbonization targets that are consistent with the overall GHG emission reduction target of minus 85 percent to ninety percent in 2050 (see Table 8.2). These targets and the phasing out of nuclear energy are treated as overarching goals supported by different subordinated goals and policy measures.[3]

In detail, the shares of renewables in both energy and electricity consumption are expected to increase to sixty percent and eighty percent respectively by 2050. In the same period, overall primary energy consumption ought to

3 Bundesministerium für Wirtschaft und Energie (BMWi), ed., *Expertenkommission zum Monitoring-Prozess „Energie der Zukunft," Stellungnahme zum ersten Fortschrittsbericht der Bundesregierung für das Bereichtsjahr 2013* (BMWi: Berlin, 2014), accessed March 30, 2017. <https://www.bmwi.de/Redaktion/DE/Publikationen/Energie/zweiter-monitoring-bericht-energie-der-zukunft.pdf?__blob=publicationFile&v=8>.

TABLE 8.1 *Gross electricity production in Germany from 1990 to 2011*

Energy Source	1990	1991	1992	1993	1994	1995	1996	1997	1998	1999	2000
						bn. kWh					
Lignite	170.9	158.3	154.5	147.5	146.1	142.6	144.3	141.7	139.4	136.0	148.3
Nuclear	152.5	147.4	158.8	153.5	151.2	154.1	161.6	170.3	161.6	170.0	169.6
Hard Coal	140.8	149.8	141.9	146.2	144.6	147.1	152.7	143.1	153.4	143.1	143.1
Natural Gas	35.9	36.3	33.0	32.8	36.1	41.1	45.6	48.1	50.7	51.8	49.2
Mineral Oil Products	10.8	14.8	13.2	10.1	10.1	9.1	8.1	7.4	6.7	6.3	5.9
Renewable Energy:	19.7	17.5	20.5	21.2	23.0	25.1	23.0	24.2	26.3	29.1	37.8
- Wind	k.A.	0.1	0.3	0.6	0.9	1.5	2.0	3.0	4.5	5.5	9.5
- Hydro[b]	19.7	15.9	18.6	19.0	20.2	21.6	18.8	19.0	19.0	20.7	24.9
- Biomass	k.A.	0.3	0.3	0.4	0.6	0.7	0.8	0.9	1.1	1.2	1.6
- Photovoltaic	k.A.	0.0	0.0	0.0	0.0	0.0	0.0	0.0	0.0	0.0	0.0
- Geothermal	k.A.	0.0	0.0	0.0	0.0	0.0	0.0	0.0	0.0	0.0	0.0
- Waste[c]	k.A.	1.2	1.3	1.3	1.3	1.3	1.3	1.4	1.6	1.7	1.8
Other Energy Carrier	19.3	16.2	16.3	15.8	17.4	17.7	17.4	17.4	19.1	20.0	22.6
Total Gross Electricity Production	*549.9*	*540.2*	*538.2*	*527.1*	*528.5*	*536.8*	*552.7*	*552.3*	*557.3*	*556.3*	*576.5*
Electricity Trade with foreign country	31.9	30.4	28.4	33.6	35.9	39.7	37.4	38.0	38.3	40.6	45.1
Electricity Trade to foreign country	31.1	31.0	33.7	32.8	33.6	34.9	42.7	40.4	38.9	39.6	42.1
Electricity Trade Difference	+ 0.8	- 0.6	- 5.3	+ 0.9	+ 2.3	+ 4.8	- 5.3	- 2.3	- 0.6	+ 1.0	+ 3.1
Gross Domestic Electr. Consumption[d]	*550.7*	*539.6*	*532.9*	*528.0*	*530.8*	*541.6*	*547.4*	*550.0*	*556.7*	*557.3*	*579.6*
Chance to previous year in %	*X*	*-2.0*	*- 1.3*	*-0.9*	*+0.5*	*+2.0*	*+1.1*	*+0.5*	*+1.2*	*+0.1*	*X*
						Structure of Electricity Production in %					
Lignite	31.1	29.4	28.7	28.0	27.6	26.6	26.1	25.7	25.0	24.4	25.7
Nuclear	27.7	27.3	29.5	29.2	28.6	28.7	29.2	30.8	29.1	30.7	29.5
Hard Coal	25.6	27.7	26.4	27.7	27.4	27.4	27.6	25.9	27.5	25.7	24.8
Natural Gas	6.5	6.7	6.1	6.2	6.8	7.7	8.3	8.7	9.1	9.3	8.5
Mineral Oil Products	2.0	2.7	2.5	1.9	1.9	1.7	1.5	1.3	1.2	1.1	1.0
Renewable Energy:	3.6	3.2	3.8	4.0	4.4	4.7	4.2	4.4	4.7	5.2	6.6
- Hydrot[b]	3.6	2.9	3.5	3.6	3.8	4.0	3.4	3.4	3.4	3.7	4.3
- Wind	k.A.	0.0	0.1	0.1	0.2	0.3	0.4	0.5	0.8	1.0	1.6
- Biomass	k.A.	0.1	0.1	0.1	0.1	0.1	0.1	0.2	0.2	0.2	0.3
- Photovoltaic	k.A.	0.0	0.0	0.0	0.0	0.0	0.0	0.0	0.0	0.0	0.0
- Geothermal	k.A.	0.0	0.0	0.0	0.0	0.0	0.0	0.0	0.0	0.0	0.0
- Waste[c]	k.A.	0.2	0.2	0.2	0.2	0.3	0.2	0.3	0.3	0.3	0.3
Other Energy Carrier	3.5	3.0	3.0	3.0	3.3	3.2	3.1	3.2	3.4	3.6	3.9
Total Gross production	*100.0*	*100.0*	*100.0*	*100.0*	*100.0*	*100.0*	*100.0*	*100.0*	*100.0*	*100.0*	*100.0*
						in %					
reporting: Share of renewable energy to meet electricity demand	X	3.2	3.8	4.0	4.6	4.9	4.4	4.7	5.0	5.7	6.6
Discrepancy by rounding											

Notes: a) interim reporting, partly estimated.- b) Production in Pump hydro Storage Plants - c) Only Production from biogenic Waste (ca. 50 %). - d) Including Grid losses and self consumption.

Sources: Statistisches Bundesamt; Bundesministerium für Wirtschaft und Technologie; BDEW Bundesverband der Energie- und Wasserwirtschaft e.V.; Statistik der Kohlenwirtschaft e.V.; AG Energiebilanzen e.V.

2001	2002	2003	2004	2005	2006	2007	2008	2009	2010	2011	2012	2013	2014[a]
						bn. kWh							
154.8	158.0	158.2	158.0	154.1	151.1	155.1	150.6	145.6	145.9	150.1	160.7	160.9	155.8
171.3	164.8	165.1	167.1	163.0	167.4	140.5	148.8	134.9	140.5	108.0	99.5	97.3	97.1
138.4	134.6	146.5	140.8	134.1	137.9	142.0	124.6	107.9	117.0	111.8	116.4	121.7	109.0
55.5	56.3	61.4	61.4	71.0	73.4	75.9	86.7	78.8	86.8	84.9	76.4	67.5	58.3
6.1	8.7	9.9	10.3	11.6	10.5	9.6	9.2	9.6	8.4	6.6	7.6	7.2	6.0
38.9	46.1	45.4	56.5	63.2	71.4	87.5	92.4	94.1	102.8	123.2	143.8	152.4	160.6
10.5	15.8	18.7	25.5	27.2	30.7	39.7	40.6	38.6	37.8	48.9	50.7	50.8	54.7
23.2	23.7	17.7	19.9	19.6	20.0	21.2	20.4	19.1	21.0	18.1		0.9	1.3
3.3	4.5	6.5	8.4	12.0	14.5	19.1	22.3	25.5	27.6	31.9	22.1	23.0	20.5
0.1	0.2	0.3	0.6	1.3	2.2	3.1	4.4	6.6	11.7	19.3	39.7	41.2	43.0
0.0	0.0	0.0	0.0	0.0	0.0	0.0	0.0	0.0	0.0	0.0	26.4	31.0	34.9
1.9	1.9	2.2	2.1	3.0	3.9	4.5	4.7	4.4	4.8	5.0	5.0	5.4	6.1
21.4	18.2	20.4	21.2	23.6	25.3	26.6	24.7	21.5	26.7	27.6	25.7	26.2	27.2
586.4	*586.7*	*606.7*	*615.3*	*620.6*	*636.9*	*637.2*	*637.1*	*592.4*	*628.1*	*612.1*	*630.1*	*633.2*	*614.0*
43.5	46.2	45.8	44.2	53.4	46.1	44.3	40.2	40.6	42.2	49.7	44.2	38.4	38.9
44.8	45.5	53.8	51.5	61.9	65.9	63.4	62.7	54.9	59.9	56.0	67.3	72.2	74.4
- 1.3	+ 0.7	- 8.1	- 7.3	- 8.5	- 19.8	- 19.1	- 22.4	- 14.3	- 17.7	- 6.3	- 23.1	- 33.8	- 35.5
585.1	*587.4*	*598.6*	*608.0*	*612.1*	*617.2*	*618.1*	*614.6*	*578.1*	*610.4*	*605.8*	*607.1*	*599.4*	*578.5*
+ 1.0	*+ 0.4*	*+ 1.9*	*+ 1.6*	*+ 0.7*	*+ 0.8*	*+ 0.2*	*- 0.6*	*- 6.0*	*+ 5.6*	*- 0.7*	*+ 0.0*	*- 1.3*	*- 3.5*
						Structure of Electricity Production in %							
26.4	26.9	26.1	25.7	24.8	23.7	24.4	23.6	24.6	23.2	24.5	25.5	25.4	25.4
29.3	28.1	27.2	27.2	26.3	26.4	22.0	23.4	22.8	22.4	17.6	15.8	15.4	15.8
23.6	22.9	24.1	22.9	21.6	21.6	22.3	19.6	18.2	18.6	18.3	18.5	19.2	17.8
9.5	9.6	10.1	10.0	11.4	11.5	11.9	13.6	13.3	13.8	13.9	12.1	10.7	9.5
1.0	1.5	1.6	1.7	1.9	1.6	1.5	1.4	1.6	1.3	1.1	1.2	1.1	1.0
6.6	7.9	7.5	9.2	10.2	11.2	13.7	14.5	15.9	16.4	20.1	22.8	24.1	26.2
4.0	4.0	2.9	3.2	3.2	3.1	3.3	3.2	3.2	3.3	3.0	8.0	8.0	8.9
1.8	2.7	3.1	4.1	4.4	4.8	6.2	6.4	6.5	6.0	8.0		0.1	0.2
0.6	0.8	1.1	1.4	1.9	2.3	3.0	3.5	4.3	4.4	5.2	3.5	3.6	3.3
0.0	0.0	0.1	0.1	0.2	0.3	0.5	0.7	1.1	1.9	3.2	6.3	6.5	7.0
0.0	0.0	0.0	0.0	0.0	0.0	0.0	0.0	0.0	0.0	0.0	4.2	4.9	5.7
0.3	0.3	0.4	0.3	0.5	0.6	0.7	0.7	0.7	0.8	0.8	0.8	0.9	1.0
3.6	3.1	3.4	3.3	3.8	4.0	4.2	3.9	3.6	4.3	4.5	4.1	4.1	4.3
100.0	*100.0*	*100.0*	*100.0*	*100.0*	*100.0*	*100.0*	*100.0*	*100.0*	*100.0*	*100.0*	*100.0*	*100.0*	*100.0*
						in %							
6.7	7.8	7.6	9.3	10.3	11.6	14.2	15.0	16.3	16.8	20.3	23.7	25.4	27.8

Date: 27. Februar 2015

TABLE 8.2 *Ambitious targets versus status quo*

	2013	2020	2030	2040	2050
GHG emissions (compared with 1990)	-22.6%	-40%	-55%	-70%	-80% - -95%
Renewable energies					
Share in gross final energy consumption	12.0%	18%	30%	45%	60%
Share in gross electricity consumption	25.3%	35%	50%	65%	80%
Energy efficiency					
Primary energy consumption (compared with 2008)	-4.0%	-20%			-50%
Energy productivity (final energy consumption)	0.26% p.a. (av. 2008-2013)	2.1% p.a.			
Gross electricity consumption (compared with 2008)	-3.3%	-10%			-25%
Thermal refurbishment of residential buildings	~1% p.a. (2012 value)	2% p.a.			
Final energy consumption of transport sector (compared with 2005)	1.0%	-10%			-40%

Source: BMWi (2014a) and Löschel et al. (2014b).

decrease substantially by fifty percent, electricity by twenty five percent respectively, while at the same time energy productivity is expected to increase by 2.1 percent per year. More precisely, until 2050 a nearly carbon-neutral building stock shall be in place. Following the first monitoring report, the government added two additional targets to the first monitoring report:[4] A twenty

4 Bundesministerium für Wirtschaft und Energie (BMWi), ed., *Erster Monitoring-Bericht: "Eniergie und Zukunft"* (Berlin: BMWi, 2012), accessed April 29, 2017. <https://www.bmwi.de/Redaktion/DE/Publikationen/Energie/erster-monitoring-bericht-energie-der-zukunft.pdf?__blob=publicationFile&v=7>.

percent reduction of heat demand in the existing building stock until 2020, and the reduction of primary energy consumption of buildings by about eighty percent until 2050.[5]

Therefore, the thermal refurbishment rate for the existing building stock is expected to rise to two percent per year. In addition, in the transport sector reductions of energy consumption of ten percent and of forty percent respectively are set for 2020 and 2050 respectively. E-mobility is considered to be an important pathway and therefore the increase of electric cars up to at least one million in 2020 and six million in 2030 is a further declared goal. Moreover, gross electricity generation by CHP-plants should increase from 17.3 percent in 2012 to 25 percent in 2020.

However, the actual energy transition is not only about withdrawal from the nuclear energy program but rather the comprehensive restructuring of German energy supply. If an average service life of forty years is taken as a basis for power plants, then approximately half of all German coal-fired plants would have to be shut down in the next ten years due to their age.

Across Germany, numerous new coal-fired power plants will either be in the planning stages or already under construction by 2017, with a total output of over ten gigawatts. Thus, in purely mathematical terms, it would be possible to replace all the nuclear reactors presently in operation. However, fundamentally, coal-fired power plants are not compatible with the concept of a sustainable energy transition: they produce significantly more polluting greenhouse gases than other energy sources—twice the volume produced by gas, for example.

Gas-fired power plants would be far more suitable for the transitional period. They not only produce fewer emissions but are also easier to combine with fluctuating renewable energy sources as the output of gas-fired plants can be quickly and easily ramped up and down. For electricity generation, particularly in combination with cogeneration, gas is the most efficient and also a climate-friendly form of energy supply. Further, for mobility, gas is a very interesting alternative to oil. Natural gas vehicles are subject to lower taxes, which makes them economically very attractive.

5 Both heat demand and primary energy consumption in this case do not include renewable energies: Löschel, Andreas et al., *Expertenkommission zum Monitoring-Prozess "Energie der Zukunft," Stellungnahme zum ersten Fortschrittsbericht der Bundesregierung für das Berichtsjahr 2013* (Berlin, Münster, Stuttgart: BMWi und BMUB, 2014. Therefore, the use of renewable energy facilitates achievement of both indicators, which is reasonable from the point of view of the GHG emissions reductions.

3 Aspects of Energy Transition

Renewables are very volatile. During periods when there is an abundant supply of renewable energy, the electricity networks often lack the capacity to feed the excess power to other parts of the country or abroad. With the expansion of renewables comes an increased need to upgrade, expand, and optimise the networks.[6] First, power grids will be needed to connect northern and southern Germany, particularly to transport electricity generated by the offshore windfarms to regions where nuclear and coal-fired power stations will be increasingly decommissioned, i.e., western and southern Germany. Second, an expansion of the European electricity grid is also necessary, particularly to exploit geological advantages in electricity production using renewable energy sources, to improve trade, and to optimise networks. Further, intelligent distribution networks that can optimise the volatile power supply and corresponding demand will also be needed. Along with enhanced energy storage, demand management plays an equally important role here. For example, even energy-intensive industries can manage their demand patterns so as to combine them more easily with the increasingly volatile power supply generated by the growing share of renewable energy sources.

The financial crisis has caused an increase in financial institutions' risk premiums. The government has already provided extensive funds for bank bailouts and could now insist that this money be dedicated to financing the energy transition. To this end, it is important to increase planning security for investors through reliable political and economic framework conditions.

The increasing share of renewable energy makes it imperative to establish more storage capacity. However, pumped-storage power plants are currently the only economical form of energy storage. As Germany has limited scope for establishing pumped-storage plants, the German electricity market is also dependent on other parts of the EU (the Alps region and Scandinavia). Also in order to be able to use these possibilities to their full potential, power supply lines and infrastructure must be expanded. Additionally, completely new types of electricity storage are currently being researched. Batteries from electric vehicles could be used for storage, for example, if they can simultaneously

6 There are various possible scenarios for the extension of the networks; see Deutsche Energy-Agentur, "Integration erneuerbarer Energien in die deutsche Stromversorgung im Zeitraum 2015-2020 mit Ausblick 2025," accessed July 14, 2015. <http://www.dena.de/publikationen/energiesysteme/dena-netzstudie-ii.html>. Different scenarios are discussed by Forum Netzintegration Erneuerbare Energien, "Das Stromnetz der Zukunft," accessed July 14, 2015. <http://www.duh.de/uploads/media/DUH_Broschuere_NetzintEE_2010.pdf>.

apply the 'vehicle-to-grid' option, i.e., feed the stored electricity back into the grid. This would require the infrastructure to be modified, however. New fuels are also good for energy storage. In periods of peak demand for renewable energy, hydrogen and methane could be produced to use for mobility, for instance. However, this method of storage is not yet economically viable. Due to increasing fossil fuel prices and decreasing renewable energy costs, these innovative technologies could come into use in 20 to 25 years' time.[7]

With the expansion of renewable energy, conventional power plants are becoming less and less economically attractive. Since renewables enjoy priority access to the grid and are producing increasing volumes of power, the production periods of conventional power plants are dwindling. Further, the growth of renewable energy has led to falling electricity prices on the stock exchange as a result of the "merit order effect".[8] Consequently, power plants used for peak load periods, such as gas-fired plants, lose their economic attractiveness.

To increase economic incentives for the construction and use of conventional power plants as bridging and reserve plants, the market system must be adapted and the market design modified. Shrewd regulation can help to foster the capacity required for a sustainable energy system. Above all, demand must successfully accommodate the volatile energy supply. In order to accelerate investment, the framework conditions, particularly with regard to financing the infrastructure as well as power storage and reserve capacity, must be adapted to enable a secure supply and dynamic development to be rapidly established. Appropriate financial incentives to develop electricity networks and storage, and also the necessary electricity capacity to balance supply and demand, are just as important as the rapid implementation of all energy conservation measures. Therefore, not only should funds for energy-efficient building refurbishment be increased as soon as possible, but any energy conservation potential should be exhausted, particularly in the mobility field.

Demand, however, is just as important as supply. The less energy is consumed, the lower the energy import and energy price dependencies—and the greater the energy cost savings. Apart from the mobility sector, the biggest

7 Jentsch, Mareike et al., "Power to Gas als Langzeitspeicher," *Energy 2.0* (2011): 46.

8 The extension of renewable energy results in an increase in supply and thus a shift of the supply curve to the right which, in turn, leads to a price drop if demand remains unchanged; cf. Fraunhofer-Institut für System- und Innovationsforschung (ISI), Deutsches Institut für Wirtschaftsforschung (DIW), Gesellschaft für Wirtschaftliche Strukturforschung mbH (GWS), and Institut für ZukunftsEnergieSysteme (IZES), "Einzel- und gesamtwirtschaftliche Analyse von Kosten- und Nutzenwirkungen des Ausbaus Erneuerbarer Energien im deutschen Strom- und Wärmemarkt: Update der quantifizierten Kosten- und Nutzenwirkungen für 2010."

potential for savings is in the real estate sector, or, more precisely, in the building shell. Approximately one-fifth of real estate energy needs could be saved simply by using more efficient insulation and air-conditioning technologies.[9] This would relieve individual citizens' financial burden. However, this requires adequate funding which is currently being provided by the German Investment and Development Bank (*Kreditanstalt für Wiederaufbau*, KfW).[10]

Apart from technical, economic, and political factors, the social dimension is another important consideration: the population's acceptance will depend on how well the situation is communicated to society, i.e., precisely what is involved, what steps are necessary, how big the financial burden will be, and how much financial support they will receive (for building refurbishment). Therefore, a high degree of transparency, supply of information, and participation is required, not only at the national and *Länder* level but particularly at the municipal level.

3.1 *Renewable Energy Sources—Problem Solver and Economic Factor*

Renewables are the building block for a sustainable energy supply: they are climate friendly, ensure supply security as a domestic energy producer, and can also increase competitiveness as a stimulus for growth and jobs. Renewable energy sources are of interest for all energy areas: for power and heat generation, and as alternative fuels in the transport sector. The renewables industry promotes expansion and innovation and has become a growth industry like no other sector in recent years[11] The majority of employees work in the wind energy and biomass power generation fields, followed by the solar and geothermal industries. In the course of the political process of phasing out nuclear

9 Neuhoff, Karsten et al. , "Energetische Sanierung: Handlungsbedarf auf vielen Ebenen," DIW *Wochenbericht* 34 (2011): 3-13; and Deutsche Energie-Agentur, "Der Dena-Gebäudereport 2012: Statistiken und Analysen zur Energieeffizienz im Gebäudebestand," accessed October 01, 2013, 28. <http://www.dena.de/projekte/gebaeude/dena-gebaeudereport.html>.

10 Kreditanstalt für Wiederaufbau (KfW), "Energieeffizient sanieren—Soziale Organisationen", accessed January 19, 2012. <http://www.kfw.de/kfw/de/Inlandsfoerderung/Programmuebersicht/Sozial_Investieren_-_Energetische_Gebaeudesanierung/index.jsp>. <https://www.kfw.de/inlandsfoerderung/%C3%96ffentliche-Einrichtungen/Energetische-Stadtsanierung/Finanzierungsangebote/Energieeffizient-Sanieren-kommunale-Unternehmen-%28219%29/>.

11 Blazejczak, Jürgen, et al., "Energiewende erfordert hohe Investitionen," *DIW Wochenbericht 26* (2013): 19-30; see also OECD, "Transition to a Low-Carbon Economy: Public Goals and Corporate Practices," published by OECD, 2010. <http://dx.doi.org/10.1787/9789264090231-en>.

energy and reducing the high CO_2 emissions from coal-fired power plants, renewable energy sources can also make an outstanding contribution to combating climate change and to improving security of supply by reducing dependence on imports. Because renewables are usually used for a decentralized energy supply comprising wind turbines, biomass plants, and combined heat and power, the use of renewable energy sources increases the security of supply. There are plenty of opportunities for expanding renewables. Depending on how global demand for renewable energy sources develops, the export potential can be enhanced considerably.

3.2 *New Market Integration of Renewables—More Problems than Solutions*

A key objective of the German government is to develop renewable energy sources at the lowest cost. In addition, the pace at which renewable energy sources grow should be kept within the expansion corridor. To date, renewables have been supported by fixed feed-in tariffs. Since the costs of renewable energy sources have plummeted, the intention is to now switch to a tender system—also in accordance with EU regulations. Tenders are to be introduced for both onshore and offshore wind turbines and for solar plants with an output of at least 1 megawatt (MW), thus allowing the tariffs and a sliding market premium to be calculated and specified. This approach should allow eighty percent of all renewable energy systems to be determined through competitive tenders in the future.

The main objectives of the tender process are therefore reduced costs through lower subsidy rates, better compliance with the expansion corridor, limited influence of interest groups, and greater acceptance of the expansion of renewable energy generation and, consequently, the energy transition. In theory, tenders provide various options for combining competitive pricing with elements of control. They would, for example, allow system compliancy, i.e., how renewables can relieve pressure on the energy system, or regional aspects, to be included in the award criteria.

As tempting as it might sound to use the tender process to automatically achieve low costs and compliance with the expansion corridor, its successful implementation entails difficulties. The devil is in the detail. Essentially, determining subsidies competitively through tendering can only work if there is adequate competition. There must be enough investors compared to the amount of land available to be used for renewable energy sources. In order to encourage sufficient bidders, the process must be comprehensible and the preliminary work and securities that bidders are expected to provide have to be affordable. The need to encourage the highest possible number of bidders to participate forces the regulator to make the tender regulations attractive for bidders.

Against this background, it is worth taking a look at the second key objective of the tender: better compliance with expansion objectives. This is to be achieved by putting potential renewable energy expansion out to tender in several processes each year. Experience from other countries shows that low requirements with regard to preparatory work and securities, and low financial penalties for non-compliance, lead to a high proportion of bidders failing to make use of the permission they have been granted to build plants[12] This would mean the expansion target would not be met. Securities and preparatory work are thus urgently required in order for development objectives to actually be achieved.[13] They do, however, increase the cost of participation and inevitably lead to a low number of participants. Little competition in the tenders also means higher funding rates, i.e., higher costs. In addition, if potential investors are willing or able to bear risks associated with the tenders to varying degrees, this will not lead to an efficient allocation of licenses because efficient bidders may decline to participate from the outset.

Consequently, there is a clear conflict of interest between the two key objectives of introducing a tender process: automatically generating low subsidy rates and thus lower costs, and meeting expansion targets. Hence, it is by no means certain that costs will fall and that expansion objectives will be achieved more efficiently. Financial risks are presumably factored in, low participation due to high barriers to entry could lead to prices being pushed up. Lowering qualification requirements and penalties, on the other hand, involves the risk that not all approved plants will actually be constructed and planned expansion corridors missed by a considerable margin.

The additional wish to address quality aspects such as demand-driven electricity generation, regional criteria, or maintaining the variety of participants would seem very difficult to implement in this context. In particular, small investors and citizens' energy cooperatives that could have stimulated the market to date could lose out in the new system. This is confirmed by the first

---

12 Grau, Thilo, "Comparison of Feed-in Tariffs and Tenders to Remunerate Solar Power Generation," *DIW Discussion Paper* 1363 (2014), accessed March 30, 2017. <https://www.diw.de/documents/publikationen/73/diw_01.c.437464.de/dp1363.pdf>; GIZ and Ecofys *Lessons for the Tendering System for Renewable Electricity in South Africa from International Experience in Brazil, Morocco and Peru* (Pretoria, 2013), accessed March 30, 2017. <http://www.ecofys.com/files/files/ecofys-giz-2013-international-experience-res-tendering.pdf>.

13 Deutscher Bundestag, *Ausschreibungsbericht nach § 99 des Erneuerbare-Energien-Gesetzes, Drucksache 18/7287*, published 2016, accessed July 17, 2016. <http://dip21.bundestag.de/dip21/btd/18/072/1807287.pdf>.

experiences with tender schemes of PV open area tenders: first tender rounds have shown that most often big bidders and almost no cooperatives participated. Although quite high levels of competition and low prices might have been reached, the actual realization rate is still unknown. First experiences with tendering schemes in Germany do not allow for any conclusion but provide first insights into potential challenges[14].

Although tenders may increase transparency and identify the most attractive provider among participating bidders, there are, nevertheless, many disadvantages: there is a great deal of red tape involved, there are conflicts between the objectives to be met by the tenders; strategic behavior in combination with low participation can be a problem and push up bid prices. Bidders' transaction costs will rise and are probably not affordable for all stakeholders. There are also bidding risks such as time and effort spent carrying out preparatory work without any certainty of ultimately being awarded the contract.

Tenders do not automatically lead to lower costs; they involve a high risk of not meeting renewable energy expansion targets. In addition, there is a danger that the variety of participants and therefore acceptance of the energy transition diminishes. Tenders can bring benefits. There are, however, still many challenges in their specific design.

4 Necessary Investment for Energy Transition Runs into Hundreds of Billions

According to estimates by the Federal Ministry for the Environment, Nature Conservation and Nuclear Safety (*Bundesministerium für Umwelt, Naturschutz und Reaktorsicherheit*, BMU), to double the share of renewable energy, a cumulative investment of approximately two hundred billion euros would be required over the next ten years (see Table 8.1).[15] Other studies point to a nec-

14 Klessmann, Corinna et al., *Ausschreibungen für PV-Freiflächenanlagen Auswertung der ersten zwei Runden. Auswertung der ersten zwei Runden* (Ecofys, ZSW, Fraunhofer ISI, Takon:, Berlin, 2015), accessed March 30, 2017. <https://www.erneuerbare-energien.de/EE/Redaktion/DE/Downloads/ausschreibungen-pv-freiflaechanlagen-auswertung-erste-zwei-runden.pdf?__blob=publicationFile&v=3>.

15 Bundesministerium Umwelt, Naturschutz und Reaktorsicherheit, *Ausbau erneuerbare Energien im Strombereich: EEG-Vergütungen, -Differenzkosten und -Umlage sowie ausgewählte Nutzeneffekte bis zum Jahr 2030*, (Berlin, 2008); and Bundesministerium Umwelt, Naturschutz und Reaktorsicherheit, *Leitstudie 2010, Langfristszenarien und Strategien für den Ausbau der erneuerbaren Energien in Deutschland bei Berücksichtigung der Entwicklung in Europa und global* (Berlin, 2010), 23.

essary cumulative investment of up to 235 billion euros over a period of several decades.[16] The Hamburg Institute of International Economics (*Hamburgisches WeltWirtschaftsInstitut*, HWWI) recently estimated the total monetary cost of the energy transition until 2030 at 335 billion euros. This calculation envisaged 250 billion euros being spent on subsidies for the development of renewable energy and 85 billion euros on additional investment, plants, power lines, storage, and power plant capacity.[17]

For the cost of grid expansion alone, the Federal Network Agency (*Bundesnetzagentur*, BNA) is working on the assumption of 20 to 25 billion euros over the next 15 years.[18] For energy-efficient building refurbishment, KfW has estimated that up to 75 billion euros will need to be invested over the next ten years.[19] Additional gas-fired power plant capacity of up to ten gigawatts will require further investment of up to 15 billion euros.[20] However, what must be

16 Prognos, *Investitionen durch den Ausbau erneuerbarer Energien in Deutschland* (Berlin: Prognos AG, 2010), accessed April 23, 2013. <https://www.bee-ev.de/fileadmin/Publikationen/Studien/100501_Prognos-Studie_Investitionen_BEE-Ausbauprognose_lang.pdf>; Prognos, "Konsequenzen eines Ausstiegs aus der Kernenergie bis 2022 für Deutschland und Bayern," accessed April 23, 2013. <http://www.prognos.com/fileadmin/pdf/publikationsdatenbank/Prognos_vbw_Studie__Ausstieg_Kernenergie_bis_2022.pdf>; and Erdmann, Georg, "Kosten des Ausbaus erneuerbarer Energien: Studie der Technischen Universität Berlin," accessed April 23, 2013. <http://www.vbw-bayern.de/Redaktion-%28importiert-aus-CS%29/04_Downloads/Downloads_2011/3_Wirtschaftspolitik/3.5_Energie/Publikationen/studie_eeg_energie_kosten_110907_final_NGa_2.pdf>.

17 Bräuninger, Michael and Sven Schulze, "Konsequenzen der Energiewende," accessed August 17, 2010. <http://www.hwwi.org/fileadmin/hwwi/Publikationen/Studien/HWWI-Studie-Energiewende-2012.pdf>. This study does not, however, differentiate between pure investment and additional costs. Therefore, it does not indicate which costs are to be attributed exclusively to the energy transition and which would be incurred even without the energy transition.

18 Bundesnetzagentur, "Monitoringbericht," accessed, Juli 14, 2015, accessed March 30, 2017. <https://www.bundesnetzagentur.de/SharedDocs/Downloads/DE/Allgemeines/Bundesnetzagentur/Publikationen/Berichte/2014/Monitoringbericht_2014_BF.pdf?__blob=publicationFile&v=4>. However, the Bundesnetzagentur itself concedes that only part of the grid expansion can be directly attributed to the energy transition since the renewal and repair of the grid would be necessary even without the energy transition.

19 Prognos, "Ermittlungen der Wachstumswirkungen der KfW-Programme zum Energieeffizienten Bauen und Sanieren," accessed July 14, 2015. <https://www.kfw.de/migration/Weiterleitung-zur-Startseite/Startseite/KfW-Konzern/KfW-Research/Economic-Research/Evaluationen/PDF-Dokumente-Evaluationen/Prognos_Wachstumseffekte_KfW_EBS_Endbericht.pdf>.

20 Traber, Thure, and Claudia Kemfert, "Die Auswirkungen des Atomausstiegs in Deutschland auf Strompreise und Klimaschutz in Deutschland und Europa," *Studie im Auftrag von*

borne in mind here is that not all the aforementioned investment can be attributed solely to the energy transition. Many new power plants would have to be built and much grid renewal and maintenance work would have had to be carried out even without the energy transition.[21] Overall an annual investment of 31 to 38 billion euros will be required.[22]

The funding of renewables is set out in the German Renewable Energy Sources Act (Erneuerbare-Energien-Gesetz, EEG), which enables guaranteed feed-in tariffs and priority access to the grid, both of which reduce the risk for investors. In recent years, the EEG surcharge has increased to 3.5 cents/kWh, primarily due to strong growth in photovoltaics. In 2012, the surcharge remained relatively constant at 3.592 cents/kWh but in 2013, it dramatically increased to 5.28 cents/kWh.[23] Various studies have predicted an increase in the surcharge of 3.5 to 5.1 cents/kWh over the next ten years.[24] However, there are increasing calls to cap the EEG surcharge, and also to abolish priority feed-in in order to be able to better optimise network bottlenecks and reduce costs. From an investment perspective, both of these proposals would have more of an inhibiting effect.[25]

However, since there is presently no change to the fundamental EEG concept in the pipeline, further investment will be relatively low risk. The lion's

---

Greenpeace e.V. (2012), accessed August 17, 2012. <https://www.greenpeace.de/sites/www.greenpeace.de/files/20120702-Auswirkungen-Atomausstieg-Gutachten.pdf>.

21 Ibid. The BNA also recently confirmed that the grid expansion outlined in the grid development plan can only be partially directly attributed to the energy transition; Übertragungsnetzbetreiber, "Netzentwicklungsplan 2012," accessed August 17, 2012. <http://data.netzausbau.de/2022/NEP/NEP2022_UENB-Entwurf2a.pdf>.

22 Blazejczak et al., "Energiewende erfordert hohe Investitionen," DIW *Wochenbericht* 26 (2013): 19-30.

23 50Hertz Transmission GmbH, Amprion GmbH, EnBW Transportnetze AG, and TenneT TSO GmbH, "Prognose der EEG-Umlage nach AusglMechV: Prognosekonzept und Berechnung der ÜNB," last modified October 14, 2011, accessed July 14, 2015. <https://www.netztransparenz.de/portals/1/Content/EEG-Umlage/EEG-Umlage%202012/111014_Prognose_EEG-Umlage-2012_final.pdf>.

24 Bundesverband Solarwirtschaft, and Prognos, Kosten der Solarstromförderung (Berlin, Basel: 2011); "Konsequenzen eines Ausstiegs aus der Kernenergie"; "Kosten des Ausbaus erneuerbarer Energien"; Traber, Thure, Claudia Kemfert, and Jochen Diekmann, "Strompreise: Künftig nur noch geringe Erhöhung durch erneuerbare Energien," *DIW Wochenbericht* 6 (2011): 2-9.

25 Kemfert, Claudia, and Dorothea Schäfer, "Finanzierung der Energiewende in Zeiten großer Finanzmarktinstabilität," *DIW Wochenbericht* 31 (2012): 3-14.

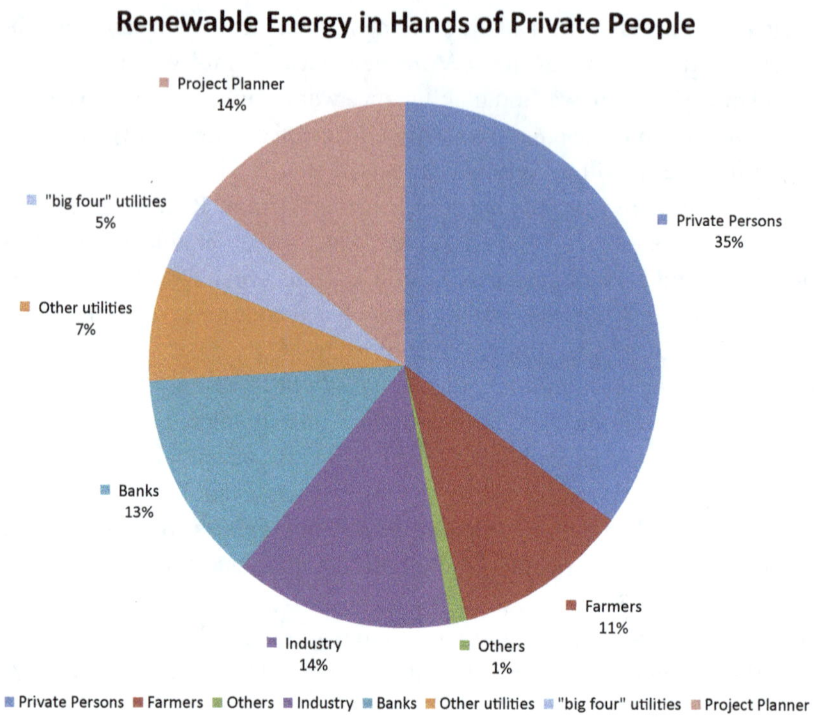

FIGURE 8.1 *Renewable energy in hands of private people*

share of investment will be made by private individuals (with forty percent of installed capacity), followed by banks (14 percent) and farmers (11 percent).[26]

Building refurbishment is mainly financed through private investment, supported by soft loans from KfW's building refurbishment program,[27] funds for which have been increased from almost 900 million euros to 1.5 billion euros per annum. This increase in funds will mainly be financed from the Energy and Climate Fund which, in turn, primarily derives from the sale of CO_2 emission certificates. However, certificate prices have plummeted as a result of the economic crisis and the current price means that it will not be possible to meet funding requirements for the building refurbishment program with emissions trading.

Investment in the electricity networks is also made by companies but this can be recovered via network fees. However, it is distinctly possible that the

26 Agentur für erneuerbare Energien, "Eigentumsverteilung erneuerbarer Energien 2011," accessed July 14, 2015. <https://www.unendlich-viel-energie.de/themen/gute-gruende/2-mehr-buergerbeteiligung>.

27 "Energieeffizient Sanieren."

current incentive regulation could actually create obstacles since investors may be worried about inadequate financing.[28]

However, warnings are increasingly being voiced: Can Germany even afford to implement an energy transition during these turbulent times of financial and debt crisis? Will companies relocate abroad? Is the entire energy transition in jeopardy? It is true that the financial crisis is currently overshadowing the necessary changes to the energy transition process. However, the transition is forging briskly ahead, above all with the extension of renewables but also with the development of new power plant capacity to replace nuclear power. The expansion of renewables is also progressing partly due to subsidies through the Renewable Energy Sources Act, which are being continued with just one or two adjustments. However, subsidy rates for photovoltaics will continue to be cut to keep costs down but, at the same time, increased financial incentives will be provided to promote the expansion of wind energy and, to a certain extent, also biomass. The financial cost to energy-intensive industries, in particular, will also be curbed, as exemptions—whether for the purchase of emissions certificates, payment of the EEG surcharge, or green tax—will continue to exist. However, the financial crisis may indeed strike the energy transition's major Achilles' heel: the urgently needed expansion of the infrastructure including the 'electricity highways', which transport electricity generated by wind power from northern to southern Germany, or the power lines transferring electricity to other European countries, and also the distribution networks for optimal and smart coupling of electricity supply and demand. If we bear in mind that constructing the necessary infrastructure can take up to ten years from approval to operation, it quickly becomes clear how urgent the need for investment is. The financial crisis could thus delay necessary investment.

The German government has initiated the energy transition. This brings with it massive challenges but, above all, tremendous opportunities. A smart energy transition could become a motor for economic growth. This makes it all the more important to create a master plan and establish a German Energy Ministry in order to advance the energy transition as quickly as possible. There are five key elements:

a) *Expansion of electricity networks and storage, and network optimisation through demand management.* The expansion and optimisation of European electricity networks is a matter of priority and an accelerated expansion is

---

28 Brunekreeft, Gert, and Roland Meyer, "Netzinvestitionen im Strommarkt: Anreiz- oder Hemmniswirkungen der deutschen Anreizregulierung?" *Energiewirtschaftliche Tagesfragen* 1-2 (2011): 2.

urgently needed. Renewable energy sources are very volatile. During periods when renewable energy supply is abundant, the electricity networks often lack the capacity to transport the excess power to other parts of the country or abroad. The expansion of 'electricity highways', both within Germany and across the entire EU, is particularly fundamental, as is the expansion of distribution networks. Electricity generation using renewables is becoming increasingly decentralised. This makes smart grids necessary to balance supply and demand better than it has previously been the case. The financial crisis increased financial institutions' risk premiums. State guarantees would at least reduce uncertainty. Further, it is also essential to create more storage capacity. However, pumped-storage plants are presently the only form of electricity storage that is, even to a certain extent, economical. Opportunities for this form of storage are, however, primarily to be found abroad, making the cross-border expansion of the network even more important. In future, innovative electricity storage methods such as batteries, methane, and hydrogen, etc. will also be possible, but they are currently completely uneconomical.

b) *Creation of suitable regulatory and framework conditions.* In order to increase incentives for the construction and use of gas-fired power plants, the market design must be modified. Shrewd regulation can help to foster the capacity necessary for a sustainable energy system. Further, demand must successfully accommodate the volatile energy supply. In order to accelerate investment, the framework conditions, particularly with regard to financing the infrastructure as well as power storage and reserve capacity, must be adapted to enable security of supply and dynamic development. Appropriate financial incentives for electricity networks and storage and also for the necessary electricity capacity to balance supply and demand are just as important as the rapid implementation of all energy conservation measures. Thus, not only should funds for energy-efficient building refurbishment be increased as quickly as possible, but all potential for energy conservation should be exhausted, particularly in the mobility field.

c) *Energy efficiency.* The German government quite rightly decided that energy conservation potential in the building sector must be exploited. However, this requires sufficient funding[29] and the planned 1.5 billion euros will barely be enough to meet the refurbishment rate.

---

29 Kreditanstalt für Wiederaufbau (KfW), "Der energetische Sanierungsbedarf und der Neubaubedarf von Gebäuden der kommunalen und sozialen Infrastruktur," Studie des Bremer Energieinstituts im Auftrag der KfW 2011, accessed July 14, 2015. <https://www.kfw.de/Download-Center/Konzernthemen/Research/PDF-Dokumente-Akzente/Akzente-Nr.-62-Mai-2012.pdf>.

d) *Social acceptance.* The population's acceptance of the measures required for the implementation of the energy transition is dependent on how well informed they are, how transparent the process is, and which opportunities for participation at the national and *Länder* levels they have.

e) *Creation of an energy ministry and implementation of a change management process.* Company managers who want to fundamentally realign their business are familiar with this essential process which includes the complete restructuring and reorientation of all key areas. Applied to the energy transition, this entails the comprehensive, decisive, and targeted implementation of all relevant areas. Although targets have already been clearly formulated, they must now be rapidly implemented. The energy transition is an incredibly complex process which, as a result, means that, in many areas, numerous loose ends need to be tied up. This needs a comprehensive overview and also faster and braver adjustments and modifications, which will only be possible with a Ministry to bring together all the different interests, develop a master plan, and take leading responsibility for the implementation of the transition. There is no room for mistakes. The system is far too fragile for that. The negative effects of inertia would also rapidly make themselves known. Thus, we can only hope that the current financial crisis does not result in nervousness and negligence with regard to the change management process needed for the energy transition. After all, the financial crisis has made one thing clear: rapid government (re)action is perfectly possible.

5 An Energy Transition "Made in Germany"

The energy transition "made in Germany" is an important contribution to global climate policy. However, Germany is not at the top of the energy transition rankings but is only mid-table, behind countries such as Sweden, Brazil, and Italy (see Figure 8.1).[30]

30 See Fraunhofer ISE, *The Energy Transformation Index*, accessed July 17, 2016. <https://www.ise.fraunhofer.de/de/downloads/pdf-files/aktuelles/ise-ises-eti.pdf>. (The Climate Change Performance Index also sees Germany only mid-table on issues such as emissions development, expansion of renewables, improving energy efficiency, and policy measures (Burck, Jan, Franziska Marten, Christoph Bals, "Climate Change Performance Index 2016," published by Climate Action Network, accessed July 17, 2016. <https://germanwatch.org/en/download/13626.pdf>. In contrast, other studies like the Global Green Economy Index (GGEI) see Germany alongside Sweden in first place on global green markets; see Tamanini et al., "The Global /Green Economy Index GGEI 2014," *Strategic Review* 2014 (Washington, New York, 2014), accessed July 17, 2016. <http://dualcitizeninc.com/

However, the German energy transition has both upsides and downsides. On the one hand, Germany has managed to considerably reduce emissions in the power sector through its expansion of renewables. On the other hand, Germany still uses too much coal for power generation,[31] which means that Germany's self-imposed climate target of a forty percent reduction by 2020 will most likely not be achievable. There are no binding climate policy measures, and emissions trading in its current form is ineffective because the CO_2 price is far too low. Therefore, accompanying measures are required, such as a carbon tax and a structured fossil-fuel phase-out. In addition, Germany has not done enough in the area of sustainable mobility; the recent "VW Diesel scandal" is a bad example of environmental protection "made in Germany." If Germany is to be celebrated as a model for climate policy and Chancellor Merkel wants to be known as the "climate chancellor," the carbon problem must be solved, and we also need to do more to save energy and improve our credibility again, particularly in the area of sustainable mobility. Consequently, policy-makers are now being called upon to also implement measures that combat climate change more effectively but run counter to economic interests. Nevertheless, the German energy transition policy has initiated a global upheaval and perhaps Germany can learn more from other countries when it come to issues like the fossil-fuel phase-out or properly measuring exhaust emissions.

The German energy transition is nevertheless setting an important precedent: thanks to investments from Germany, the rising demand and associated economies of scale have caused the costs of renewable energy sources to fall massively worldwide. For the first time, more global investment is being made in renewables than in fossil fuels.

More and more countries will follow the German example and prefer to invest in renewable energy sources than fossil fuels or atomic energy: more opportunities, fewer risks.

6 Conclusion

A smart energy transition could become a motor for economic growth. Massive investment will be made which, in turn, will create added value and jobs. There will be investment in renewable energy sources, new power plants, energy

GGEI-Report2014.pdf>). A study by the Handelsblatt Research Institute (HRI) sees Germany in eighth place out of 24 countries.

31 Kemfert, Claudia, *Battle about Electricity, Myths, Power and Monopolies* (Hamburg: Murmann, 2013)

efficiency, and sustainable buildings and mobility. The German economy is in a better position than any other to profit from the boom, since German manufacturing, in particular, can benefit tremendously from the expansion of plants, infrastructure, and power plant construction. As several studies have shown, there are great chances to create hundreds of thousands of new jobs. A smart energy transition undoubtedly presents more economic opportunities than risks. This is why Germany can continue to be the role model for a sustainable global energy transition in the world.

CHAPTER 9

Changing climates, changing spaces, changing times: Adaptation and conflict on the West Frisian Island of Ameland

*Lea Schmitt**

Abstract

Under the premise "Working Together with Water" the national management of water in the Netherlands is going through a fundamental process of change. Water is supposed to be more tightly integrated with spatial structures, no longer using exclusively technological measures, but ones that are sustainable, too. The political sphere's attempts to counter climate change with changes in spatial structure raise questions about processes surrounding legitimation. As the successful application of adaptation strategies is dependent upon a certain level of consensus at the local level, my work concerns the sphere of governance processes relating to space. This article presents the results of ethnographical field research that I conducted on the West Frisian island of Ameland. Based on an example of spatial adaptation that the local population resisted, I examine the significance for actors and people affected by strategies that involve taking measures to adjust spatially and allow natural dynamics a greater degree of influence. The field research showed that the resistance of some inhabitants of Ameland against spatial adaptation to the effects of climate change touched upon competing understandings of space and time, and also of reality. Physical changes in the material surroundings go hand in hand with a change in existing social power relations, against which one defends oneself. Furthermore, the case study of Ameland makes clear that the situation concerning on Ameland was glocal in nature: There was an overlapping of discourses, actors and power relations that relate to various spatial levels.

* This chapter is a translated excerpt from my dissertation: Schmitt, Lea, *Klimawandel, Raumwandel, Zeitwandel: Adaptationen und Konflikte auf der westfriesischen Insel Ameland* (PhD diss., University of Duisburg-Essen, 2013).

 | DOI 10.1163/9789004356825_011

1 Introduction

In a project entitled "Samen werken met water" (Working Together with Water),[1] which has been running now for some years, the national management of water in the Netherlands is going through a fundamental process of change. Water is supposed to be more tightly integrated with spatial structures, no longer using exclusively technological measures, but ones that are sustainable too.[2] The Second or New Delta Commission, among others, had suggested these explicitly integrative approaches in 2008.[3] This state commission was set up to formulate long-term recommendations for the nationwide management of water up until the year 2100—paying particular attention to the continuous sinking of the ground in the Netherlands and the accelerated rise in sea level as a result of climate change.[4] The Second Delta Commission's recommendations were also incorporated into the first National Water Plan, that is, the vision for the nationwide structuring of space.[5] The Ministry of Infrastructure and Environment (Ministerie van Infrastructuur en Milieu),[6] declared: "The water

1 The author's translation.

2 Cf. Rijswick, Marleen van, and Herman Havekes, *European and Dutch Water Law* (Groningen: Europa Law Publishing, 2012), 259f; Kabat, Pavel et al., "Dutch Coasts in Transition," *Nature Geoscience* 2.7 (2009): 450-452; Dijk, Judith M. van, "Water Assessment in the Netherlands," *Impact Assessment and Project Appraisal* 24.3 (2006): 199-209; Brugge, Rutger van der, Jan Rotmans, and Derk Loorbach, "The Transition in Dutch Water Management," *Regional Environmental Change* 5.4 (2005): 164-176.

3 Cf. Deltacommissie, *Samen werken met Water: Een Land dat Leeft, Bouwt aan zijn Toekomst, Bevindingen van de Deltacommissie 2008*, accessed January 25, 2014. <http://www.deltacommissie.com/doc/2008-09-03%20Advies%20Deltacommissie.pdf>. The official title is Commission for Sustainable Coastal Development, customarily referred to as the Second or New Delta Commission.

4 While the Intergovernmental Panel on Climate Change's (IPCC) Fourth Assessment Report forecasts a rise in sea level of between 18 and 38 centimetre by the year 2100 (cf. Intergovernmental Panel on Climate Change (IPCC), ed., *Climate Change 2007: Synthesis Report*, published 2007, accessed January 25, 2014, 45. <http://www.ipcc.ch/pdf/assessment-report/ar4/syr/ar4_syr.pdf>), the Delta Commission works on the assumption that there will be a rise in the Netherlands of between 65 and 130 cm (Deltacommissie, *Samen werken met Water*, 10).

5 Cf. Ministerie van Verkeer en Waterstaat, Ministerie van Volkshuisvesting, Ruimtelijke Ordening en Milieubeheer, Ministerie van LNV, ed. *Nationaal Water Plan 2009-2015*, published 2009, accessed January 25, 2014. <http://www.rijksoverheid.nl/documenten-en-publicaties/rapporten/2009/12/01/nationaal-waterplan-2009-2015.html>.

6 Referred to henceforth as Ministerie van I&M.

system changes in accordance with natural processes where it can and resists them where it has to."[7]

As such, the approach to water in the Netherlands is in transition, for the recommendations of the Second Delta Commission clearly distinguish themselves from those of the First Commission. These were applied after the devastating storm surge of 1953, in which over 1,800 people lost their lives. At the time, recommendations referred exclusively to the technical measures of storm surge barriers. The Delta Works came to symbolise this battle against the water. Then the Second Delta Commission recommended "Working Together with Water", an approach with which it even associated opportunities and new perspectives.[8]

The way in which views on water management have changed is also to be seen in the national programme "Adaptatie Ruimte en Klimaat" (Adaptation in Space and Climate), which the ministry established in cooperation with the provinces, local authorities, and regional water associations in 2006. Its stated, ambitious goal is to become a 'climate-proof spatial institution' for the long term in the Netherlands.[9] A year later, from this programme emerged the national strategy for adaptation,[10] entitled "Maak Ruimte Voor Klimaat" (Make Room for the Climate). Climate change was to be met with a change in spatial structure.

Beyond spatial adaptation in the face of climate change, the New Delta Commission, as well as the national policy programme, represents a *normative* change in water management:

---

7 Ministerie van Infrastructuur en Milieu, ed., *Structuurvisie Infrastructuur en Ruimte: Nederland Concurrerend, Bereikbaar, Leefbaar en Veilig*, published 2012, accessed January 25, 2014, 51. <http://www.rijksoverheid.nl/documenten-en-publicaties/rapporten/2012/03/13/structuurvisie-infrastructuur-en-ruimte.html>. It would be an exaggeration to contend that the Dutch political sphere had never taken into consideration recommendations of the kind formulated in 2008 by the "Second Delta Commission" concerning water. However, an approach to the integrated management of water embracing multiple sectors was first conceived of in 2009 (cf. Rijswick, and Havekes, *European and Dutch Water Law*, 91f).

8 Cf. Deltacommissie, *Samen werken met Water*, 21ff.

9 Ministerie van Volkshuisvesting, Ruimtelijke Ordening en Milieubeheer, Ministerie van Verkeer en Waterstaat, Ministerie van LNV, Ministerie van EZ, ed., *Nationaal Programme Adaptatie Ruimte voor Klimaat.*

10 Adaptation to the effects of climate change is referred to here as "adaptation" and not as "adaption" as found in the works of certain authors. My reasons for this are etymological: the verb stem of the Latin verb is "adaptare", "adapta-" and not "adapt-".

> Working with nature implies a gradual paradigm shift moving from a perspective of water (nature) as an enemy toward a perspective of water (nature) as an ally, a reappreciation of aspects of an accommodation strategy as an alternative to a strict protection strategy (although safety remains a fundamental part of the overall strategy).[11]

This kind of approach has been particularly well realised in the Wadden area in the context of (inter)nationally recognised areas in which natural habitats and the littoral zone are protected and a pronounced natural dynamic is utterly essential. This is the opposite of, for instance, the low lying and densely populated Randstad. That nature's self-organisation enjoys a principally scientific and political legitimacy raises the question as to what social consequences result on a local level from the extended leeway given to nature and whether those people affected accept this new leeway. In the remainder of this chapter, I present the results of ethnographical field research that I conducted on the West Frisian island of Ameland. Based on an example of spatial adaptation that the local population resisted, I examine the significance for respective actors and those people affected of strategies that involve taking measures to adjust spatially and allow natural dynamics a greater degree of influence.

The following questions guide my analysis:

1. The manner in which actors identify with space and time and associated changes: Which of the various groups of actors come to the fore in relation to which conceptions of space and time? What sort of resistances and concerns are prompted by spatial transformations in a given location? How does this relate to locally established ways of identifying with space and time?
2. Coalitions and power over interpretation: Which constellations of actors, networks and power structures emerge? Which bodies of knowledge are (not) appealed to?
3. Local-global connections: To what extent does the global scope of the topic of climate change play a role for the respective actors and their perception of space and time? How is this displayed at a local level and in everyday life?

11 Koningsveld, Mark van et al., "Living with Sea-Level Rise and Climate Change: A Case Study of the Netherlands," *Journal of Coastal Research* 24.2 (2008): 376.

2 Theoretical Basis and Methodical Approach

Drawing upon Peter Berger and Thomas Luckmann's *The Social Construction of Reality*,[12] the phenomenological sociology of knowledge and recent social constructionism provide the theoretical basis of this study. It is from this point of view that the conflicts to which adaptation measures lead on Ameland will be investigated. I work on the assumption that the recommendations made on the basis of scientific expertise with a view to extending the space given over to natural dynamics have met with local resistance because they are formulated in terms of a competing view of reality. On this basis, I turn to the "unproblematic sector of everyday life",[13] in order to better understand the problematic against the background of the unproblematic. As a result, the life-world[14] of the everyday on Ameland becomes central to the aim of my study. It is therefore meaningful to adopt the broadly conceived concept of knowledge that the sociology of knowledge provides: that is, knowledge is what counts as knowledge.

The study therefore reconstructs various realities relating to space and time and draws attention to processes and practices, in accordance with which these realities are socially established, defended or reformulated.[15]

However, what recent social constructionism fails satisfactorily to illustrate is the aspect of collective or institutional action. It pays just as little attention to the materiality of the everyday world and its significance for processes involved in the social construction of reality. In these respects, the theoretical and conceptual reflections of the sociologist of space and of knowledge Gabriela Christmann prove to be helpful. In the context of spatial structural change, Christmann is concerned with communicative processes of negotiation between actors in possession of differing knowledge of space.[16] As such, in

12 Berger, Peter L., and Thomas Luckmann, *Die gesellschaftliche Konstruktion der Wirklichkeit: Eine Theorie der Wissenssoziologie* (Frankfurt a.M.: Fischer, 2009) (first published 1966).

13 Berger & Luckmann, *Die gesllschaftliche Konstruktion*, 27.

14 This involves festivals and work just as it involves institutions, objects and events; cf. Knoblauch, Hubert, *Kommunikationskultur: Die kommunikative Konstruktion kultureller Kontexte* (Berlin: de Gruyter, 1995), 12.

15 Social constructionism does not merely mean to pursue different realities but, at the same time, to look at "the processes by which any body of 'knowledge' comes to be socially established as 'reality'" (Berger, and Luckmann, *Die gesellschaftliche Konstruktion der Wirklichkeit*, 3).

16 Christmann, Gabriela B., "Kommunikative Raumkonstruktionen als (Proto-) Governance," in *Governance und Raum*, ed. Heiderose Kilper, 27-48 (Baden-Baden: Nomos,

addition to individual actors, she also considers collective and institutional actors (groups of actors, networks, political actors). Further, she looks at aspects of governance relating to changing power relations between actors that can lead to amalgamations or to rifts—depending on whether existing interpretations of spatial realities are conducive to one another and the success of these interpretations.[17]

Further, my approach is that of a social constructionism that extends to the issue of materiality.[18] In particular, I consider the reflections of the philosopher and ethnographer Bruno Latour a source of inspiration in this regard. He critiques sociology for ignoring material objects or viewing them as being the responsibility of other disciplines.[19] With a view to integrating an enhanced degree of realism, he wishes to switch from *realpolitik* to *dingpolitik*[20] and

---

2010), 29; Christmann, Gabriela B., and Kerstin Büttner, "Raumpioniere, Raumwissen, Kommunikation—zum Konzept kommunikativer Raumkonstruktion," *Berichte zur deutschen Landeskunde* 85 (2011): 361-378. Christmann therefore combines the approach of communicative social constructionism with concepts of relational ideas of space—cf. Löw, Martina, *Raumsoziologie* (Frankfurt a.M.: Suhrkamp), 2001; Läpple, Dieter, "Essay über den Raum: Für ein gesellschaftswissenschaftliches Raumkonzept," in *Stadt und Raum: Soziologische Analysen*, ed. Hartmut Häußermann and Walter Siebel (Pfaffenweiler: Centaurus, 1991), 157-207—and considerations of discourse theory—cf. Keller, Reiner, "Wissenssoziologische Diskursanalyse als interpretative Analytik," in *Die diskursive Konstruktion von Wirklichkeit: Zum Verhältnis von Wissenssoziologie und Diskursforschung*, ed. Reiner Keller et al. (Konstanz: UVK, 2005), 49-75, as well as the spatial dimensions of identity and governance; cf. Kilper, Heiderose, "Governance und die soziale Konstruktion von Räumen: Eine Einführung," in *Governance und Raum*, ed. Heiderose Kilper (Baden-Baden: Nomos, 2010), 11-13, after Mayntz, Renate, "Von der Steuerungstheorie zu Global Governance," in *Governance in einer sich wandelnden Welt*, ed. Gunnar Folke Schuppert and Michael Zürn (Wiesbaden: VS Verlag für Sozialwissenschaften, 2008).

17 Cf. Christmann, "Kommunikative Raumkonstruktionen," 44ff.

18 Cf. Christmann, Gabriela B. et al., *Vulnerabilität und Resilienz in sozio-räumlicher Perspektive: Begriffliche Klärungen und theoretischer Rahmen*, Working Paper, 44 (Erkner: Leibniz-Institut für Regionalentwicklung und Strukturplanung, 2011), accessed January 25, 2014. <http://www.irs-net.de/download/wp_vulnerabilitaet.pdf>; Christmann, Gabriela B., and Oliver Ibert, "Vulnerability and Resilience in a Socio-spatial Perspective," *Raumforschung und Raumordnung* 70 (2012): 259-272.

19 Cf. Latour, Bruno, *We Have Never Been Modern*, trans. Catherine Porter (Cambridge, MA: Harvard University Press, 1993), 103.

20 Latour, Bruno, "From Realpolitik to Dingpolitik or How to Make Things Public," in *Making Things Public: Atmospheres of Democracy. Exhibition at ZKM, Center for Art and Media Karlsruhe 20.03.–3.10.2005*, ed. Bruno Latour and Peter Weibel (Cambridge: MIT Press, 2005), 14f, 41.

argues for a project of "symmetrical anthropology",[21] in the framework of which he deems things to be "actants".[22] Such a "sociology of associations" analyses networks that vary in scope and consist of a wealth of entities that are only mistakenly conceived of as incommensurable. I too consider objects to have a certain degree of agency but no intentionality, the latter being a purely human property. Nonetheless, I do take the issue of their effect[23] into consideration.[24] My analysis therefore draws attention to how the natural phenomenon of a sandbank is incorporated into processes of the construction of reality. It is thereby possible to observe those interactions between actors that relate to space as political actions: nature "remains mobilisable, humanisable and socialisable".[25] Accordingly, it is productive to look at how certain actors position themselves as speakers for certain natural objects, since such practices touch upon far-reaching legitimation processes and constructions of reality.

The fieldwork was methodologically conceptualised as "strategically situated single-site ethnography":[26] though I remained on one and the same site for the overwhelming majority of the duration of my fieldwork on the island of Ameland, I understood this location as globalised or glocalised.[27] The background to this conceptualisation is the assumption that ethnographic field research must do justice to the fact that it takes place in a globalised world. In

21 Latour, *We Have Never Been Modern*, 106.

22 Latour wishes to overcome what he considers modernistic distinctions between natural and social beings, since he conceives of the everyday as being increasingly shaped by "hybrids" or "quasi-objects", these being neither entirely material nor social (cf. Latour, *We Have never Been Modern*, 55), though just as capable as human actors of bringing about change.

23 Holzinger, Marcus, "Welcher Realismus? Welcher Sozialkonstruktivismus? Ein Kommentar zu Georg Kneers Verteidigung des Sozialkonstruktivismus und zu Bruno Latours Akteur-Netzwerk-Theorie," *Zeitschrift für Soziologie* 38.6 (2009): 532.

24 Here I depart from Latour, who would go one step further as regards the emergence of action. In his view, the approach of social constructionism, which touches upon the attribution of meaning, maintains the separation of nature and culture instead of disposing with it (cf. Christmann, et al., *Vulnerabilität und Resilienz in sozio-räumlicher Perspektive*, 23).

25 Latour, *We Have never Been Modern*, 37.

26 Marcus, George E., "Ethnography in/of the World System: The Emergence of Multi-Sited Ethnography," *Annual Review of Anthropology* 24 (1995): 110.

27 Following Roland Robertson's concept of "glocalisation", the local and the supra-local levels interfere with one another. As such, local and supra-local knowledge do not represent separate spheres but spheres that are connected to one another—cf. Robertson, Roland, "Glocalization: Time-Space and Homogeneity–Heterogeneity," in *Global Modernities*, ed. Mike Featherstone (London: Sage Publications, 1995), 25-44.

accordance with which, the local and the social can no longer be considered in a restrictive, essentialist manner: the concept of the field must be revised. A given location must therefore be understood as reaching out beyond itself and intertwined in a wealth of different spatial models, constellations of actors and power structures.[28]

The fact that concepts of locality and community always involve material and social components has consequences for the methodological considerations behind this investigation: given that the local field can no longer be represented in an exclusively descriptive manner, attention must be paid to the perceptions of actors and the way in which they position themselves in order to draw out what exactly these are based on, against what they are defined and what kind of locality they interactively bring forth. This impacts on the ethnologist's research, which is not conceived of as the work of an annalist of a place taken for granted. Rather, the point is to observe the social as "place-making projects",[29] that is, to discover the complex social ways in which places are produced.[30] Thus the field research was plurilocal not so much in a geographical but rather in a conceptual sense. It is in this way that ethnography does justice to the complexity of the research situation, in that it avoids isomorphism, that is to say, the equivalence of different geographical scales and levels of analysis.[31] With the help of a "strategically situated single-site ethnography", it is possible to successfully produce a network-like description of social reality in a fragmented world in which different spatial scales, power relations and constellations of actors cross over with one another.

---

28 Cf. Appadurai, Arjun, "Theory in Anthropology: Center and Periphery," *Comparative Studies in Society and History* 28.1 (1986): 356-361; Gille, Zsuzsa, and Seán Ó Riain, "Global Ethnography," *Annual Review of Sociology* 28 (2002): 271-295; Gupta, Akhil, and James Ferguson, "Beyond 'Culture': Space, Identity, and the Politics of Difference," *Cultural Anthropology* 7.1 (1992): 6-23.

29 The concept of "place-making" hails originally from Akhil Gupta and James Ferguson (Gupta, and Ferguson, "Beyond 'Culture'"). The authors draw a parallel between the concept of "place-making" and that of "friction" as used by Anna Lowenhaupt Tsing: Tsing, Anna Lowenhaupt, *Friction: An Ethnography of Global Connection* (Princeton: Princeton University Press, 2005)—the point being to draw attention to the mutual determination and connection of global and local processes.

30 Gille, and Ó Riain, "Global Ethnography," 278.

31 Cf. "Beyond 'Culture'." 11f. See also: Wimmer, Andreas, and Nina Glick-Schiller, "Methodological Nationalism and Beyond: Nation-State Building, Migration and the Social Sciences," *Global Networks* 2.4 (2002): 301-334.

The study is based on two types of data collection. (In-)direct or (non-)participant observations were applied alongside narrative interviews[32] or ethnographic conversations.[33] Observing "lived cultural orders"[34] was the focus of my work; that is, the experience of relevant events and discourse and their contextualisation over time. A "follow the conflict"[35] strategy was central to the materialisation of the field. Further important materialisation techniques were the "follow the metaphor", "follow the people" and "follow the thing" strategy.[36] In this way, networks of actors as well as relevant semantic fields could be highlighted. A sandbank that lay before the coast on which the village of Ballum was located and had led to significant conflict made an outstanding ethnographic point of departure insofar as it led to the crossing of paths of different actors. Indeed, the following is about this particular sandbank, the so-called "Zandhaaken" off the Dutch island of Ameland.

3 Build with Nature or Be Built Over by Nature? The Zandhaaken on Ballum Beach

For years, the coast of the village of Ballum on Ameland has undergone substantial change: On the North Sea side of the island, a sandbank known as the "Bornrif" has slowly emerged, "moving" towards the coast of Ameland and, in a hook-like form (which is why it is spoken of as the "Zandhaaken" or "sand hook"), has finally reached the same height as the village of Ballum.

The situation in Ballum is largely determined by the recommendations of the New Delta Commission, which are geared toward allowing nature a greater degree of agency in times of climate change, in order to affect spatial structures that are climate resilient in the long term. Thus, the Netherlands had already thoroughly incorporated dynamic coastal protection measures into its coastal maintenance programme as early as 1990: in order to counter structural coastal erosion and to provide the coastal foundation with more sand, the Ministry for Transport and Water Resource Management made the decision to

32 Cf. Przyborski, Aglaja, and Monika Wohlrab-Sah, *Qualitative Sozialforschung: Ein Arbeitsbuch* (München: Oldenbourg 2008), 92-94.

33 Spradley, James P., *The Ethnographic Interview* (Belmont: Wadsworth Thomson Learning, 2008).

34 Amann, Klaus, and Stefan Hirschauer, "Die Befremdung der eigenen Kultur: Ein Programm," in *Die Befremdung der eigenen Kultur: Zur ethnographischen Herausforderung soziologischer Empirie*, ed. Stefan Hirschauer (Frankfurt a.M.: Suhrkamp, 1997), 21.

35 Marcus, "Ethnography in/of the World System," 110.

36 Ibid., 106ff.

stop stabilising the sand and instead to allow greater fluctuation of sediment. The premise of "leaving nature to run its course" had been adopted from the 1990s onward:[37] Natural processes or the power of wind and water had been given greater leeway ever since.[38] Given that dynamic coastal maintenance leads not only to an increase in erosion but also in sedimentary processes, the ministry has made long-term provisions for an increased volume of sand in the coastal foundation.[39] With the introduction of a considerably altered strategy, a concept of coastal protection was established that is considered to provide for the "sustainable maintenance of security and sustainable upkeep of the functions and values of the dune area" and,[40] already at this time, for a suitable way of dealing with rising sea levels.[41] The dynamic coastal protection programme introduced in the 1990s can in this respect be considered a resilience strategy insofar as it is (among other things) a spatially oriented adaptation measure taken with the anticipated effects of climate change in mind. That is

---

37 Arens, Bas, M.A.M. Löffler, and E.M. Nuijen, *Evaluatie Dynamisch Kustbeheer Friese Waddeneilanden*, published 2007, accessed January 25, 2014, v. <https://publicwiki.deltares.nl/download/attachments/17334280/Arens+ea+%282007%29+Dynamisch+Kustbeheer+Friese+Waddeneilanden.pdf>.

38 However, this dynamic was only provided for within the limits of the "basiskustlijn" ("coastal baseline") and "primaire waterkering" ("primary coastal defence"): The coastal baseline relates to the coastline established in 1990. It guarantees that the coast will not recede beyond a given mark, independent of dynamic maintenance measures. Should the volume of sand at a given section of coast fall below a certain level, more sand will be deposited. However, the coastline is permitted to fluctuate freely at the uninhabited, easterly ends of the West Frisian islands of Ameland, Terschelling and Schiermonnikoog. In contrast to the coastal baseline, the primary coastal defence is located on the land. It constitutes a system of dyke and dune rings that were constructed after the storm surge catastrophe of 1953. These dyke rings benefit from the so-called height of the delta. Within such dyke rings, the unconditional protection of people and assets is guaranteed by law. Thus everything possible is done to stop so-called external waters from the North Sea, the Ijsselmeer or large rivers entering inland territories. Just as on other islands in Wadden, this ring is located on the edge of the innermost dune. While the low-lying part of the island is protected, the dune area lies outside of the dykes (cf. Ministerie van Verkeer en Waterstaat, ed., *Derde Kustnota: Traditie, Trends en Toekoms,* published 2000, accessed January 25, 2014. <http://stive51.home.xs4all.nl/deltalinks/3ekustnota.pdf>.

39 Cf. Ministerie van Verkeer en Waterstaat, *Kustverdediging na 1990: Beleidskeuze voor de Kustlijnzorg. Tweede Kamer, Vergaderjaar 1989-1990, 21 136, nrs. 5-6*, published 1990, accessed January 25, 2014. <http://repository.tudelft.nl/view/hydro/uuid:a4c7d579-d528-4f26-9364-1346817230ba/>.

40 Ibid., 7.

41 Ibid., 38.

to say, the political foundation for current recommendations hailing from the "Working Together with Water" programme was already laid in 1990.

The ministry had already mentioned the Zandhaaken on Ameland in connection with dynamic coastal maintenance in the mid-1990s and billed the overall process as being characterised by a "significant natural dynamic".[42] It argued that the coast of Ameland had significantly increased in size during 1960 and 1980 but that this trend had started to reverse by the mid-1980s and that, after a process running its course over a total period of between fifty and sixty years, one could expect to see a "stable or even slightly extended coastline".[43] Some years later, the ministry observed in phenomena such as the Zandhaaken one of several possibilities for growing with the sea level, that is, for a "solution that looked to the sea". What is meant here is that the coastline can be extended in the direction of the sea with the help of a large "sand puffer".[44] The Zandhaaken at Ballum presented just such a "sand puffer", and one that was considered in the framework of RWS (Rijkswaterstaat) expertise as a "stock of sand" that would provide the coastal foundation [Küstenfundament] with more sand on a long-term basis than sedimentation processes allowed to break away.[45]

The dynamics of the processes at Ballum led to large parts of the sand beach being broken off, particularly during the course of the last past decade, and an inland lake forming on the surface of the beach. The resident population is particularly angry that, for years, it has not been possible to use the beach at Ballumer Strand as a conventional beach for bathing. In the summer of 2010, some residents and companies came together to form a citizen's association entitled Friends of Ballum's Bathing Beach, which mobilised in order to achieve the goal of reconstructing the beach previously used for bathing.[46] Though seasonal summer tourism doubtless features among the most important sources of income on Ameland, the course of the conflict over the bathing beach on Ameland showed that not everything hinged on purely economic questions. Rather, the spatial politics of the approach to adaptation led to a clash of different understandings of reality and took its toll on power relations between the groups of actors involved: the Dutch coastal protection

42 Ministerie van Verkeer en Waterstaat, ed., *Tweede Kustnota: Kustbalans 1995*, published 1996, accessed January 25, 2014, 12. <http://de.scribd.com/doc/76339527/Kustbalans-1995-de-tweede-kustnota>.

43 Loc. cit.

44 Ministerie van Verkeer en Waterstaat, *Derde Kustnota*, 55.

45 Cf. Arens, Löffler, and Nuijen, *Evaluatie Dynamisch Kustbeheer Friese Waddeneilanden*, 47.

46 This was not registered as a formal society but constituted a group of engaged citizens in Ballum who understood their work as being to represent the village.

authorities—the "Rijkswaterstaat"—and its expertise, as well as on the Friends of Ballum's Bathing Beach established in protest against the Zandhaaken.

Rijkswaterstaat took the position already sketched out above, that the Zandhaaken constituted part of a process belonging to a cycle lasting approximately sixty years that now had to be seen through. They saw no need for negotiations since the sand bank that had been washed ashore provided more sediment than was to be found previously in the coastal foundation—without relying on expensive sand spraying using ships. RWS demanded in no uncertain terms that natural processes were allowed time at Ballum—something that it had continued to do over a period of many years. It had already described the Zandhaaken at Ballum in 2001 in the regional newspaper as a leisurely process and commented on the situation as follows:

> It is a question of time. Ballum will receive a large bathing beach as a result of the play of nature. [...] The inland lake continuously fills itself with sand and, with time, fills the tidal creek with sand. Nature itself restores certain things.[47]

According to RWS, the "course that natural processes take" must be accepted since, after some years, the Zandhaaken would finally join with the land and Ballum would once again have access to a bathing beach.

This kind of argumentation shows how the coastal protection authority drew on geomorphological expertise for its legitimation and thus points to cultural change within the institution: up until the 1960s, RWS was the dominant actor with respect to public infrastructure works on land and water, possessing engineering expertise, money and authority, but by the 1970s this prestige had come under pressure: "The heroes of the Delta Works were quickly demoted to the status of short-sighted engineers, authoritarian specialists, asphalt freaks, environmental polluters and landscape destroyers."[48] It was not only massive cuts in finances and personnel that defined the 1970s and 1980s.[49] In addition, many biologists, planners and administrative experts begun work and, in this way, introduced a new kind of expertise, even if it was some time before these

47 "Wind Water en Zand Spelen hun Spel bij Strand Ballum," *Leeuwarder Courant*, August 11, 2000, 9.

48 Lintsen, Harry, "Two Centuries of Central Water Management in the Netherlands," *Technology and Culture* 43.3 (2002): 551; Kaijser, Arne, "System Building from Below: Institutional Change in Dutch Water Control Systems," *Technology and Culture* 43.3 (2002): 566.

49 Ibid., 567.

experts assumed leading roles within the service.[50] In the new era of RWS, the focus was on the environment as well as on "integral" and "interactive" approaches: Economy, security and environment were thought of collectively as part of related processes involving different social groups.[51]

Civil society pitted a rhetoric of urgency against the RWS's line of argument. The phenomenon of the emergent reefs that the RWS perceived as a process were registered by Ballum's citizens' association as a permanent state. The Friends of Ballum's Bathing Beach had grown weary of waiting. They saw themselves as being robbed of their immediate future and helplessly facing instead a future promised by RWS that was in effect a kind of "future perfect", that is, an act or event to be completed in the future: Only once the reef had emerged from the sea and the inland lake filled in with sand would they once again have access to a conventional sand beach. The washed-up Zandhaaken involved an "anticipated retrospection".[52]

The example of the Zandhaaken at Ballum shows how natural processes are not only allowed more space but more time, too. Further, it clearly shows how each group of actors orients themselves according to different conceptions of time. RWS orients itself to processes involving natural phenomena that take place in real time, as well as to recollected futures. This causes uncertainty among the 'bathing friends' who coordinate their activities in accordance with the rhythms of the tourist season and with much shorter time frames, namely everyday and biographical ones. The field research data shows the attitude that the citizens' association adopts to the dynamic coastal protection that the RWS cite as the most important argument in favour of the Zandhaaken. The citizens' association considers this a form of power politics in two respects: not only does it lead to and legitimise a specific spatial structure but, at the same time, it establishes a particular temporal logic and norm in the coastal area. The protected process materialises in no uncertain terms on the Ballum coastal landscape itself and makes the local area dynamic and processual, too. Illustrative of which is that the people of Ballum in no way recognised a

50 Kaijser, "System Building from Below: Institutional Change in Dutch Water Control Systems".

51 Ibid., 567-568. For example, RWS has signed a treaty with nature protection parties that supports scientific monitoring of sand spraying with a view to avoiding disadvantageous impacts on the natural environment (cf. Ministerie van Verkeer en Waterstaat, Waddenvereniging et al. 2009).

52 Schütz, Alfred, "Tiresias oder unser Wissen von zukünftigen Ereignissen (1959)," in *Studien zur soziologischen Theorie*, ed. Alfred Schütz (Den Hag: Nijhoff, 1972), 259-278. Schütz refers here to Edmund Husserl and his concept of Vorerinnerung or "protention," with which Schütz describes the anticipation of temporally remote events (Schütz, "Tiresias": 268).

bathing beach, but rather an intertidal beach [Wattstrand]—that is, a peculiarly ambiguous landscape, half water, half sand and silt. With regard to dynamic coastal protection, one can also speak of Dutch society as a "geomorphology-based society":[53] Geomorphology is supposed to deliver the technological lever that makes the complex problem of the accelerated rise in sea level controllable, such that this technology changes society, too. Accordingly, a "geomorphology-based society" has elected to adopt a geomorphological temporal logic in the form of real-time processes and recollected futures and, wherever possible, legitimised their application.

The situation in Ballum is further complicated by the fact that it is not only the premises on which dynamic coastal protection is founded that rhythmically effect the landscape on the island but the parameters of nature protection, too—above all in the form of the European species protection programme Natura2000.[54] Its implementation has prompted heated debate on Ameland that has continued for years, just as it has done on other West Frisian Wadden Sea islands. More than a few among Ameland's population refer to Natura2000 satirically as "nature accounting", an expression of their scepticism concerning a course of action aimed at halting the reduction in species diversity in a bureaucratic manner: Certain types of habitats are associated with certain species and these are then counted per habitat area in order to learn about these areas' qualitative condition. In this way, Natura2000 areas such as the dunes and the North Sea zones form a further temporal framework oriented toward

---

53 This turn of phrase draws upon one that Bruno Latour uses, namely the "geometry-based society" (Latour, *We Have Never Been Modern*, 111). This follows Latour's reflections on the pulley narrative of Plutarch concerning the way in which Archimedes effortlessly moved a three-masted merchantman using a system of compound pulleys and thus convinced King Hiero of the significance of science and technology. Latour comments that not only did Archimedes "overturn power relations through the intermediary of the compound pulley, he also reversed political relations by offering the king a real mechanism for making one man physically stronger than the multitude" (ibid., 110). By adding the "lever of technology" to the "play of political representation", according to Latour a "new collective" comes into existence: a "geometry-based society" (ibid., 111ff).

54 The institution has set aside a network of related natural areas enhancing biodiversity. The vision of the biotope network relates back to the 1992 Rio Conference and the Habitats Directive. Cf. Radkau, Joachim, *Natur und Macht: Eine Weltgeschichte der Umwelt* (München: C.H. Beck, 2002). In the Netherlands, the Ministry for Economy, Agriculture and Innovation has earmarked around 160 Natura2000 areas, which together constitute a surface area of about 1.1 million hectares, among which around 2000 hectares are located on the island of Ameland, where the dunes and the North Sea zones have been included in the programme from 2009 onward. Cf. Regiegroep Natura2000, accessed January 27, 2014. <http://natura2000.nl>.

ecological data. Thus, the counting of species in a given habitat area amounts to an assessment on the basis of which to create an ecological system of reference or an ideal ecological state—a fixed point in time that, from the point of view of biology, can be considered good in terms of the diversity of species. The politics of biodiversity attempts to return to this ideal state or at least to approach similar levels of biodiversity. While dynamic coastal protection seeks to protect certain processes, conservation measures attempt to reconstruct specific points in time from an ecological perspective.

It was striking that the relation with nature was often spoken about on Ameland as a lawful relation—both with regard to the Zandhaaken and to the conservation measures. Specialist turns of expression and contexts such as "red list varieties"[55] or "Article 20" were widely used in everyday conversations. This means that the people of Ameland were generally familiar with the way in which conservation shaped their surroundings and that this was a part of everyday experience, especially where the observance of restrictions relating to nature protection was concerned. The so-called "Article 20", for instance, was a paragraph in nature protection law that empowered "regional government to limit access to a natural monument [...], a Natura2000 area or part of such an area insofar as this is necessary to the protection of natural properties."[56] Nature is thus perceived as a powerful contractual partner: Temporal structures that are preserved through dynamic coastal maintenance and nature protection oppose those of the actors affected, such as members of the citizens' association, with a real decisiveness based in law and, therefore, with a "solid facticity".[57]

On Ameland and on the other marine islands in Wadden, nature is perceived as a powerful contractual partner, something related in large part to the region's natural spatial pre-conditions. As such, the Wadden area is highly valued on an international level as a littoral zone rich in species, as reflected at numerous conventions on animal and plant protection as well as in the awarding of the title of Natural World Heritage site in 2009. At the level of the national political sphere, the Wadden area was made a nature protection area in the

55 A list of endangered animal and plant species created by the Ministry for Farming, Nature and Food Quality. Birds receive special attention on Ameland. There are two bird protection institutions to be found on the island: Vogelwacht Nes-Buren and Hollum-Ballum.

56 Environmental Protection Law of the year 1998 (<http://wetten.overheid.nl/BWBR0000641/geldigheidsdatum_28-02-2013#HoofdstukIII>).

57 Rosa, Hartmut, *Beschleunigung: Die Veränderung der Zeitstrukturen in der Moderne* (Frankfurt a.M.: Suhrkamp, 2005), 15.

1970s.[58] Today, the region is considered *the* nature reserve of the Netherlands. On this basis, demands are made on the region that are no longer realisable in the highly urbanised and densely populated western provinces, such as those relating to open spaces, absolute darkness and an open horizon. The economically highly lucrative exploitation of gas is also subject to the primacy of nature. This is noticeable locally in "compensation" provided in the form of nature restoration projects and associated finance and research programmes. The history of water management in the delta region of the Netherlands is to a great extent related to intensive human intervention in the natural spatial surroundings, something that has been reflected upon more critically in the contemporary Dutch public sphere. As one of the few remaining areas that has yet to be densely populated, the Wadden area constitutes, especially in the view of a network of conservationists and scientists, a kind of natural spatial instance of the memory of a landscape that has remained dynamic.

Field research further shows how the approach to dynamic, integrated coastal maintenance can be received in principle in a populistic manner, too. Thus, during the provincial elections taking place at the time, an Ameland member of the populist rightwing Party for Freedom (Partij voor de Vrijheid, PVV) attempted to use for his own purposes the situation surrounding the Zandhaaken or, rather, the general atmosphere as determined by anger, disappointment, and insecurity. In his opinion, this situation was "a very important theme for the PVV", confirming in the process his commitment to Ballum's "economic development" and not to the "nature people" and their "leftist hobbies". In the same way, he also tried to win over the 'Ballum bathing friends' to his cause, even though they would never have considered co-operation with the local PVV. It could be observed how the local PVV member tailored his rhetoric and arguments to the PVV on the state level. At this particular time, the party had already shifted its focus away from the threat that Islamism posed to that posed by Europe. Eurosceptic statements are thus related to an antipathy felt toward so-called leftist greens and the rhetoric of social inequality between the north and west of the country. In a Wadden community such as Ameland, this leads to a scenario in which European nature protection regulations like Natura2000 or a case like that of the Ballum Zandhaaken can potentially be reconstrued as Brussels imposing its will. Nonetheless, during parliamentary elections in 2010, Geert Wilders' party received only 12.6 percent of the votes

58 Oosterveld, Hendrik R., "Een Wereld van Verschill. Veertig Jaar Beleid in het Waddengebied" (paper presented at the fourth symposium "Waddenacademie KNAW," Leeuwarden, June 6, 2010), accessed April 5, 2017. <http://www.waddenacademie.nl/file admin/inhoud/pdf/04-bibliotheek/Een_wereld_van_verschil_web.pdf>.

cast on Ameland (countrywide 15.4 percent) and during the provincial elections in 2011, only 12.4 percent (countrywide 12.4 percent).[59]

Hardly a year after being grounded, the 'bathing friends' were to their great surprise permitted by the RWS at a regional level to alter Ballum's beach as "natural processes there would have indicated". The second condition was to establish a consensus on a local level among nature protection actors and to receive permission from the Ministry for Economy, Agriculture and Innovation. Given that the citizens' association feared that permission could be retracted—a fear indicative of the way in which power relations were perceived—they immediately sought a meeting with local representatives of both bird protection societies and the leader of the Environmental Working Group of Ameland. Two measures were of paramount importance to citizens' associations: Firstly, the section of beach between Ballum beach pavilion and the inland lake should be topped up with clean sand so as to provide a clean beach surface. Secondly, a dam should be built extending from the newly created beach surface to the Zandhaaken so that tourists could reach the beach without getting their feet wet. In view of the white sand and the surf, the conditions at the location were in their view optimal for the 'bathers'.

The project was in fact realised only a few days after the RWS agreeing in principle and establishing consensus on a local level, such that the three-day alteration of Ballum resembled a folk festival. Around thirty volunteers from the village of Ballum, with a population of 400, spent their entire weekend on dump trucks, diggers, cranes, tractors, and bulldozers. Thus, within the shortest of timespans, thousands of cubic metres of sand were loaded, deposited in layers and rolled out. The number of volunteers increased constantly—at least one hobby builder was to be in found in every family on Ameland, along with one or the other agricultural vehicle. In communication, the Ballum actors thus represented the work on the beach as "civil disobedience" in the face of "the green mafia". In their opinion, it was in particular the scientific group of geomorphologists, as well as nature protection groups such as the Wadden Society,[60] that were responsible for the strained relations concerning their beach. Accordingly, their spatial restructuring work makes clear that they

59 All election results are taken from the webpages of the "Kiesraad" (<http://www.verkiezingsuitslagen.nl/>).

60 A non-governmental organisation, Wadden Society is inseparable from the development of the Wadden area as a nature reserve, since it was born of the protest movement against the agricultural diking plan. With around 45,000 society members, it is one of the few conservation organisations in the Netherlands that can demonstrate a rise in membership.

invested their efforts in overwriting with a different spatial aesthetic the geomorphological logic of "leaving nature to take its course", which had materialised in connection with Ballum beach in the form of the Zandhaaken. The comparison of aerial photographs of the Ballum section of beach *before* and *after* the reconstruction is, in this respect, particularly suggestive. The direct comparison suggests a kind of iconoclasm. While previously curved lines dominate—in the form of the meandering tidal creek and the edges of the inland lake—the visual impression *after* the intervention is by contrast quite different in that it is now *straight* lines that come to the fore: the dam construction can be easily recognised, which follows a perfectly straight line through the inland lake. In turn, its banks were straightened on one side and made into the outer limit of a perpendicular area of the sandy beach created (Figures 9.1[61] and 9.2[62]):

The images make visible two different understandings of nature: One that is not interventionist and in which *nature* is represented as the formative influence, and another that is interventionist, where the *human* emerges as the shaping actor.

Even if the intervention of volunteers on Ballum beach was represented in communication as "civil disobedience," this is not an assertion that can be accepted unconditionally: One has to speak of a moderate oppositional action, if one understands the intervention as resistance against the primacy of a geomorphological conception of time situated in space. In addition, the volunteers did not await official permission from the ministry responsible. The rhetorical staging of "civil disobedience" against the "geomorphologists", those wearing "green-tinted glasses" and "nature clubs such as the Wadden Society" implies a perceived power divide between the volunteers and those conservation-sector actors in possession of geomorphological, biological expertise.[63] Resistance

61 Photo © Ina Nagtegaal.

62 Photo © Ina Nagtegaal.

63 Chapter 4 of my dissertation describes how scientific and conservation actors in the Wadden area can look back on a decade's worth of institutionalised cooperation and how compatible their expertise is. The network makes use of the possibilities that (discourse-) specific temporal and local structures have to offer, both where responding to the accelerated rise in sea-level with a sustainable level of security in mind and where the Wadden area as a dynamic area is concerned. This in order to invest their initial concerns with new momentum through the redynamisation of nature. Conditions allow the actor network to relate concepts of coastal security and nature development to one another. However, the extent of the influence that such networks carry is of a moral rather than a political nature. At this point in time, the Netherlands found itself not only in an economic but also a domestic political crisis, which considerably lessened the chances of nature

FIGURE 9.1
Ballum beach before the intervention

FIGURE 9.2 *Ballum beach after the intervention*

was exercised in the face of a politics of time received as power politics, such that one attempted to overwrite space with an alternative concept of time. Thus, the Ballum volunteers succeeded in overwriting the processual nature of the area (the "intertidal beach" "Wattstrand") and, with regard to its material make-up, completing the transition to an ensemble of conditions (concerning the beach surface, the dam, etc.). The alteration works on the beach thus effected the acceleration of natural processes. The machines fulfilled a

protection and economic actors turning their ideas into reality. Cf. Schmitt, *Klimawandel, Raumwandel, Zeitwandel*.

function of time, in that they altered space so that it was as if the sand-fill and sedimentation processes of the Zandhaaken had already taken effect. The acceleration of geomorphological time served to fill the semantic vacuum that the citizens of Ballum had associated with the former condition of the landscape.[64] However, even if a "beach reconstruction" was spoken of, the bathing beach that had previously existed was not entirely reconstructed—to which end the inland lake would have had to have been replaced by sand too—but only worked upon for so long as it took to reach the point at which it once again became meaningful to the people of Ballum. Thus the alterations were spoken of as the building of "Ballum adventure beach", on which the activities typical of a sand beach would once again be possible, just as one would also be able to catch shrimps in an inland lake or tidal creek.[65] From a grammatical point of view, with the "adventure beach", the temporal mood of the future tense (as opposed to the future perfect) entered into the picture once again: A temporal dimension that was temporally less remote from the lifeworld of the people of Ballum and that was more conducive to concrete negotiations and planning than the recollected future based on prognoses concerning geomorphological processes.

Again, observed from another perspective, it will not do to speak of resistance. For the people of Ballum were only able to wind time on temporarily. Just a few days after the reconstruction of the "bathing beach", storms led to the dam breaking and the majority of the sand distributed on the beach being blown away. The brown, damp sand once again appeared on the surface. The temporal regime was not broken but only temporarily disrupted. That none of this disrupted the positive atmosphere among the people of Ballum following the completion of the alteration works and that the citizens' association took this development in their stride shows that the symbolic act of reconstructing the beach was clearly more important than its material results and their maintenance. Further, a closer look at the course that events took shows that the material restructuring of Ballum beach did not constitute a breach but, paradoxically, confirmation of prevailing temporal norms: the alterations were carried out in accordance with a framework of design that natural processes had already indicated. Thus, at the point at which the dam was built, a process of sand filling could already be observed. As a result, the "reconstruction of the bathing beach" did not sideline the contours of natural processes so much as it highlighted them. As such, the opposition paradoxically effected some degree

64 An example of this is the reference to the inland lake as a "hole" or "stinking hole", which expresses in a visually striking way a lack of meaning.

65 See N.N., *Amelander Persbureau*, 07 May 2011, which exemplifies this phenomena.

of corroboration of existing power relations.[66] The communicative and performative staging of the beach alterations as opposition and civil disobedience in the face of the "geomorphologists" and "nature clubs" as well as the haste with which the volunteers settled on the day point to the Ballum residents' consciousness of being seen: they showed themselves conscious of the circumstances in which the island and the surrounding Wadden region found itself, that is, under the observation and the influence of a variety of influential actors, even if these were not physically there, on location. The RWS group of expert geomorphologists thus played an outstanding role, since, with the help of this specialist knowledge, the coastal protection authorities made and legitimated decisions concerning temporal order (real time processes and recollected futures) that resulted in associated spatial situations.

The conflict surrounding the Zandhaaken cannot therefore be considered in isolation from events in the small town of Ter Heijde on the coast of southern Holland. Here, at the same time, RWS carried out a globally unique pilot project involving a so-called "sand motor" and promoted this as an innovative form of spatial adaptation to climate change.[67] To this end, the coastal protection authorities allowed around twenty million cubic meters of sand to be distributed on the coast before the tiny town. This artificially created reef was, with the effects of water and wind currents over the years, supposed to break up along the coast and provide the coastal foundation with more sand.[68] The "sand motor" technique distinguished itself from the customary sand spraying in that a large volume of sand was not distributed along the coast by the use of a ship but through the power of wind and wave. The World Wide Fund For Nature (WWF) considered an eco-engineering technology such as that of the "sand motor" as a "natural approach to providing a solution" and, as such, as a suitable spatial adaptation strategy in the face of climate change, particularly in the delta region.[69]

66 This also explains why the Wadden Society did not comment all that much on the course that events on Ameland took. Of course, one would have much preferred to see "nature be given free rein"; on the other hand, the alterations were not considered to have had much impact on "natural processes". The so-called "green mafia" did not therefore campaign for the alterations to be reversed—as the people of Ballum had feared—since they could see no cause for doing so.

67 Cf. Provincie Zuid-Holland, *De Zandmotor: Van Zand naar Land. The Sand Engine: From Sand to Land*, published 2009, accessed January 27, 2014. <http://www.dezandmotor.nl/uploads/2011/03/090525-brochure-zandmotor-1.pdf>.

68 Cf. ibid., 9.

69 Cf. ibid., 13.

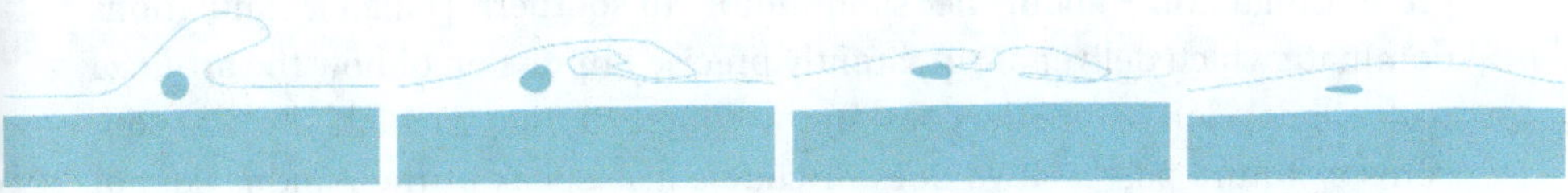

FIGURE 9.3 *The phases of "sand motors", as forecast by Rijkswaterstaat in the beginning, after five, after ten, and after twenty years*

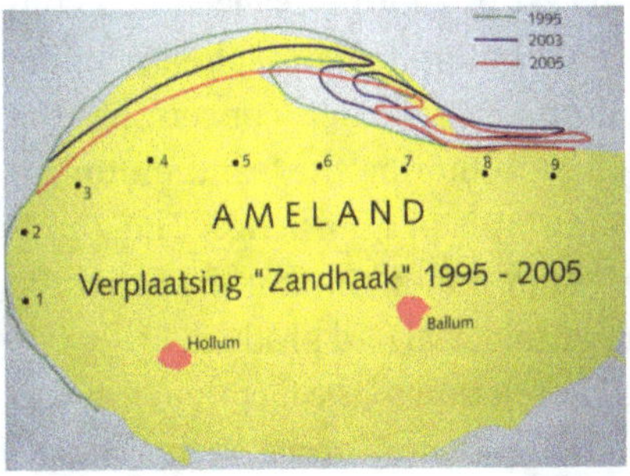

FIGURE 9.4
Comparative representation of the development of the "Zandhaaken" at Ballum from 1995 to 2005

While the "sand motor" at Ter Heijde did, like an artificially created reef, resemble "natural processes", the measures taken on Ameland with the Zandhaaken already achieved the results shown in the following diagrammatic representations, which Rijkswaterstaat used to explain the course that the "sand motor" and the Zandhaaken took (figures 9.3[70] and 9.4[71]).

The central difference between the "sand motor" in southern Holland and the Zandhaaken at Ballum lies in the fact that on Ameland it was a case of a self-initiated mechanism that could be controlled only to a limited extent, whereas at Ter Heijde, it was Rijkswaterstaat that determined the situation of the motor. A further difference can be noted with regard to communications on the part of RWS: while the coastal protection authorities used the term "sand motor" at Ter Heijde, and described this as an innovative spatial adaptation measure, they spoke of the Zandhaaken of Ameland as a "natural phenomenon". The ways in which the situations were communicated at Ballum and at Ter Heijde also clearly distinguish themselves in another respect: In a

70 Courtesy of: Provincie Zuid-Holland, and Ministerie van Verkeer en Waterstaat, *De Zandmotor Delta Duin*, published 2011, accessed January 27, 2014, 1. <http://www.wshvh.nl/downloads/21-4_factsheetsZandmotor.pdf>.

71 Courtesy of: Ministerie van Verkeer en Waterstaat, ed. *Het Natuurfenomeen van Ballum*, Informationsblatt, published 2006a.

promotional film[72] about the "sand motor" in southern Holland, simulations dominate which deliver an apparently precise impression of how the artificial reef will develop before the coast in the course of time. As such, the film concerns a "future past".[73] RWS does of course also speak of the Zandhaaken of Ameland as if it were a form of such a recollected future, but thus emphasises above all the *openness* of the situation and the future, as conveyed in 2006 in an associated information leaflet it distributed on the island.[74] The first case concerns the mistaken certainty of the forecast of a future realized in a controlled manner; the second, a description of a geomorphological phenomenon. The central difference between the information leaflet concerning the Zandhaaken of Ameland (2006) and the film promoting the "sand motor" in southern Holland (2009) is thus the varying status of knowledge. The Zandhaaken represents an *earlier* phase of the "sand motor," in that the scientific expertise is far less advanced than in the technological end product.

Such an observation must be considered in relation to the fact that, from an economic perspective, water is considered a "top sector" in the Netherlands[75] and innovative eco-engineering pilot projects such as that involving the "sand motor" touch in turn upon fundamental research. Phenomena such as the Zandhaaken of Ballum are the subject of ground-breaking experiments that are conducted on an (inter)national level as part of essential research in the area of eco-engineering in order to establish trends that can provide solutions and, where relevant—as in the case of the "sand motor"—turn these into lucrative articles for export worldwide.[76] In this context, the Wadden region, because of its natural spatial characteristics as a dynamic and large-scale tidal area, presents a highly attractive laboratory. The knowledge generated there is, for the Netherlands, an important economic good. The geomorphological research concerning the sand bank washed up at Ballum is an essential precondition for the development of a delta technology such as the "sand motor"

72 (cf. Rijksoverheid 2011)

73 Welzer, Harald, "Vergangene Zukünfte und zukünftige Vergangenheiten," in *Erzählte Zukunft: Zur inter- und intragenerationellen Aushandlung von Erwartungen*, ed. Jens Kroh and Sophie Neuenkirch (Göttingen: Wallstein, 2011), 19-26.

74 Ministerie van Verkeer en Waterstaat, *Het Natuurfenomeen van Ballum*.

75 Regarding which, more can be found here: <http://www.rijksoverheid.nl/onderwerpen/ondernemersklimaat-en-innovatie/investeren-in-topsectoren/water>.

76 Thus, Vietnam is a more important trading partner of the Netherlands in the highly lucrative water sector, which, among others, concerns the areas of delta technology, water management and climate (<http://www.rijksoverheid.nl/ministeries/eleni/nieuws/2011/03/28/nederlandse-regering-opent-deuren-in-vietnam-voor-nederlands-bedrijfsleven.html>).

to be at all possible. For it *touches* on information that was *generated* with the help of natural phenomena.

From the perspective of the economy of knowledge, both are lucrative: valuable, fundamental research can be run on the basis of the Zandhaaken, whereas the "sand motor" represents knowledge that has been refined in a product for export. The various phases or states of knowledge are, drawing on Michel Foucault, commensurable as concerns the "capitalisation of time" and efforts to "organise profitable durations" as well as to "[d]ivide duration into successive or parallel segments, each of which must end at a specific time".[77] The temporal characteristics of both the "sand motor" as knowledge and the Zandhaaken as research assume their place in a value chain within Dutch knowledge society. Ameland turns out to be an important part of this chain, in that it represents for RWS a natural space that is also a large laboratory and for scientific actors a natural space that provides a structure for the development of adaptation technologies.[78]

4 Concluding Observations

My work concerns the sphere of governance processes relating to space, since the Dutch government bureaucracy envisages that the country's spatial structure should be made "climate-proof", to which end it considers cooperation between various actors essential. The point of departure of this investigation was to analyse realities relating to (cultural) space, since I assume that adaptation necessarily implies negotiations concerning various realities relating to space and that, in this respect, connections between governance, knowledge, power, and social dynamics as well as asymmetries concerning knowledge and

77 Foucault, Michel, *Discipline and Punish: The Birth of the Prison*, trans. Alan Sheridan (London: Vintage Books, 1995), 202f.

78 My dissertation shows how Ameland as a natural space represents in several respects a significant open laboratory for the Dutch in providing the basis not only for resilience research but also for vulnerability research. The latter is possible since on the east of the island, gas exploitation takes place leading to subsidence. This is of advantage to researchers looking at the simulated effects of the accelerated rise in sea level that is anticipated in the future. For researchers, eastern Ameland thus represents a "future memory" (Zukunftsgedächtnis)—cf. Welzer, "Vergangene Zukünfte und zukünftige Vergangenheiten," with the help of which information can already be gathered today on the effects of a future rise in sea level of up to 1.3 meters every century. This accords approximately with the negative scenario, on the basis of which the Delta Commision is currently working (cf. Deltacommissie, *Samen werken met Water*, 25).

power become apparent, such that the disruption of a shared perspective on actions taken can be explained. In this respect, it is worth treating the conflict on Ameland concerning the Ballum Zandhaaken not as a local trifle but emphasising instead the exemplary and structural aspects of this case as well as their import for local governance processes relating to climate change that are dependent on co-operation between various actors:

The case study of Ameland makes clear that both the situation concerning the Zandhaaken and the constellation of actors were glocal in nature: There was an overlapping of discourses, actors and power relations that relate to various spatial levels. The challenge for an analysis of the general cooperative process lies in having to observe the problem's setting and the constellations locked in conflict as being spatially grounded yet, at the same time, embedded in multiple complex structures. This becomes particularly pressing in the Wadden Sea region, in which a particularly diverse set of actors is a permanent characteristic.

In order to be able to make statements concerning the significance of spatial adaptation measures taken in response to climate change and of the greater emphasis placed on the role of nature's self-organization, it is essential to contextualise the technologies involved together with the topic of climate change in relation to the sociocultural and socio-spatial history of the island and the Wadden Sea region, as well as that of the actors and associated conflicts. Thus the citizens of the Wadden Sea region gained their first experience of conflict with the network of science and conservation that emerged since the rise of the environmental movement and the accompanying rules of conservation legislation. The relations between the actors are thus marked by the way in which the conflict unfolds, as became clear in the controversy surrounding the Zandhaaken. Added to which, historical factors relating to the region must also be taken into consideration—the Wadden area as a nature reserve and research area of (inter)national significance—as well as the historical background concerning the actors: Including the fact that the science and conservation network in the Wadden area can draw on a long period of cooperation and that the associated expertise connects very well. The suggestion for adaptation from the New Delta Commission, namely to work together with water, means the region around the Wadden Sea islands relying to a greater extent on geomorphological expertise. So, the existing controversy surrounding dynamic coastal protection will continue. Adaptations can thus lead to the re-opening of existing controversies—on the level of content as well as on that of relations between actors.

In addition to such a locally and regionally specific contextualisation, attention must at the same time be paid to an embedding in supra-local networks and context. The case study made the relevance of the knowledge economy clear in relation to the dynamic littoral landscape. This landscape is significant for science as a large open laboratory in the form of a natural space, in which important fundamental research can be conducted that generates innovative adaptation technologies, which can then be exported worldwide. With the help of an ethnographic approach, it is possible to represent the discourse on climate change as it is refracted at a local level: The Ballum case study shows how the topic of climate change transforms, that is, how it is expressed, whether explicitly or only very indirectly, according to the perspective of each actor(s' group). In this regard, it is of central importance to establish which concrete life realities and socially shared knowledge systems are touched upon by an approach to adaptation policy in the spirit of "working together with water". On Ameland, it was shown that anti-scientific reflexes rooted in a longer history of conflict (scientifically informed nature and coastal protection that had led to constraints and the loss of jobs) could in principle connect with the populist political currents of a party such as the PVV, which increasingly wishes to promote itself as assuming an anti-European and anti-Brussels position.

Furthermore, the case study of the Zandhaaken at Ballum makes clear that cultural transition does not come about as a result of the shift from fighting against the water to "working together with water", but in the accompanying change in power relations between the actor groups involved. On Ameland, a temporal confusion among the actor groups bears commenting upon, which stems from the various frameworks that underpin their expectations and actions. While the citizens' association sees itself as being robbed of its future, adaptations to climate change mean security in the face of the effects of climate change when considered from the point of view of science and policy. As regards the latter, we are however concerned here with a logic in the spirit of a future past—in accordance with the adage, "We will *have* built with nature and, thus, *have* adapted to climate change." This is about a temporal mood in accordance with which, at first, the immediate future is paradoxically allowed to become somewhat vague and the present is relegated to a state of transition that, in turn, can have a negative effect on the courses of action available to those affected by these measures. Thus it feels as if Ballum is caught up in a long-lasting, real-time process. This problem becomes more complex against the background of the existing temporal structure of the Wadden and North Sea landscape that has resulted from conservational regulations. The temporal logic (allowing natural processes their *time*) established with the spatial logic (allowing natural processes their *space*) is perceived locally as power politics. As for adaptations and resilient constructions, these therefore have to do with decisions concerning normative spatial and temporal orders, in accordance with which those affected must coordinate themselves and suppress other ways of relating to reality. The field research showed that the resistance of some Ameland people to spatial adaptation to the effects of climate change touched upon competing understandings of space and time, and of reality. Physical changes in the material surroundings go hand in hand with change in existing social power relations, against which one defends oneself. As such, the question as to whose interpretation reigns over the area is central for the actors involved.

The political sphere's attempts to counter climate change with changes in spatial structure raise questions about the processes surrounding legitimation. The successful application of adaptation strategies is dependent upon a certain level of consensus at a local level. Dutch political culture traditionally allows civil society actors considerable influence.[79] The "consensus in the polder"[80] is understood today as a principle that is supportive of the state. As far as political attempts at spatial transformation are concerned, certain aspects of governance play a major role, that is, "the forms and methods of coordination among more or less autonomous actors, whose actions are interdependent, are thus capable of mutually influencing or supporting one another."[81] As such, I understand "governance" not as a normative but as an analytical and descriptive concept that contributes to the manner in which hierarchical and statist forms of political control (government) are increasingly dependent on the participation of non-state actors (governance).[82] Since the Dutch ministerial bureaucracy proposes that the country's spatial structures should be made more "climate-proof," and therefore considers the cooperation of various actors necessary, I investigated the processes of "local

79 Papadopoulos, Yannis, "Governance und Demokratie," in *Governance – Regieren in komplexen Regelsystemen: Eine Einführung*, ed. Arthur Benz (Wiesbaden: vs Verlag für Sozialwissenschaften, 2004), 217.

80 Wielenga, Friso, "Konsens im Polder: Politik und politische Kultur in den Niederlanden nach 1945," in *Länderbericht Niederlande: Geschichte – Wirtschaft – Gesellschaft*, ed. Friso Wielenga und Ilona Taute (Bonn: Bundeszentrale für Politische Bildung, 2004), 13-130.

81 Benz, Arthur et al., "Einleitung," in *Handbuch Governance: Theoretische Grundlagen und empirische Anwendungsfelder*, ed. Arthur Benz et al. (Wiesbaden: vs Verlag für Sozialwissenschaften, 2007), 9.

82 Cf. Kilper, "Governance und die soziale Konstruktion von Räumen," 43-60; see also Benz, Arthur et al., "Einleitung," 14f; Nohlen, Dieter, and Rainer-Olaf Schultze, ed., *Lexikon der Politikwissenschaft* (München: C.H. Beck, 2005), 324.

climate change governance"[83] or, in other words, governance processes to which spatial considerations are central.[84] As has been seen, reality constructs are based on both spatial constructs and, to a significant degree, on conceptions of time: not only do they emerge in a given span of time but they are also based to a large extent on temporal constructs that are both normative and relative.

That the level of trust in the political authorities among Dutch citizens has in recent years declined further impedes the search for acceptance and legitimation.[85] "Political discontent" is widespread in Dutch society.[86] These kinds of developments and prognoses must in turn be interpreted against the background of a continuous decline in citizens' trust in the political institutions of western democracies ever since the 1970s.[87]

Additionally, climate change is effectively the "wicked [problem] *par excellence*"[88]—that is, a problem which is, both scientifically and politically, not only peculiarly difficult to penetrate and get a handle on but, in addition, highly contingent on developments in all sectors and at all levels. Given the

83 Cf. Pohlmann, Angela, "Local Climate Change Governance," in *Global Transformations Towards a Low Carbon Society* (Working Paper), ed. Anita Engels (Hamburg: University of Hamburg/KlimaCampus, 2011).

84 Cf. Kilper, "Governance und die soziale Konstruktion von Räumen." Angela Pohlmann offers an overview of the literature in the social sciences concerning local governance processes in the context of climate change (cf. Pohlmann, "Local Climate Change Governance"). She defines "climate change governance" as a "process including different stakeholders who, in some form of participatory agreement, deal with the causes and consequences of climate change" (ibid., 5).

85 Cf. Wielenga, Friso, "Grundzüge niederländischer Politik im 20. Jahrhundert," in *Nachbar Niederlande: Eine landeskundliche Einführung*, ed. Friso Wielenga (Münster: Aschendorff, 2007), 58.

86 Dekker, Paul, and Josje den Ridder, "Politisches Unbehagen in den Niederlanden: Wie sich das kleine liberale und tolerante Land verändert hat," *INDES – Zeitschrift für Politik und Gesellschaft* (2011): 46-55.

87 Cf. Newton, Kenneth, and Pippa Norris, "Confidence in Public Institutions: Faith Culture or Performance?" in *Disaffected Democracies: What's Troubling the Trilateral Countries?*, ed. Susan J. Pharr and Robert D. Putnam (Princeton: Princeton University Press, 2000); Dalton, Russell J., *Democratic Challenges, Democratic Choices: The Erosion of Political Support in Advanced Industrial Democracies* (Oxford: Oxford University Press, 2004).

88 Jordan, Andrew, Dave Huitema, and Harro van Asselt, "Climate Change Policy in the European Union: An Introduction," in *Climate Change Policy in the European Union: Confronting the Dilemmas of Mitigation and Adaptation?*, ed. Andrew Jordan, Dave Huitema, Harro van Asselt, Tim Rayner and Frans Berkhout (Cambridge: Cambridge University Press, 2010), 4.

great instability and complexity of the problem, more and more expertise from the natural sciences has to be taken into account during the course of political decision-making processes, in the hope of reducing diverse uncertainties. The topic of climate change is thus a particularly fertile case of the "scientification of politics",[89] which is ambivalent in terms of democratic legitimacy: from the perspective of an understanding of democracy oriented toward outputs, experts can indeed be considered legitimate actors since they can, on account of their knowledge, contribute in certain circumstances to enhanced decision-making. However—and now comes the input-oriented perspective—these experts cannot be considered democratically legitimate, that is, they are not elected actors who can be called to account for their actions by, for example, being voted out of office.[90] The policy of dealing with climate change through spatial change is exemplary of an orientation and an understanding of legitimation based on output. "The government [takes] responsibility for its citizens through its good policy that, far from having to be related back to democratic processes, is in fact often divorced from these."[91] Drawing upon Colin Crouch's concept of "post-democracy",[92] political scientist Gary Schaal argues that normative ways of dealing with citizens' preferences have changed too: Bringing the "post-democratic actors", that is, political and economic elites, into play has increased in importance. This is justified with the argument that "in the context of a political reality perceived as hyper-complex, citizens may well possess opinions on political matters but lack a sufficient level of knowledge in order to be able to respond adequately to the situation".[93] As such, post-democratic actors can also be considered scientific experts, without whom a political response to climate change (such as the tighter integration of planning as concerns water and spatial planning) would not be thinkable. Modern politics is no longer conceivable without the incorporation of such expert knowledge. Whereas it was primarily engineers who were represented in the First Delta

89 Kopfmüller, Jürgen, "Einführung," in *Risiko Klima: Der Treibhauseffekt als Herausforderung für Wissenschaft und Politik*, ed. Jürgen Kopfmüller und Reinhard Coenen (Frankfurt a.M.: Campus, 1997), 12.

90 Cf. Benz, Arthur, and Yannis Papadopoulos, "Actors, Institutions and Democratic Governance: Comparing Across Levels," in *Governance and Democracy: Comparing National, European and International Experiences*, ed. Arthur Benz und Yannis Papadopoulos (London: Routledge, 2006), 277.

91 Schaal, Gary P., and and Claudia Ritzi, "Politische Führung in der 'Postdemokratie'," *Aus Politik und Zeitgeschichte* 2-3 (2010): 11.

92 Cf. Crouch, Colin, *Post-Democracy* (Cambridge: Polity, 2004).

93 Schaal, Gary P., and Claudia Ritzi, "Politische Führung in der 'Postdemokratie'," *Aus Politik und Zeitgeschichte* 2-3 (2010): 9-14, 13.

Commission in 1953, it was ecologists, economists, and expert planners in particular who found themselves in the new commission of 2007. The incorporation of non-state actors such as experts[94] is far from unproblematic from the point of view of representative democracy and can exasperate efforts to establish consensus at a local level, since citizenry is confronted with a policy based to a large degree on scientific findings that it cannot fully grasp. Those affected by associated measures may well understand this as the expression of an elite form of rule. At a local level, it is not only acceptance of the political project *per se* that has to be found but, at the same time, also of the scientific background from which the project emerged, and the implications (sometimes in competition with one another) that this scientific background has for understanding the reality of the situation.

All in all, adaptations must be considered multi-level processes: The actors' understanding of the overall situation is based on different frameworks as regards value, perception and interpretation. This can cause difficulties in reaching agreement as well as for the "governance of adaptation",[95] the latter being of particular relevance in this context.[96] An approach to governance that takes into consideration cognitive theory and that is oriented towards processes can aid the analysis of resulting conflicts, particularly where such an approach treats knowledge "as both the object and the precondition of governance".[97] In following this logic, Benz et al. observe that:

> Analyses [would have to] pay far more attention to the cognitive dimension and establish whether and how the world pictures and interpretative frameworks of actors can be reconciled with one another. This would also suggest a stronger focus on the question of how a problem is perceived in

94 On which see the contribution by Papadopoulos, "Governance und Demokratie," who considers the democratic legitimacy of governance.

95 Biesbroeck, Robbert G., et al., "Europe Adapts to Climate Change: Comparing National Adaptation Strategies," *Global Environmental Change* 20.3 (2010): 440-450.

96 Termeer, Dewulf et al. also refer to "governance of wicked climate adaptation problems": Termeer, Katrien, Art Dewulf, and Gerard Breeman, "Governance of Wicked Climate Adaptation Problems," in *Climate Change Governance*, ed. Jörg Knieling and Walter Leal Filho (Berlin: Springer, 2013), 27-40.

97 Botzem, Sebastian et al., "Die Dynamik des Governance-Ansatzes: Vier Dimensionen von Wandel," in *Governance von und durch Wissen: Interdisziplinäre Tagung "Governance von und durch Wissen" (Berlin)*, ed. Gunnar Folke Schuppert and Andreas Voßkuhle (Baden-Baden: Nomos, 2008), 22. See also in this respect: Schuppert, Gunnar Folke, and Andreas Voßkuhle, ed., *Governance von und durch Wissen: Interdisziplinäre Tagung "Governance von und durch Wissen" (Berlin)* (Baden-Baden: Nomos, 2008).

> governance analyses—if everyone involved has very different interpretations of the situation in which the problem arises and thus the need for coordination, then it is not surprising that attempts at collective regulation are not successful or do not even come to fruition in the first place.[98]

Since the case of Ameland concerns plans for spatial restructuring, in my analysis I paid special attention to space and to associated frameworks that the actors involved in the process of coordination draw upon as regards perception and interpretation, particularly where temporality is concerned. An analysis of actors' conceptions and actions as regards space leads to the recognition of how governance, knowledge, power, and social dynamics, as well as symmetries of knowledge and power interrelate, such that a shared perspective on the matter in hand is put to an end. I attended to processes that both require a coordinated, cooperative plan of action and that were prerequisite for the same. In this respect Gabriela Christmann uses the concept of "proto-governance":

> Proto-governance concerns the fundamental and much more wide-ranging, communicative processes from which commonly held interpretations of space emerge among groups, networks, and publics, along with dynamic spaces of culture and identity, in all of their ideal and material manifestations.[99]

Thus, a common knowledge of space or, in other words, shared life realities, can have a decisive influence on the course that collective attempts at regulation take. Here the point of departure was the consideration that a governance of adaptation cannot succeed if the associated policies fail to connect to a given lifeworld. The topic of climate change does not imply that actors share the same perception of associated problems. Instead, it would be worthwhile to take a step back and look at how knowledge of space is embedded in the lifeworld as well as culturally and temporally. The study takes the realities of a given (cultural) area into consideration when assuming that spatial adaptation involves the negotiation of various realities relating to space and that collective negotiations between actors can break off if the concrete realities of people's lives and the systemisation of shared knowledge conflict.

98 Benz et al., "Einleitung," 19.

99 Christmann, Gabriela B., "Kommunikative Raumkonstruktionen," 29.

5 Outlook

The Netherlands has assumed a leading position in the EU with respect to adaptation policy, as a result of their spatial approach to adaptation to the effects of climate change. In order to have an effect, however, this approach must also be applied. The example described here of the Zandhaaken at Ballum, which represents the raw mechanism of an eco-engineering technology, is but one example of a relevant process of governance which has caused conflict between the groups of actors involved and, from this point of view, is far from being a precondition suited to a cooperative course of action.

The case study of Ballum makes clear that the local citizens must come to terms with the far-reaching consequences of a phenomena such as the Zandhaaken: Rijkswaterstaat legitimises the measures taken off the coast of Ballum on the basis of the prevailing regime of coastal protection, the corollary of which was that the citizens of Ballum were faced with the challenge of having to cope in principle with long-lasting geomorphological real-time processes. However, that these extend considerably beyond the temporal horizons of lifetimes is cause enough for implicitly driving the local population to insecurity, anger, and resistance.

RWS, as well as actors in administrative and political spheres, dealt with the situation in Ballum in a thoroughly uncreative manner. Instead of rather shame-facedly granting the citizens' association special powers to alter the beach, it would have been a viable alternative to make a precedent of the situation at Ballum that would be relevant beyond the concrete, local situation.[100] In order for processes of adaptation to be successfully applied, a *political* solution would have been necessary in addition to the specifics of the technocratic solution—which, in this case, entailed allowing the Zandhaaken the necessary space and the necessary time to provide additional sediment for the coastal system. Here, a democratic and discursive form of the relevant adaptation is required in order to give due consideration to the different forms of legitimate knowledge that the actors involved possessed and to bring these actors into deliberative negotiations with one another.

100 To write down ethnographic observations also always means to opt for the description of a part of a process. Of course, the various subplots continue to unfold. Thus, in June 2012—when I was no longer on Ameland, but in Berlin—the "bathing friends" informed me that the work at Ballum would resume once more. However, it was only a 'small solution' that took effect in the end, since their temporal and spatial perspectives on the concrete situation in Ballum remained limited.

To this end, it would have been necessary to take an initial step toward transparently conveying the glocality of the situation, that is, its complexity: Thus, a natural phenomenon like the Zandhaaken, as observed in a dynamic littoral zone like the Wadden area, is of great significance for the Netherlands in several respects. It can assist in the generation of knowledge concerning the development of innovative adaptation technologies that are of service to the cultivation of resilience as well as a meaningful export product in the water sector, a central area for the Netherlands. It is at the same time clear that the raw natural mechanism of the Zandhaaken leads to difficulties for those affected locally. The local and supra-local as well as the positive and negative consequences of this natural phenomenon must therefore be considered in context in order to convey that one finds oneself in a specific dilemma: that which is beneficial for a country's wellbeing and future proves to be disadvantageous to a group of people affected on the local level.

A second step would have to be for the actors from political, institutional, and civil society spheres to deliberate with respect to the situation, such that learning processes take effect that can lead to the transformation of individual preferences oriented to the present into preferences relating to the common good that are oriented toward the future. Such a process would ideally lead to the development of a positive attitude to the situation on the part of those affected locally, that is, to the recognition that the Zandhaaken constituted a phenomenon that contributed to the common future of life in the Netherlands and elsewhere in the world. In this way, an enhanced consciousness could be inspired as to just what climate policy measures mean in terms of long-term thinking. At the same time, an associated way of going about things in the here and now could also enter into effect.

As for political and institutional actors, deliberative processes could, by contrast, create a requisite consciousness of the kinds of difficulties that can be experienced locally. Ideally, this would result in common efforts to find a provisional solution that would guarantee those affected locally a certain leeway without however losing sight of supra-local and future-oriented dimensions, given that the future has its beginning in the present.

A situation such as that found in Ballum could be used in order to enhance cooperation between the various groups of actors instead of leading to ever more entrenched conflict between them. As regards the approach to adaptation, it is necessary to complement a technocratic approach to problem-solving with suitable political formats. A successful 'governance of adaptation' implies not only the need for a technically specialised solution tailored to the inertia of a given location or region but also a flexible manner of implementation that does justice to the socio-cultural specificities of the situation.

PART 5

Prospects

CHAPTER 10

The Age of Uncertainty: the Challenges of Climate Change for the Insurance Business

Franz Mauelshagen

Abstract

We are living in an age of uncertainty: the dynamics of modern societies and their impacts on the global environment—anthropogenic climate change in particular—have changed our relationship with the future and the conditions for planning it. Uncertainty has undermined traditional prognostics based on the future of the past, with "nature" as the static scenery, towards which the drama of history was unfolding. The history of scenario planning after World War II indicates that the age of uncertainty began some time around 1970. It is a symptom of the Great Acceleration. This context helps us see the broader picture behind the new challenges climate change poses to the insurance business in Europe and elsewhere. Expanding insurance globally has been recommended as a crucial strategy of climate change adaptation. However, the European case shows that climate change poses unprecedented adaptive challenges to the insurance business itself in a part of the world, where it has been well established for a long time. The frequency and severity of meteorological and climatological hazards are changing. Thus, the uncertainties of climate change create new uncertainties for insurance, as the example of crop insurance in Switzerland will illustrate.

1 Introduction

Climate change has made our future uncertain in many ways. In the most extreme of imaginations, climate change even features as a threat to the survival of humanity. Though disliked by many, such apocalyptic scenarios may not be completely out of proportion; in fact, some such scenarios must be seen in context with a return of catastrophism in geology that is scientifically well founded in recent advances in geological knowledge that have changed our understanding of earth history.[1] Criticism that climate change advocates rely

1 See Mauelshagen, Franz, "Modernity's Decline? Climate Change, the Anthropocene and the Return of Catastrophism," in *Decline*, ed. William O'Reilly et al. (Budapest, New York NJ: forth-

 | DOI 10.1163/9789004356825_012

on apocalyptic scenarios to help their case and have little else to offer with regard to the future is definitely unjustified. As we shall see, early concerns about climate change and its potential effects were framed, almost from the beginning of the early 1970s, in terms of strategic planning. Faced with a bunch of uncertainties that surrounded the future of climate change, climate scientists started applying and advancing future scenarios to handle the unpredictable socio-economic development of societies and the emissions of greenhouse gases that would result from their use of various energy resources.

The post-World-War II era marks the beginning of a shift from traditional forms of forecasting based on probability calculations to future scenarios as a new tool to approach the future from the point of view of strategic decision making. It is a shift from risk to uncertainty, which is meaningful as a diagnosis of our time and its future, and in which rapid socio-economic change is, and will continue to be, paralleled by rapid environmental change. The first two sections of this chapter will explain this in greater detail, before insurance moves into the focus of attention. Insurance has been suggested as a means of adapting to climate change. However, in Europe, where insurance has a long history and is already well established, exploring the adaptive challenges climate change poses to that business seems just as important as considering new insurance schemes as a potential future path for climate change adaptation. To start with, I will consider the uncertainties climate change brings about for the insurance business in the context of a wider spectrum of uncertainties relevant to the insurance business in general. In part four I will then look at the case of crop insurance in Switzerland, and discuss the challenges climate change is already posing here, in the wider context of the history of adaptation that business went through over a period of a hundred and fifty years—adaptation that is to both its socio-political environment as well as to the meteorological hazard of hail.

---

coming 2017). In another article, I have explored the interferences between popular science fiction and scenarios of rapid climate change: see Mauelshagen, Franz, "Climate Catastrophism: The History of the Future of Climate Change," in *Historical Disasters in Context: Science, Religion, and Politics*, ed. Andrea Janku, Gerrit Jasper, and Franz Mauelshagen (New York, London: 2012). The excellent book Horn, Eva, *Zukunft als Katastrophe* (Frankfurt a.M.: S. Fischer, 2014), provides a much more elaborate analysis in terms of the narrative contents of science fiction.

2 Future's Past

It is a commonplace that the future is open and cannot be predicted. Historians tend to be satisfied with that translation of openness into unpredictability and base their scepticism towards any attempt to plan the future on such general grounds. The history of the twentieth century, in particular the fate of ideologically founded societal utopias within it, seems to support such scepticism. However, even beyond ideologies and utopias, modern societies have never been satisfied with leaving the future to fate. For them, recognising the general openness of the future as an essential part of the modern human condition did not at all mean that individual or collective desires for security and plannability had been rendered obsolete.

Quite the contrary! Future's fundamental openness created a new space for planning and organization—in other words: provision, prevention, preparedness, and other approaches to planning the future only emerged from it.[2] They were born from the expectation that certain threats, hazards, or risks were identifiable prior to their actual occurrence, and that it made sense to meet them proactively. In doing so, public as well as private institutions have learnt to apply a variety of predictive procedures founded on probability calculation. Return periods or frequencies of risk occurrence in the areas of public health or natural hazards provide ample examples for this.

The history of risk calculation goes back several centuries.[3] Its methods are nowadays so common, publicly and scientifically acknowledged, that we are little aware of the fact that they have developed into powerful tools for projecting our individual and collective futures. Healthcare systems, retirement provision, disaster control, and weather forecasts are all based on quantitative projections extending no more than a couple of decades into the future, often much less. Probability calculations play a key role in all of them, and in any

---

2 Koselleck, Reinhart, *Futures Past: On the Semantics of Historical Time* (Cambridge, Mass.: MIT Press, 1985), already made that very clear. According to Koselleck, the modern era saw the "reoccupation of a prophesied future by a predicted future," 21. Future visions and future management are recurring themes in Hölscher, Lucian, *Die Entdeckung der Zukunft* (Frankfurt a.M.: Fischer-Taschenbuch-Verlag, 1999).

3 The term risk (or Italian: *rischio*) was used in marine insurance already back in the Renaissance, mostly in Italy. Probability calculation is a later invention, from the seventeenth century. See Daston, Lorraine, *Classical Probability in the Enlightenment* (Princeton: Princeton University Press, 1988). The most extensive historical account on probability and the uses of statistics is Hald, Anders, *A History of Probability and Statistics and Their Applications before 1750* (New York: Wiley, 1990) and Hald, Anders, *A History of Mathematical Statistics from 1750 to 1930* (New York: Wiley, 1998).

case, they are based on statistical information on the past, most fundamentally: the frequency of a certain event or occurrence, such as an earthquake or a disease. Only historical information enables reliable projections of the expected occurrence of the same events in the future. In few brief words: Expectations about the future are generated from the past. Or in Reinhard Koselleck's words: "The experience of history, founded as it was on repeatability, bound prospective futures to the past".[4] In modernity, the rationality of probability calculations replaced astrological or prophetic schemes of predicting the future. Probability became the most powerful forecasting tool, applied almost universally if possible.

However, the future of the past appears to be guaranteed only in the context of metaphysical assumptions about cyclic models of history that are essentially pre-modern. This assumption is inherited from the circumstance that "risk" and "probability" circulated as ideas before modern nation states and their security systems emerged. That is, neither "risk" nor "probability" were genuine tools of the modern nation state, nor were modern institutions such as insurance companies (in fact, insurance was another early modern invention); they only established a special relationship with modernity due to the exclusion of other forms of prognosis.[5] At the same time, ideas of progress radically broke with the old metaphysics of historical cycles and, thus, produced conflict with the rationale underlying predicting the future based on probability. Koselleck wrote: "The future contained in this progress is characterized by two main features: first, the increasing speed with which it

4 See Koselleck, *Futures Past*, 21. Koselleck preferred examples of political decision making to illustrate that point. But insurance history provides many more examples. For references, see the next footnote.

5 The genealogy of both, insurance and reinsurance, reaches far beyond modernity. See, in particular, Zwierlein, Cornel, *Der gezähmte Prometheus. Feuer und Sicherheit zwischen Früher Neuzeit und Moderne* (Göttingen: Vandenhoeck & Ruprecht, 2011), 24-39 (state of research summary) and 45-72 (part B) on premodern insurance. Lengwiler, Martin, *Risikopolitik im Sozialstaat. Die Schweizerische Unfallversicherung 1870-1970* (Köln: Böhlau, 2006) is a seminal book in social security and insurance in the case of the Swiss modern state. More recently, insurance history has entered the stage of environmental history. See Rohland, Eleonora, *Sharing the Risk: Fire, Climate, and Disaster: Swiss Re 1864-1906* (Lancaster: Crucible Books, 2011) on reinsurance, Lübken, Uwe, *Die Natur Der Gefahr. Überschwemmungen am Ohio River im 19. und 20. Jahrhundert* (Göttingen: Vandenhoeck & Ruprecht, 2014), 278-281, on the U.S. *Flood Insurance Program*, and Mauelshagen, Franz, "Sharing the Risk of Hail: Insurance, Reinsurance and the Variability of Hailstorms in Switzerland, 1880-1932," *Environment & History* 17.1 (2011): 171-191, as well as Mauelshagen, Franz, "Die Vergesellschaftung Des Hagelrisikos: Zur Geschichte der Landwirtschaftlichen Hagelversicherung in der Schweiz, 1818-1950, " *Traverse* 2014.3 (2014), 60-71, on hail insurance in Switzerland.

approaches us, its unknown quality. 'Unknown' because this accelerated time, i.e., our history, abbreviated the space of experiences, robbed them of their constancy, and continually brought into play new, unknown factors, so that even the actuality or complexity of these unknown quantities could not be ascertained."[6]

In other words, the dynamics of change that characterize modern societies undermine the future of the past, at least potentially. Seen from the perspective of the dynamics that unfolded with industrialization, the incommensurability of historical processes is no longer a given, quasi-metaphysical truth, but itself the result of a historical process of acceleration that escapes from the historical cycles produced by previous agrarian civilizations. Another consequence of that acceleration is a contraction of what Koselleck called the "space of experiences", which generates expectations of the future. To describe the effect, one could speak of a de-historicization of the future, which also implies a de-historicization of provision, care, and prevention. When outpaced by change, the past loses its relevance for the future, and that is also the moment when future planning requires reorientation.

3 The Age of Uncertainty

Change undermining the validity of futures past is more than just a good theory. It helps to describe the circumstances that gave birth to scenario planning after World War II. Scenarios were first used to inform military strategic planning in the Cold War. Then, from the 1970s, it was applied in two other emerging areas of serious future concern: energy security and climate change. In all three contexts, the developers and users of scenario planning felt that new uncertainties questioned the applicability of traditional calculations of probability and risk.[7]

The threat of nuclear war and a potential self-annihilation of humanity was clearly beyond any historical experience of conventional warfare. Assessing the consequences of nuclear conflict required dealing with uncertainties about the enemy's nuclear arsenal and strategy, as well as the impact the actual use of nuclear weapons would have had on populations and on the

---

6 Koselleck, *Futures Past*, 22.

7 This section is based on a more comprehensive account of the history of future scenarios after World War II: Mauelshagen, Franz, "Das Zeitalter Der Ungewissheit: Zukunftsszenarien Und Globale Bedrohung Nach Dem Zweiten Weltkrieg," in *Vorsorgen in der Moderne*, ed. Nicolai Hanning, (München: 2016), with detailed references.

environment.[8] The innovation of scenario planning was introduced to the energy business by Royal Dutch Shell strategic planners,[9] who also described the situation to which they responded as one characterized by uncertainty. In retrospect, Shell's chief scenario planner, Pierre Wack (1922-1997), remembered: "Traditional planning was based on forecasts, which worked reasonably well in the relatively stable 1950s and 1960s. Since the early 1970s, however, forecasting errors have become more frequent and occasionally of dramatic and unprecedented magnitude."[10] The business environment had become the main source for new uncertainties, particularly due to changes in the geopolitical world order following the decolonization of the Middle East and its oil producing countries. The oil crisis of 1973 worked as a catalyst with regard to the public perception of these changes in Western societies.

In the case of climate change, uncertainties ruled from the beginning. While the greenhouse effect and the potential for changing the climate by changing the chemical composition of the atmosphere had been known since the nineteenth century, still, in the middle of the twentieth century, biogeochemical feedbacks in the earth system, in particular the absorption of CO_2 by the oceans, had been studied too little to assess how CO_2 released from the burning of fossil fuels would accumulate in the atmosphere. In fact, it was even unclear if it was accumulating at all or to any extent that would indeed affect average temperatures on a global scale, before Charles David Keeling (1928-2005) developed a reliable method of measuring atmospheric CO_2 and started his famous series from the Mauna Loa observatory on Hawaii in 1958.[11]

8 Hamblin, Jacob D., *Arming Mother Nature: The Birth of Catastrophic Environmentalism* (New York: Oxford University Press, 2013) deals extensively with the many links between the nuclear age of warfare and environmental science(s). For the role of futurism in the context of U.S. mililtary strategic planning after WW II, see Andersson, Jenny, "The Great Future Debate and the Struggle for the World," *American Historical Review* 117.5 (2012): 1411-1430.

9 See in particular Wilkinson, Angela, and Roland Kupers, *The Essence of Scenarios: Learning from the Shell Experience* (Amsterdam: Amsterdam University Press, 2014), and Wilkinson, Angela, and Roland Kupers, "Living in the Futures: How Scenario Planning Changed Corporate Strategy," *Harvard Business Review* 91.5 (2013): 3-11.

10 Wack, Pierre, "Scenarios. Uncharted Waters Ahead," *Harvard Business Review* 63.5 (1985): 73.

11 For a more extensive account of the story of the "Keeling Curve" see Hamblin, *Arming Mother Nature*, 93-96; Howe, Joshua P., *Behind the Curve: Science and the Politics of Global Warming* (Seattle: University of Washington Press, 2014), 16-43; and Weart, Spencer, *The Discovery of Global Warming* (online), chapter "Money for Keeling: Monitoring CO_2 Levels," accessed April 5, 2017. <http://history.aip.org/climate/Kfunds.htm>.

Anthropogenic warming of the global climate became a subject of serious concern only after the issue of measuring atmospheric CO_2 had been settled.

Today, climate models can simulate the effects of different CO_2 emission pathways on global average temperature with some degree of reliability. Uncertainties remain, partly due to the state of climate research, partly due to the character of certain models and the non-linear character of the climate system. However, the greatest source of uncertainty is the future socio-economic development of nearly two hundred states around the world.

Like all scenarios, climate scenarios are not forecasts or prognoses. They apply where classical projections and probability calculations have been undermined by uncertainties. They are used as tools for assessing the consequences of possible future paths of society for climate change—though not so much for climate assessment, but for strategic thinking. IPCC climate scenarios open future horizons until the end of the 21st century. Other scenarios opt for longer time periods. Generally speaking, evaluating changes in greenhouse gas cycles requires long-term, sometimes even geological timescales. A century may appear long by the standards of political and social change. But a century is not a long time by the standards of biogeophysical processes that are relevant to the climate system. The divergence between processes in society on the one hand and in the earth system on the other is the origin of problems of scale that raise some of the most serious questions to climate policy, because they require making provisions with regard to an unusually long-term future horizon, often reaching far beyond any time periods in which political decision makers are accustomed to think.

The problem of scale—resulting from the interference between historical time and geological time—defines one of the key problems to environmental politics in the 21st century. Abstract as this may sound, problems of scale are underlying an enormous spectrum of sustainability issues. Today, human activities impact all parts of the earth system on a level that supports the idea that, in sum total, humanity has become the equivalent of a geophysical force and that we are now living in a new geological era, for which Paul Crutzen and Eugene Stoermer suggested the term "Anthropocene", now under review for formalisation by the International Commission on Stratigraphy.[12] Quantitative

12 The "Anthropocene" was first suggested in the following article: Crutzen, Paul J. and Eugene F. Stoermer, "The 'Anthropocene'," *Global Change Newsletter 41* (2000): 17-18. See also Steffen, Will et al., "The Anthropocene: Conceptual and Historical Perspectives," *Philosophical Transactions*, Series A 369.1938 (2011): 842-867. The International Commission on Stratigraphy started debating the formalisation of the term much more recently, setting up the Anthropocene Working Group.

indicators of change, both in society (where economic indicators dominate over others) and the earth system (climate change, decline in biodiversity etc.) since the beginning of industrialization, have helped in identifying what is now discussed as the Great Acceleration from circa 1950.[13] More than anything else, the Great Acceleration undermines the relative stability of the conditions the Holocene provided for life on earth as well as for the development of complex societies. The massive disturbances human activities have caused in many chemical cycles, such as the carbon, nitrogen, and phosphorous cycles, have produced serious uncertainty about the continuation of Holocene stability in the future.[14] Rapid environmental change means that "futures past" is also questioned on the level of geological timescales.

The coincidence of the Great Acceleration with the emergence of future scenarios after World War II is symptomatic for an era characterised by rapid change in society *and* nature. In this era, knowledge of the past loses its relevance for forecasting the future based on probability—in other words, we have entered an era of transition from risk to uncertainty. That transition can barely be described in the categories provided by sociologies of risk, which hardly ever distinguish between risk and uncertainty.[15] There may be good reasons for this from the sociological point of view. However, this distinction is essential to understanding the history and future of forecasting and the emergence of scenario planning in our time.

4 Uncertainty in the Insurance Business

The diagnosis that we are now living in an age of uncertainty seems abstract at first glance. But nevertheless, it has very concrete implications for all levels of futures politics, including the politics of climate change, which is with regard to both, mitigation and adaptation. In the following, I will focus on a very specific form of adaptation: Some influential studies have suggested that an expansion of insurance might be a way to bolster the expected impacts of

13 For the latest update on quantitative indicators, see Steffen, Will et al., "The Trajectory of the Anthropocene: The Great Acceleration," *The Anthropocene Review* 2.1 (2015): 81-98.

14 That is the point behind the so-called planetary boundaries as introduced by Rockström, Johan, et al., "Planetary Boundaries: Exploring the Safe Operating Space for Humanity," *Ecology and Society* 14.2 (2009); for the latest update see Steffen et al., "Planetary Boundaries."

15 To name just two of the most prolific sociological theorists of risk in modern society, both Anthony Giddens and Ulrich Beck have used broad definitions of "risk" that do not distinguish between "risk" and "uncertainty."

global warming. Regarding insurance, Nicholas Stern stated that "A substantial part of any strategy must [...] be to facilitate recovery from damage after the act." And Anthony Giddens assessed "Insurance is likely to be crucial in adaptation".[16]

However, it must not be overlooked that climate change also poses new challenges to the insurance business. Most climate scientists expect higher frequencies and/or greater severity of meteorological and climatological hazards as a consequence of climate change. These changes already affect many branches of the insurance business. Greater risks of drought and fire, for example, are a challenge to the business of fire insurance. Climate change challenges "business as usual" strategies across the entire insurance business. Thus, while expanding insurance schemes may be seen by society as an adaptation strategy for climate change risks, it must not be overlooked that changing climatological and meteorological hazards also pose challenges for insurance to adapt. Actually, related concerns are already quite prominent within the insurance business. Reinsurance companies in particular have warned about the potential destabilising effects of climate change, particularly in parts of the world where insurance has been traditionally well established and covers a high degree of potential risks. This applies to Europe. But before I will look at an example, in the final section of this chapter, I will discuss the problem of uncertainty within the broader context of insurance risk to define the place that new uncertainties caused by climate change have in it.

Generally, insurance risk is determined by probability calculations, which combine two time series, one representing the frequency of a damaging event (which may result from a natural hazard or another kind of hazard), and the other representing the damage resulting from its impact. These time series form the knowledge base of insurance. Under normal circumstances, this knowledge base expands over time, allowing greater statistical precision.[17]

Beyond these statistical calculations, insurance is always built on the law of large numbers, meaning that insurance seeks to share risk among as many policyholders as possible.[18] The logic of the law of large numbers is, by the way,

16 Stern, Nicholas H., *A Blueprint for a Safer Planet: How to Manage Climate Change and Create a New Era of Progress and Prosperity* (London: Bodley Head, 2009), 66; and Giddens, Anthony, *The Politics of Climate Change* (Cambridge: Polity Press, 2009), 165. For a broader explanation of the theoretical approach taken in this section, see Mauelshagen, Franz, "Ungewissheit in der Soziosphäre. Risiko und Versicherung im Klimawandel," in *Unberechenbare Umwelt. Zum Umgang mit Unsicherheit und Nicht-Wissen*, ed. von Detten, Roderich, Fenn Faber, and Martin Bemmann (Wiesbaden: Springer, 2013).

17 Zweifel, Peter, and Roland Eisen. *Versicherungsökonomie* (Berlin: Springer, 2003), 34.

18 Zweifel, and Eisen, *Versicherungsökonomie*, 240-243.

a powerful tool to reduce social complexity in risk management. However, in practice many social factors, or even specific features of policyholders, be they individuals, groups, or organisations, might lead to disturbances in the functioning of insurance, as they may—directly or indirectly—influence the distribution of insured risk. Changes in the insurance business environment often affect the conditions under which insurance operates, particularly when that environment is changing rapidly. In this specific context, the environment is defined, from the perspective of the insurance business, as the entirety of what surrounds that business, while not directly belonging to it. That could be the legal framework of the insurance business as defined by a state or an association of states, or political circumstances, or the natural environment.

Calculations of insurance risk are prone to uncertainties from various sources: Statistical calculations of the probability of damaging events and the damage expected from them cannot anticipate when exactly a damaging event will occur. Given a return period of, say, a hundred years for such an event, nobody expects it to return precisely after a hundred years. Thus, probability does not mean *predicting* an event. The remaining uncertainty about the exact time of occurrence is, in fact, essential to insurance, as perfect forecasts would make it unnecessary to accumulate capital from insurance premiums to prepare for the damage. While the insurance business somehow builds on the uncertainty of forecasting a damaging event, the lack of knowledge about the occurrence of a damaging event may still have serious consequences for insurance companies as well. Proof from insurance history shows that an accumulation of loss by coincidence, which may be highly improbable from the point of view of probability calculation, can lead to insolvency. While insurance companies are social organisations, which reduce the vulnerability of policyholders to (economic) loss and potentially strengthen the resilience of social systems, they are of course vulnerable to economic loss themselves. The vulnerability of an insurance company depends on the number of contracts, the exclusion of moral hazard and adverse selection, as well as on the distribution of risks in a market that varies in time and space. Insurance companies are particularly vulnerable in the early stages of their existence, when the number of policyholders is small and the law of large numbers therefore does not work for them.[19] Also, at an early point in the evolution of a company it often happens that the time to generate financial reserves from surpluses has not been sufficient to survive a major damaging event affecting the property insured by that respective insurance company.

---

19 Examples for early business vulnerability: Mauelshagen, "Sharing the Risk," 178-181, and Rohland, *Sharing the Risk*, 30-37.

There is a more significant element of uncertainty inherent to the repetitive occurrence of loss accumulation over short periods of time. Such anomalies in the frequency of damaging events—also known as catastrophic loss—often raise the question of whether the statistical baseline for probability calculations is reliable or not. Anomalies in the frequency of natural extremes can be random, or they may be unexpected due to a lack of reliable information over certain periods of time, or they may result from changing natural conditions, such as climate change.

Let me illustrate this point using an example: After much hesitation, the Swiss Reinsurance Company (in the following: SwissRe) in 1927 decided to extend their risk pool in crop reinsurance (called hail reinsurance in the German speaking countries), i.e. the insurance of crop insurance. That very year, 1927, had produced major losses for hail insurance companies in Switzerland, which is why the reinsurers counted on a period of modest weather conditions for many years to come. Yet SwissRe's hail branch produced deficits again in 1928, 1930, and 1931, and faced record loss in 1929 (which almost went unnoticed because of the global financial crisis of that same year). Probability calculations were based on statistics provided by insurance companies and weather stations. After that series of unprecedented losses, risk managers declared these statistics unreliable, because the underlying observation periods were either too short or devalued by climate change. Indeed, climate change became the official argument when SwissRe cancelled policies to reduce the crop insurance risk pool as early as 1932.[20]

Insurance companies have two textbook reactions to changing frequencies and severities of extremes: either they raise premiums or they shrink (or completely cancel) their business in branches likely to become unprofitable. Another option, which is not in the textbooks, is to build lobby groups and pressure for change, i.e., reducing the vulnerability of built environments and agriculture to natural hazards. We know that such lobby or pressure groups were formed after the fire in Sundsvall (Sweden) in 1888 and after the earthquake and fire in San Francisco in 1906.[21] In both cases, existing social networks among leading insurance managers helped forming such groups. In the case of Sundsvall, they finally succeeded in transforming the risk of fire by pushing towards more fire-resilient construction in the rebuilding of the city.

A third source of uncertainty is social contingency. Societies have adapted to natural hazards in many ways and, depending on various circumstances,

20 Mauelshagen, "Sharing the Risk of Hail", 185-190.

21 I take both these examples from Rohland, *Sharing the Risk*, 87-123, where the historical circumstances are explained in much greater detail.

their adaptive efforts to changing risks are sometimes more, sometimes less successful. It is a complex, non-deterministic evolutionary process from which societies emerge somewhat modified. In order for adaptation strategies to be successful it is important that the organizational efforts of a society match the complexity of the challenge. In the case of global warming this complexity is obviously enormous, because various relevant climate elements—temperature, precipitation, atmospheric circulation patterns (like the El Niño Southern and North Atlantic Oscillations), extremes, meteorological and climatological hazards etc.—will change in a non-linear fashion, and the scale of change varies locally.[22] Adding to this is that the course of global warming, due to its predominantly anthropogenic character, largely depends on greenhouse gas emissions produced from within the social sphere. Thus, natural and social forces of climate change become intertwined—which is really the signature of the Anthropocene—creating new challenges for insurance companies to manage changing risk patterns.

However, that change creates uncertainties in the first instance, as it is beyond predictability at which point in time probability calculations can be based on a new relatively stable state of the climate system, which allows the defining of return periods. In this situation, the future of the past is no longer guaranteed, as was explained in more abstract terms in the first section of this chapter. In the case of insurance, climate change undermines probability calculations based on the historical record of natural hazards and damaging events resulting from them. This situation is equivalent to a journey into the past, to the early days of the insurance business, when meteorological statistics and other relevant information still needed to be collected. Compared with that earlier stage, it may be seen as an advantage that insurance has already matured in terms of the distribution of risk among large numbers of policyholders—that is, in places like Europe, where insurance has a long tradition. However, the global dimension of climate change raises doubts that growing out of the problem will be an option for the insurance business. If the frequencies and severities of meteorological and climatological hazards rise globally, balancing more damaging events in one place with less in another may no longer be possible, particularly if the increase in natural hazards and extremes will occur across their entire spectrum.

22 Intergovernmental Panel on Climate Change, *Climate Change 2013: The Physical Science Basis,* Fifth Assessment Report (Cambridge, UK: Cambridge University Press, 2013).

5 Hail and Crop Insurance in Switzerland

To illustrate the effects of anthropogenic climate change and the uncertainties emerging with it on the insurance business, I will finally look at a relatively simple example: crop insurance (or agricultural hail insurance) in Switzerland. It is "simple" compared with more complex markets as well as compared with more complex insurance structures. Crop insurance emerged as a distinct branch of insurance in Switzerland—distinct even from insurance against hail damage in building insurance and car insurance. The relative simplicity of the example makes it easier to embed climate change, its present and future effects, in a brief historical account of adaptation of crop insurance to the hazard of hail in Switzerland. Choosing the example of crop insurance in a small Alpine country may raise the question of representativity. However, it must be said that hail is by no means marginal compared with other meteorological hazards in Europe. In fact, it has been one of the most damaging hazards, particularly in parts of Europe surrounding the Alps.[23]

Hailstorms are among the costliest natural hazards in Switzerland—a fact that has been recognised for a long time. Already in 1888, the Swiss federal government (Bundesrat) stated: "Compared with other countries, Switzerland is exposed to the damaging effects of hail fall to a very high degree."[24] Any assessment of hail risk, like this one, is based on historical knowledge about the severity and frequency of hailstorms, their fluctuation, and change in time. However, there is also a social dimension to the assessment of hail risk: the vulnerability of people and the material values that are exposed to potentially destructive physical forces. Cultural changes—e.g. trends in house construction—often severely influence the potential for property damage. Also, the invention and proliferation of automobiles has created a new category of exposed property that played no role prior to the twentieth century.[25]

---

23 Other high-risk regions for hail occurrence in Europe are Southern France, North Italy, Southern Germany, Austria, and the Czech Republic. Vinet, Freddy, "Climatology of Hail in France," *Atmospheric Research* 56 (2001); Bertato, Mirko et al., "The Climatology of Hail in the Plain of Friuli Venezia Giulia," *Atmospheric Research* 67-68 (2003); Chroma, Kateřina, Ruldolf Brázdil, and Radim Tolasz, "Spatio-temporal Variability of Hailstorms for Moravia and Silesia in the Summer Half-Year of the Period 1961-2000," *Meteorologický časopis* 8 (2005); Brázdil, Ruldolf and Karel Kirchner, *Vybrané přírodní extrémy a jelich dopady na Moravě a ve Sleszku/Selected Natural Extremes and their Impacts in Moravia and Silesia* (Brno, Praha, Ostrava 2007), 217-237.

24 *Bundesblatt*, 40, vol. IV, No. 53, 8 December 1888, 795.

25 See also for the following: Mauelshagen, "Die Vergesellschaftung des Hagelrisikos," 60-62.

The invention of crop insurance in Switzerland was initiated by the economic society in Bern early in the 19th century. However, through the entire century the "ancient regime" of charitable aid and state support after damaging hail events continued. After hail insurance proliferated rapidly in the 1820s, companies operating on a Cantonal level proved too vulnerable. Around 1850 several foreign private insurance companies began supplying the unregulated Swiss market. However, competition on a small but highly risky territory led to failure, because premiums were never fully adjusted to the real costs for damage compensation.

In 1880 the Swiss Hail Insurance Company (in the following: Swiss Hail) was founded. This also marks the beginning of a monopolistic market solution, which privileged a non-profit oriented mutual company on the national level. This new system was politically implemented by Cantonal and Federal subventions beginning in 1890: while crop insurance was not made mandatory, as had been demanded in previous debates, political institutions subsidised crop insurance exclusively if taken deliberately with a company operating exclusively on Swiss territory. This was the end of 19th-century laissez-faire business and led to the exclusion of foreign crop insurance companies from the Swiss market—a process praised as the establishment of a national "community of hazard".[26] This new national community of hazard interfered with cantonal insurance organisations until 1942, when the last cantonal crop insurance company went bankrupt. The risk distribution that was established after 1880 helped in stabilizing hail insurance in Switzerland to some degree. It developed from a complex interplay between the risk geography of Switzerland, its political organization, the specific national framework for insurance markets, and the preferences of the largest group of (potential) policyholders, Swiss Hail's clientele: farmers. A great majority of them preferred a national, non-profit-oriented, mutual form of company organization, which is how Swiss Hail was set up as a company.[27]

This very same clientele of farmers was also sceptical about reinsurance companies operating on international markets, which is why Swiss Hail did not take reinsurance before 1927. Thus, nationalism and even regionalism (i.e. with regard to cantonal insurance companies operating in some places until 1942) clearly influenced the evolution of risk sharing in the framework of Swiss Hail. When Swiss Hail suffered great loss in 1927, the company finally decided

---

26 Fäh, Emil, *Die Hagelversicherung in der Schweiz in Vergangenheit und Gegenwart. Zum 75jährigen Bestehen der Schweizerischen Hagel-Versicherungs-Gesellschaft 1880-1954* (Zürich: Schweizerische Hagel-Versicherungs-Gesellschaft, 1954), 39.

27 Mauelshagen, "Die Vergesellschaftung des Hagelrisikos," 64.

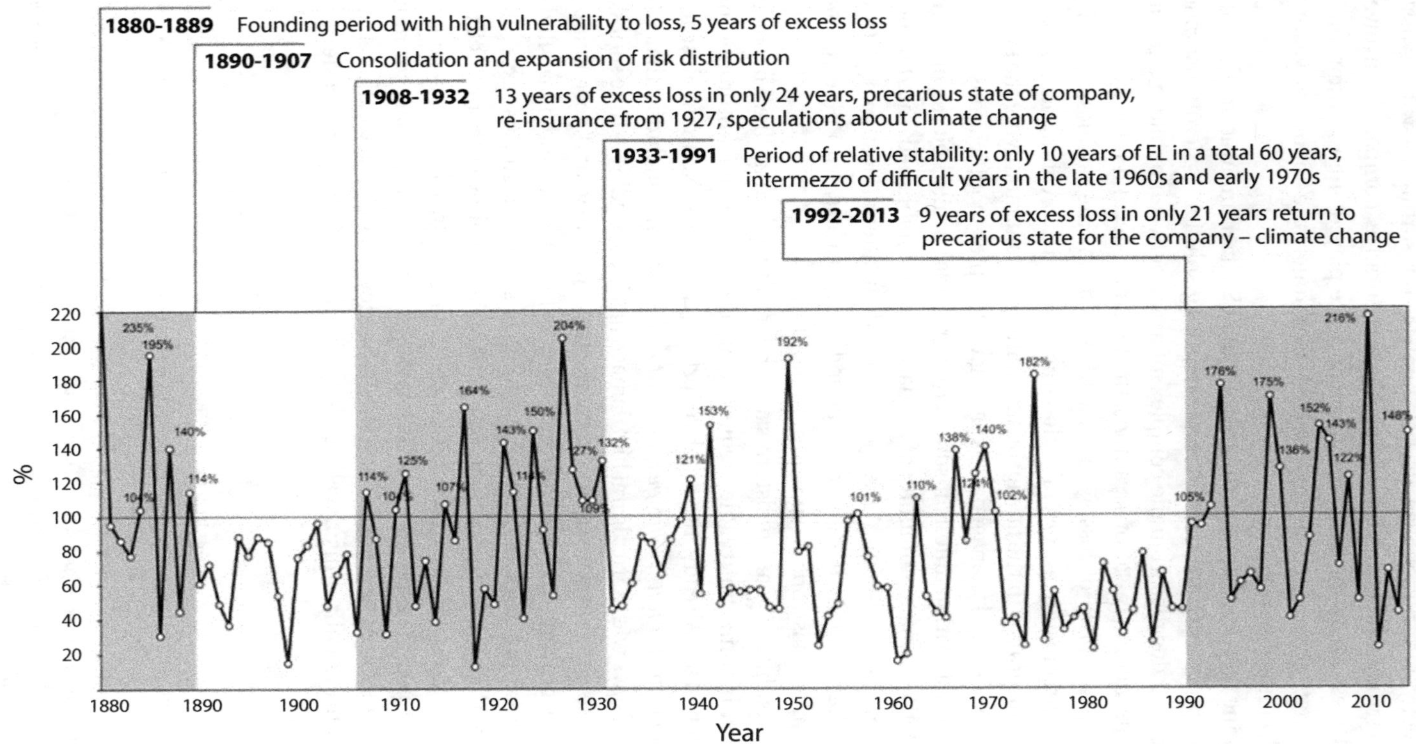

FIGURE 10.1 *Swiss Hail Insurance Company, 1880-2013 (crop insurance only), damage compensation (paid to policyholders) as percentage of annual total premium incomes; periods when business was difficult for Swiss Hail are marked in gray; data: Swiss Hail Insurance Company (Annual Reports)*

to indirectly internationalize its risk pool by reinsuring it—with SwissRe accepting the greatest share among the reinsurance companies involved. Political pressure by the Federal Insurance Office played an important role in making that step, which helped Swiss Hail through the following years of unusual, continuous loss.[28]

Adapting crop insurance to the risk of hail in Switzerland was anything but a passive process. Quite the contrary: a working solution for crop insurance emerged over decades from the complex interplay between public and private forces—farmer organisations, cantons, the federal state, and reinsurers—and from the experience of loss as an incentive to adapt the business model to changing circumstances. Overall, after the somewhat revolutionary step of seeking reinsurance in the late 1920s, adaptations meant modifying the system gradually. Swiss Hail managed to get along with minor modifications, such as greater diversification among insured crops (pricing), while holding on to the established national community of solidarity. Once established, that business model remained stable until the 1990s. Since then, after almost sixty years of stability, the company has entered a period of severe loss, with nine years of accumulative catastrophic damage from hailstorms.

In the last fifteen years, hailstorms and the resulting damage have been discussed, with some noticeable frequency, in the context of anthropogenic climate change. Already the final report of the Swiss National Research Programme 31, *Climatic Changes and Natural Hazards*, concluded that "the number of days with agricultural hail damage has increased".[29] A time series of days with severe hail occurrence in Switzerland from 1920 and 2005 demonstrates the growing trend of extreme hail between 1980 and 1994. Radar observations provide evidence for a doubling of severe hailstorms (extending over more than a hundred kilometers) within the twenty-year period 1983 to 2003.[30] The latest OcCC-report on Climate Change in Switzerland 2050 concludes that peasants, property owners, and insurers should prepare for more extreme hailstorms to come in case the frequency of synoptic weather situations favouring hailstorms develops along the trend of the last two decades (figure 10.2[31]).

---

28 Mauelshagen, "Sharing the Risk of Hail", 182-185.

29 Schiesser, Hans-Heinrich, *Klimatologie der Stürme und Sturmsysteme anhand von Radar- und Schadendaten* (Zürich: Vdf, 1997), 11.

30 Schiesser, Hans-Heinrich, *Hagelstürme in der Schweiz: Wiederkehrperioden von schadenbringenden Hagelkorngrössen— Eine Abschätzung* (Zürich, 2006).

31 OcCC et al., ed., *Klimaänderung und die Schweiz 2050. Erwartete Auswirkungen auf Umwelt, Gesellschaft und Wirtschaft* (Bern, 2007): 145.

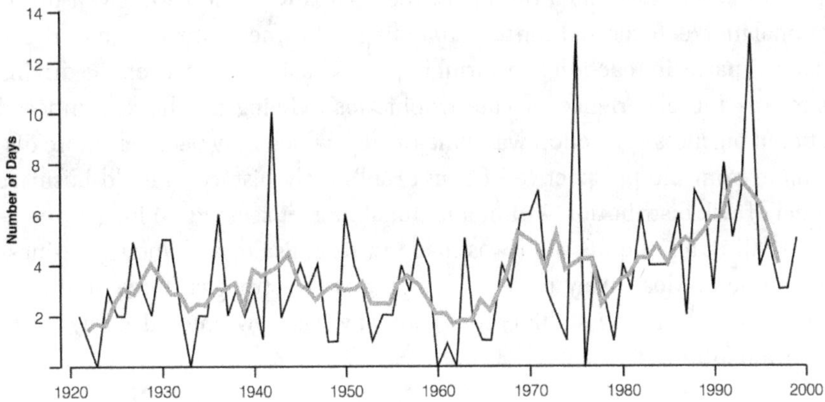

FIGURE 10.2 *Number of days with hail affecting one hundred or more municipalities in Switzerland, 1920-2005; grey line: five-year running mean*

However, the same report argues that existing climate models cannot project the future of hailstorms, for which the most important reason is that hail occurrence is locally a highly variable meteorological phenomenon.[32] In other words, existing models of global warming cannot predict the effect global change is likely to have on hailstorm patterns (frequency, severity etc.), which is also, to some extent, due to the limits of existing time series on hailstorm occurrence. Projections drawn from a regional climate model representing conditions for northern Switzerland in 2071-2100 indicated a "pronounced increase in year-to-year variability [of temperature and precipitation] in response to greenhouse-gas forcing",[33] which may lead to higher frequency of synoptic situations leading to hailstorms.

In sum, there are uncertainties (1) about how climate change will affect hail occurrence in Switzerland and therefore also (2) about the relation between recent statistical evidence for increasing numbers of hailstorms and climate change; and (3) there are the general uncertainties about how climate change will develop in the long term (depending on emission pathways of societies in the future) and how that may affect hailstorm occurrence and damages resulting from them on local and regional levels. Though it is logical and plausible to assume that climate change will lead to even more severe changes in the frequency and severity of hailstorms in the future, no insurance company, such as Swiss Hail, can wait for the evidence before starting to adapt. Experiencing a

32 ibid.

33 Schär, Christoph et al., "The Role of Increasing Temperature Variability in European Summer Heatwaves," *Nature* 427 (2004): 335.

streak of severe loss since the 1990s, the company has already given up its national market focus and started expanding its business into foreign territory. That was partly in reaction to a shrinking Swiss business. However, the driving force was the experience of catastrophic loss. Giving up the self-imposed national business restriction was a major step, which only becomes more discernible from the perspective of Swiss Hail's early history. The old business model of risk distribution within a national market seemed no longer operational. All in all, the decision was made in response to a changing business environment. Most likely climate change was the strongest force behind it. After a period of long stability, the company has now entered its new and uncertain future.

6 Conclusions

To judge the potential role of insurance in dealing with natural hazards and their changing patterns in times of global warming, we need a deeper understanding of the capability of insurance to strengthen social resilience. Historical experience is obviously relevant in this context, as we learn about decision making and path dependence. But the value and role that historical knowledge and experience have for the future are changing in our time of accelerated societal and environmental change. That applies to the insurance business, particularly where meteorological and climatological hazards affect that business. The uncertainties of climate change are already calling into question business models based on risk calculations and statistical projections, which are no longer fully reliable.

Insurance may reduce social vulnerability; yet insurance companies themselves are vulnerable to the uncertainties produced by climate change. Some policymakers today turn scientific uncertainty about the relationship between meteorological hazards, such as hailstorms, and climate change into an argument for inaction. However, the impetus of the insurance industry cleary diverges from this. While politicians are concerned about state budgets and costs of prevention, insurers look at uncertainty from the perspective of potential loss. Politicians want to make sure that investment in mitigation or adaptation will not be wasted. From the point of view of risk management, however, the implications of scientific uncertainty are entirely different. In the face of loss there is less time to wait for more certainty to be delivered by scientific research. Delaying decision making involves the danger of a decline in the social capacity to cope with meteorological hazards and damage or loss resulting from them, which would be the case if insurance were to withdraw from business because of failure to adapt to climatic changes and other changes in the business environment. Proactive and robust decision making is called for, because the most important capacity to preserve is that of decision making itself. Running out of alternatives might already be a clear indication of future collapse. Such are the challenges of the age of uncertainty.

CHAPTER 11

Climate Change and Future Pasts: Preliminary Considerations on a Historiography of the Future

Claus Leggewie

Abstract

After decades of a rather bright and optimistic futurology "future" in our days is mostly evaluated in much more sceptical and pessimistic terms. This is true for science-based scenarios (e.g. on the consequences of climate change or the degradation of biodiversity or the carrying capacity of the planet Earth in general) as well as for fiction (and science fiction). The reaction of political decision-makers was what is called in this article "futurisation"; the concept of boundaries (to economic growth or greenhouse gas emissions or the amount of public and private debt) almost naturally introduces a political approach focusing on the chances of future generations. Climate change has brought a cultural reorientation regarding the dominant time consciousness of postmodern societies, as is shown mainly in reference to the German case.

•••

> "Living backwards!" Alice repeated in great astonishment.
> "I never heard of such a thing!"— "but there's one great advantage in it, that one's memory works both ways."
> "I'm sure mine only works one way," Alice remarked.
> "I can't remember things before they happen."
> "It's a poor sort of memory that only works backwards," the Queen remarked.
> "What sort of things do you remember best?" Alice ventured to ask.
> "Oh, things that happened the week after next," the Queen replied in a careless tone.[1]

∴

1 Carroll, Lewis, *Through the Looking-Glass, and What Alice Found There* (1871), ch. 5.

 | DOI 10.1163/9789004356825_013

1 Introduction

There is a growing consensus in the scientific community, in public opinion, and among political elites that the past releases of greenhouse gases into the Earth's atmosphere have had negative effects on the climate and that further releases, above all of carbon dioxide and methane, must be reduced in the medium term and stopped altogether in the long term (around 2070) to avoid a level of global warming that may have dangerous consequences for humanity's future living conditions. Socio-ecological research has provided growing evidence relating to these global risks, showing that "planetary boundaries" must be respected in order to avoid jeopardizing the Earth system.

The validity of these hypotheses, of the consequent concept of "boundaries" (or of "guard rails"[2]) and of the political consequences of related global cooperation and contracts will not be discussed here.[3] Rather, climate change shall be considered as an outstanding example of a driving force of what we might call the "futurisation" of politics—that is, shifting the political frame of reference from today's goals (distribution of welfare, securing continuous growth, crisis management etc.) to the assumed interests of coming generations.[4] Political life *per se* is, of course, oriented toward the shaping and planning of possible futures, which are consciously anticipated in decisions taken in the 'here and now'.[5] Such programs are then put to the vote within democratic polities; at the same time, it has to be remembered that every intentional act has its unintended consequences, which lead to the modification, perversion, and distortion of original intentions.[6] As such, it is trivial to say that political life is oriented toward the future. However, the hypothesis of planetary bound-

---

2 The German Advisory Council on Global Change (WBGU) has proposed the following definition: Guard rails are "quantitatively definable damage thresholds whose transgression either today or in future would have such intolerable consequences that even large-scale benefits in other areas could not compensate these". Cf. German Advisory Council on Global Change (WBGU), *World in Transition—A Social Contract for Sustainability* (Berlin: WBGU, 2011): 32.

3 Cf. German Advisory Council on Global Change (WBGU), *World in Transition: Ways Towards Global Environmental Solutions* (Berlin: Springer Verlag, 1996); German Advisory Council on Global Change (WBGU), *Climate Protection as a World Citizen Movement* (Berlin: WBGU, 2014).

4 The sociological concept of generation was elaborated by Karl Mannheim (1952) and Joachim Mathes.

5 Parsons, Talcott, *The Structure of Social Action* (New York: MacGraw Hill, 1937); Arendt, Hannah, *The Human Condition* (Chicago: University of Chicago Press, 1958).

6 Merton, Robert, "The Sociology of Knowledge," *Isis* 27 (1937): 493-503; Adriana, Mica, Arkadiusz Peisert, Jan Winczorek, ed. *Sociology and the Unintended. Robert Merton Revisited* (Frankfurt a.M.: Peter Lang, 2011).

aries conveys a message that is fundamentally critical of "traditional" (i.e. modernist) concepts of progress and growth, which opposes a general "less is more" to the *plus ultra*[7] of modernity and, to a certain extent, erects a clearly visible stop sign upon what had until now been thought of as wide open horizons. The probable consequences of dangerous climate change also challenge the presentism of current politics. Orientation to the present was an accomplishment of modern-era democracies, which do neither bind people to tradition nor to utopias; but the focus on 'here and now", under the auspices of planetary boundaries, may ignore responsibilities and destroy future options.

This is to postulate nothing less than a new *Sattelzeit* ("saddle period"), as the historian Reinhart Koselleck described the period between 1750 and 1850.[8] It was then, according to Koselleck, that a fundamental shift in mental perceptions and epistemic concepts took place in the wake of the social and economic upheaval of the Industrial Revolution and the French Revolution. This allowed for a new perception of time to develop and a semantic shift in political and philosophical terms, neither of which has lost their global significance until today. However, with this paradigm in mind, to make planetary boundaries the frame of reference for political action is a fundamentally provocative move.

Therefore, the following presents some empirical evidence and normative considerations relating to policy concerns that put the putative interests of future generations' front and centre.[9] The first question concerns the task of shaping the future under the considerable time pressure that is linked to planetary boundaries; and how to conceive of this task in a way that it becomes the 'objective' target of political decisions. The second question relates to how, under these premises, a political decision-making process can be reorganized

7 Peter Sloterdijk sees in this phrase that of the last Holy Roman Emperor Charles V, "the call for anarchic increase in the capitalist world, which one calls growth and progress", Sloterdijk, Peter, *Zeilen und Tage: Notizen 2008-2011* (Frankfurt a.M.: Suhrkamp, 2012), 86. iae chief attraction, in terms,ur ines neuen Erdzeiralters iae chief attraction, in terms,ur ines neuen Erdzeiralters

8 The outstanding oeuvre of this distinguished historian should be more visible in the international academic community. Cf. Koselleck, Reinhart, *Futures Past: On the Semantics of Historical Time* (Cambridge: MIT Press, 1985); Koselleck, Reinhart, *Critique and Crisis: Enlightenment and the Pathogenesis of Modern Society* (Cambridge: MIT Press, 1988).

9 My concept of 'generation' as a sociological category of time is distinct from trendy "generationalism". Cf. White, Jonathan, "Thinking Generations," *British Journal of Sociology* 64.2 (2013): 216-247.

such that it stands in the way of both delay tactics and a technocratic decisionism.[10]

2 Saddle Period 1 (1770-1850)

In modernity, time is predominantly perceived as linear and the future as open.[11] This perception of time supersedes an understanding of time as cyclical, or an indifference to time, that had prevailed previously. Once a techno-scientific civilization was established, thanks not least to the chronometer and methodological quantification in the natural sciences, it became 'natural' to think of time as chronological and as 'marching on relentlessly'; that is, to perceive a time regime that applied to every area of life. In parallel to which, a consciousness of "historical time" established itself, the modes of which are—a recent invention!—past, present, and future. Thus, historical time undermined symbolic and hermeneutical interpretations of the passage of time conceived of in non-linear terms and as being, in a certain sense, reversible.

Thus, modernity is characterised by a dualism of heterogeneous, often conflicting concepts of time. Philosophers like Kant, Heidegger, and Rorty are interpreted as saying that

> the reification of time as read on our watches and calendars, and which strikes us as a reality independent of ourselves as subjects, follows [...] from the temporal processuality of our self-constitution. [...] As a rule, we enter into a future, the content of which we determine according to our needs and plans; a future from which we exclude the ultimate horizon of death.[12]

---

10 With Habermas, decisionism can be described as a type of political decision-making based on the authority of the scientifically informed decision-makers and the legitimacy of instrumental rationality. Cf. Habermas, Jürgen, *The Philosophical Discourse of Modernity* (Oxford: Polity, 1987), quoted in Elgar, Andrew, *Habermas, The Key Concepts* (Oxford; New York: Routledge, 2006), 36-38.

11 Assmann, Aleida, Ist die Zeit aus den Fugen? Aufstieg und Fall des Zeitregimes der Moderne (München: Hanser, 2013).

12 Sandbothe, Mike, "Die Verzeitlichung der Zeit in der modernen Philosophie," in *Die Wiederentdeckung der Zeit: Reflexionen – Analysen – Konzepte*, ed. Antje Gimmler, Mike Sandbothe, and Walther Ch. Zimmerli (Darmstadt: PRIMUS Verlag und Wissenschaftliche Buchgesellschaft, 1997), 52.

In a programmatic sociological paper entitled "Entwurf einer Kritischen Theorie spätmoderner Zeitlichkeit" ("Towards a Critical Theory of Late Modern Temporality"), Hartmut Rosa takes Heidegger's critique of the objectified power of time as the basis for diagnosing how an accelerated pace of life hollows out the sovereignty of time for the individual.[13] From this, Rosa derives the counterfactual postulate of "deceleration" ("Entschleunigung").[14]

In the Euro-American and, more general, Pan-Western grand narrative, "Progress" (as distinct from individual instances of progress) is a "collective singular"[15] that emerged during the eighteenth century with authors such as French philosophers Turgot[16] (1973) and Condorcet[17] (1795), and animated the French Revolution.[18] The separation of (past) experience and (future) expectation, that which is characteristic of this narrative, means that time itself must be reconfigured in the modern era, as it is no longer possible to deduce from past experience what has to be done in the 'here and now' (and tomorrow). Human consciousness itself becomes historical; people live according to expectations that exceed the limits of the individual.

The drivers of Progress are science and technology (including liberal und later socialist movements), thanks to which the future will not only be *different* but also *better*. The perception of time as cyclical, which stems from the passing of seasons, the repetition of events and of religious rites pertaining to the finite nature of human existence, gives way to a perception of time as linear. Instead of the cycle, in which humans formerly imagined themselves to be, there is a line with a single destination: perpetual improvement (aka

---

13 Rosa, Hartmut, *Beschleunigung. Die Veränderung der Zeitstrukturen in der Moderne* (Frankfurt a.M.: Suhrkamp, 2005).

14 Rosa, Hartmut, Alienation and Acceleration: Towards a Critical Theory of Late-Modern Temporality (Aarhus: Aarhus Universitetsforlag, 2010); Rosa, Hartmut, Social Acceleration: A New Theory of Modernity (New York: Columbia University Press, 2013).

15 Koselleck, Reinhart, *Kritik und Krise: Eine Studie zur Pathogenese der bürgerlichen Welt* (Frankfurt a.M.: Suhrkamp, 1973); Koselleck, Reinhart, *Vergangene Zukunft: Zur Semantik geschichtlicher Zeiten* (Frankfurt a.M.: Suhrkamp, 1979); Koselleck, Reinhart, *Begriffsgeschichten* (Frankfurt a.M.: Suhrkamp, 2006); Koselleck, Reinhart, *Vom Sinn und Unsinn der Geschichte: Aufsätze und Vorträge aus vier Jahrzehnten* (Berlin: Suhrkamp, 2010).

16 Turgot, Anne-Robert-Jacques, *Turgot on Progress, Sociology and Economics: A Philosophical Review of the Successive Advances of the Human Mind, on Universal History* [*and*] *Reflections on the Formation and the Distribution of Wealth, translated, edited and with an introduction by Ronald L. Meek* (Cambridge: Cambridge University Press, 1973).

17 Condorcet, Marquis de, Esquisse d'un tableau historique des progrès de l'esprit humain (Paris: Agasse, 1795).

18 Manuel, Frank E., *The Prophets of Paris* (Cambridge MA: Harvard University Press 1962).

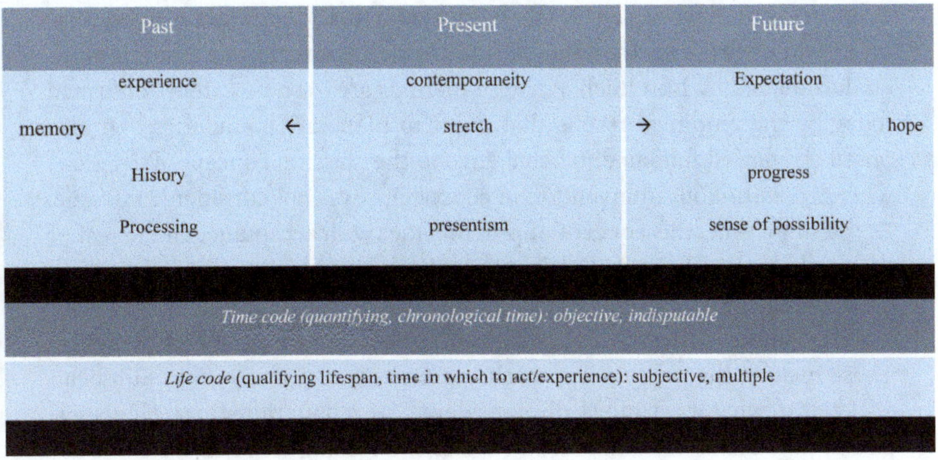

FIGURE 11.1 *(Post-)modern perceptions of time*

plus ultra). This became a self-justifying norm that has released enormous energies.

2.1 *Utopian Programmes*

Since the beginning of the Industrial Revolution, there have always been counter-currents and dystopias that contrasted with the optimism surrounding progress: Luddites and the *Lebensreform* movement[19] either turned their backs on received notions of Progress or proclaimed alternative forms of progress "on a human scale", "in harmony with nature", that reflected a "respect for divine creation" and the like. Since the 1960s and 1970s, with the new social movements and the emergence of Green parties, this revitalised kind of critique provided the basis for a political and social fault line that also loosened up the Right/Left dualism of industrial societies. In 1979, Koselleck, for example, tried to convince the Christian Democratic Union (CDU), West Germany's leading advocates of economic growth at the time that the ecologists and Greens (then mocked as the "enemies of Progress" had formulated a *general* interest of far more importance than the short-sighted outlook, which predominated in the political sphere.[20] Modern ecology drew attention to a "limit

19 Fox, Nicols, *Against the Machine: The Hidden Luddite History in Literature, Art, and Individual Lives* (Washington DC: Island Press, 2003); Linse, Ulrich, "Das 'natürliche' Leben: Die Lebensreform," in Die Erfindung des Menschen: Schöpfungsträume und Körperbilder 1500-2000, ed. Richard van Dülmen, (Böhlau: Wien, 1998), 435-458.

20 Jeismann, Michael, "Grüne Gegenwart. Die Grenze der Ökologie: Reinhart Kosellecks Beg riffsgeschichten,"*Frankfurter Allgemeine Zeitung*, November 12, 2006.

to growth" not to be exceeded. This coincided with the approaches to political ecology of André Gorz[21] and the Club of Rome[22] or notions of a convivial society formulated by Ivan Illich,[23] whose findings are to some extent confirmed today by the empirical results that point to planetary boundaries.[24] It also posed a renewed fundamental challenge to the classical concept of progress, whereby continuous intervention in ecosystems was not considered an issue. At the same time, this concept of progress met with acceptance in the rest of the world, particularly in the global South, thus supplying proof of its both dynamic and destructive potential.

Other modes of temporal consciousness also deserve brief consideration. These include the political approach in Germany and many other European societies to what was supposedly considered "zero hour" after 1945, characterised by post-dictatorial and post-war societies engaging in the *Aufarbeitung der Vergangenheit*,[25] that is, "coming to terms with the past", and in transitional justice, wherein responsibility for the mistaken behaviour and misdeeds of grandparents and parents was handed on to descendants, who provided material and non-material compensation wherever possible. This, by the way, unfolded an interesting pattern of generational interaction, as it encompasses prescriptions for future generations as well.

A period of "muddling through" (*Durchwurschteln*) preceded, characterised by a fixation on the present frequently attributed to a "sceptical generation"[26]; problems to which there were no short-term solutions were kicked into the long grass and the cost of externalities discounted on the basis of forecasts of future performance. Implicit to which was the belief that the then dominant patterns of problem solving (connected to the continuous rise in gross national product, and to technological development and adaptive strategies) would also render descendants sufficiently resilient and competent. The utilitarian

21 Gorz, André, *Écologie et Liberté* (Paris: Galilée, 1977).

22 Meadows, Dennis et al., ed. *Die Grenzen des Wachstums – Bericht des Club of Rome zur Lage der Menschheit* (Stuttgart: DVA, Stuttgart 1972).

23 Illich, Ivan, *Tools for Convivability* (New York: Harper & Row, 1973).

24 Rockström, Johan et al., "A Safe Operating Space for Humanity," *Nature* 461 (2009), 472-475. doi:10.1038/461472a; Steffen, Will et al., "Planetary Boundaries: Guiding Human Development on a Changing Planet," *Science* (2015),.doi:10.1126/science.1259855.

25 Adorno, Theordor W., "Was bedeutet Aufarbeitung der Vergangenheit? [1959]" in *Gesammelte Schriften* 10.2 (Frankfurt a.M.: Suhrkamp 1977): 555-572; Fischer, Torben, and Matthias N. Lorenz, "Lexikon der "Vergangenheitsbewältigung," in *Deutschland: Debatten- und Diskursgeschichte des Nationalsozialismus nach 1945* (Bielefeld: Transcript, 2007).

26 Schelsky, Helmut, Die skeptische Generation: Eine Soziologie der deutschen Jugend (München: Diedrichs, 1957).

conventions of taxation and planning also determined the horizons of expectation in regard to state welfare programmes, which have since hit an objective limit following demographic change, the deregulation of labour markets and the explosion of healthcare costs. This was presentism taken to the extreme.[27]

Finally, religious notions of paradise also deserve mention, where only salvation in the hereafter counts.[28] Millenarian and chiliastic currents have, above all in monotheistic religions, a long tradition among those who have turned their backs on this world (Max Weber's *Weltabgewandtheit*: "unworldliness/otherworldliness"); today, they once again exercise a powerful influence on secular political regimes, via Evangelical communities and Pentecostalism in the United States and via the influence of Salafist fundamentalists on Sunni governments in the Middle East, South Asia and Africa.[29]

If, by way of conclusion, *Vergangenheitsbewältigung* ("coming to terms with the past", see also above), *Weltabgewandtheit*, and political theology can be considered to provide alternative temporal horizons, then they share a certain indifference to the passing of time in this world or an overemphasis of the future, which renders them politically sterile or emphatically anti-political.

3 'Themes of the Future'

What are then 'themes of the future', i.e. themes that can or should be considered particularly relevant to the future? The criteria for the following selection include the medium- and long-term relevance of problems that require a timespan almost certain to exceed that of two or three legislative periods. A prime example of which is (as mentioned) the as yet unsolved storage of highly radioactive atomic waste, which is produced daily in large quantities. Secondly, there is a preponderance of socio-economic trends that have special relevance for the future and that, on the basis of previous experience, appear difficult to reverse. Here, demographic change in most OECD countries suggests itself as an obvious example that will probably lead to a rapidly growing age gap

27 During this period, the "peaceful utilisation" of nuclear energy was decided by parliaments and bureaucracies without any clue about a safe deposit of nuclear waste. Cf. Radkau, Joachim, and Lothar Hahn, *Aufstieg und Fall der deutschen Atomwirtschaft* (München: Oekom, 2013).

28 Kippenberg, Hans G., and Tilman Seidensticker, ed., The 9/11 Handbook: Annotated Translation and Interpretation of the Attackers' Spiritual Manual (London: Equinox, 2006).

29 Landes, Richard, *Heaven on Earth: The Varieties of the Millennial Experience* (Oxford: Oxford University Press, 2011).

between stagnating and shrinking societies in the global North, and fast- or rapidly growing societies in the global South; and, in both regions, to variation in the demand for intervention by the welfare state. Linked to this is an urbanisation process that will take cities close to being the standard format for life on Earth by the end of the twenty-first century.[30] A third characteristic of futurisation is the uncertain basis of knowledge—the many 'unknowns'—that associated processes provide for diagnoses and decision-making to the scientific community, decision-makers and the general public.

Decarbonisation (i.e. a low fossil-fuel or low a carbon-economy and society with a minimal output of carbon dioxide) offers a key example in this respect, a project deemed to be urgent from the perspective of the global eco-balance and climate crisis. Decarbonisation is accepted by G7 countries as the leading economies as well as by many UN institutions as a policy target. When it comes to specific details, there is much uncertainty surrounding decarbonisation, a process fraught with some risk for economic systems. A fourth characteristic is to be found in the fact that future problems call for considerable intervention in the levels of resources and of freedom available to future generations. This gives rise to a preventative and anticipatory style of politics, in which the "voice of future generations" is virtually rendered audible.

In the current socio-political discourse, dates in the near future are attached to certain political themes and decisions. From the German "Agenda 2010" (see below), through the *Schuldenbremse* ("debt brake") prescribed in some OECD countries, to the (Sustainable) Millennium Goals (SMGs) of the UN and other international organisations; all such policies seem to come with a "date of completion" and, as such, must be considered more or less concretely in relation to a horizon of realisation in the near or distant future. This applies just as much to focussing policy on desired reductions and the themes raised during the course of political power play (the political dimension) as to attempts at the normative encoding of the possible interests of future generations in constitutions and programmatic statements (the polity dimension).

Let me briefly go through some examples of soft and strong futurisation, mostly taken from the German political arena:

a) *Agenda 2010* was a concept that political leaders, the administration and think tanks prepared for the reform of the German welfare state and labour market, and which was carried out from 2003 by former Chancellor Gerhard

30 German Advisory Council on Global Change (WBGU), ed., *Humanity on the Move: Unlocking the Transformative Power of cities. Flagship Report* (Berlin: WBGU 2016:).

Schröder.[31] The description circulating among the Chancellor's inner circle referred back to the resolution of Europe's heads of state and government leaders in 2000 to make the European Union—within the framework of the Lisbon Strategy—the most competitive and dynamic knowledge-based economy in the world. *Agenda 2010* was supposed to function as a basis for reorganising and trimming back the welfare state, introducing the "Hartz plan" (a set of recommendations submitted by a commission on reforms to the German labour market in 2002) and a formula for sustainable pension schemes through flexibilisation.

Chancellor Schröder intended a paradigm shift. The original aim of welfare was to provide for risks such as accidents at work and prolonged illness, to bridge periods during which individuals are unemployed and to provide the long-term unemployed with a certain level of income. Up until this point, the welfare state's compensatory functions related overwhelmingly to short-term risks and the reorientation of the "intergenerational contract" such that provisions for the elderly were generated not directly by family members but by guaranteed state pension schemes. In the wake of the protracted worldwide economic crisis, which began in the mid-1970s, welfare provisions have not only been extended to other risk groups and relief measures. In most OECD states, they account for a growing portion of state spending and have contributed in a major way to public debt. *Agenda 2010* and analogous attempts by British and US governments "to end welfare as we knew it" (Bill Clinton)[32] all aim to alleviate the burden on the state through reintegrating those who are capable of working into the labour market and to hinder a "culture of dependence" on benefits. In fact, they went hand in hand with the deregulation of the labour market and a wage differential that meant millions of workers subject to social insurance contributions became dependent on benefits and that state spending was reduced at the cost of a growing social divide. We cannot discuss the pros and cons of this policy project but have to confine ourselves to pointing out the attempt to place time-stamps (the year 2010) at particular reform projects (ironically neoliberal ones) as usually observed in centrally planned economies.

---

31 Schöllgen, Gregor, *Gerhard Schröder: Die Biographie* (München: DVA Sachbuch, 2015); Deutsche Welle, "A Quick Guide to 'Agenda 2010'," published October 17, 2003, accessed November 9, 2015. <http://www.dw.com/en/a-quick-guide-to-agenda-2010/a-988374-1>.

32 Clinton, Bill, "How We Ended Welfare, Together," *The New York Times*, August 22, 2006, accessed November 18, 2015. <http://www.nytimes.com/2006/08/22/opinion/22clinton.html?_r=0>.

b) *Schuldenbremse* ("debt brake") is a term that emerged from the constitutional law ruling proposed in 2009 by a specially convened Federal? Commission. The ruling was intended to limit Germany's public debt, in that Federal Government and the *Länder* (German states) would create binding guidelines for the reduction of the budgetary deficit. As a result of which, the federal budget and the budgets of the *Länder* were in principle to be balanced without recourse to credit. This accords with the medium-term sustainable average-limit for structural deficit as defined in the European Stability and Growth Pact. So here, too, the impetus for the domestic German measures was European, as Germany's deficit was significantly above the Maastricht Treaty ceiling of sixty percent of gross domestic product (GDP). As a result of the government's *Schuldenbremse*, revenues from borrowing at a federal level in 2016 must not exceed 0.35 percent of GDP. As of 2020, the *Länder* are no longer permitted to include revenues from borrowing at all when balancing their budgets. The general idea is to slash public debt from a 'decadent' percentage of GDP (Japan 230 percent, Greece 177 percent, Italy 132 percent, major OECD countries around eighty percent) and "never again to stray from the path of virtue".[33]

This measure is just as controversial as Agenda 2010 and analogous welfare state reforms. It has been argued that the cost of the politics of austerity may fall upon the most impoverished section of the population; whole states, such as Greece, have hit rock bottom as a result of the austerity programmes imposed upon them by the institutions of the former Troika (aka "The Institutions", namely the International Monetary Fond (IMF), the European Commission and the European Central Bank). On the other hand, it remains the case that, if the repayment of government debt is no fiction, then it can scarcely be predicted whether the massive amounts now channelled into servicing that debt can ever be reduced—these having themselves been financed as a rule through the creation of further debt. The Keynesian inspiration of deficit spending was perverted into a constant mechanism of irreversible long-term indebtedness. This policy measure shall not be discussed here either, but again its normative legitimation was under the premise of protecting "future generations" and breaking a vicious cycle. Germany's constitutional *Schuldenbremse* of 2009 is likely to be the model for OECD countries that do not yet have one.[34]

---

33 "Tie Your Hands, Please: Is Germany's Fiscal Straitjacket an Example for Others?" *The Economist*, December 10, 2011, accessed November 9, 2015. <http://www.economist.com/node/21541459>.

34 As Michael Hüther, a German economist close to the business sector, puts it: "There's a good side to any crisis. It offers an opportunity for collective learning, although this

Let us now turn to more strongly "future-oriented" policies with much more structural and global impact.

c) *The Sustainable Development Goals* (SDGs) were to succeed the United Nation's Millennium Development Goals (MDG), once these expire at the end of 2015. The MDG were formulated in 2001 by a working group consisting of representatives of the United Nations, the World Bank, the IMF, and the OECD's Development Assistance Committee. There were the following six goals: reducing global poverty by half and eradicating hunger; achieving universal primary education and gender equality; reducing child mortality by two thirds and death rates among mothers by three quarters; reversing trends in HIV/AIDS and malaria; significantly slowing down the rate of species extinctions; and securing sustainable access to safe drinking-water. As the UN stated in 2015, there were some noticeable achievements, yet also many unfulfilled promises.

With the SDG, sustainability and development partnerships will be added to the overarching norms of this ambitious programme. The SDG should not be reduced to poverty eradication, but must address all dimensions of sustainable development. In particular, global environmental change must be incorporated, otherwise even poverty eradication will become impossible. Up to now, insufficient attention has been paid to this link in the on-going discourse on the SDG.

In particular, scientific advisors say that guard rails will have to be incorporated into its framework. They regard global anthropogenic climate change (2°C); ocean acidification, loss of biodiversity and ecosystem services, land and soil degradation, risks posed by persistent anthropogenic substances and products (plastic and atomic waste), and finite, non-renewable strategic resources (e.g., phosphorus).[35] In this regard the WBGU argues as follows:

> (1) guard rails do not restrict development for the poorest. (2) Rather, compliance with the guard rails is a necessary pre-requisite for poverty eradication and development. (3) The consumption decisions and lifestyles of the middle and upper classes are currently making the biggest contribution to the causes of global environmental problems. Their

requires a careful analysis of the causes. The alarmingly high national debt many countries now hold, as compared to gross domestic product (GDP), is by no means simply a result of fiscal countermeasures taken after the global financial crisis of 2008-2009. It is just as much a product of a policy, carried out for decades and widely accepted, of perpetually borrowing money for public budgets." (Cf. "Tie Your Hands, Please.")

35 WGBU, *World in Transition*, 2011

consumption decisions and lifestyles should be transformed in a sustainable direction. (4) Furthermore, guard rails highlight the need for international cooperation and coordination by means of global governance to protect global common goods. Humanity must therefore find ways to achieve human progress within the planetary guard rails.[36]

4 Time Pressure

The decisive impetus for futurisation is time pressure, particularly caused by climate change. The Kyoto Protocol, an international treaty adopted in 1997 based on the 1992 United Nations Framework Convention on Climate Change (UNFCCC) has committed state parties to reduce greenhouse gas emissions, based on the premise that man-made global warming exists. 192 parties had signed the protocol. In several Conferences of Parties (COP), the parties have agreed upon that if nothing is done within a window of opportunity of between one and two decades, then any attempt at mitigation will become obsolete. Instead, the prospect of another, dystopian world will have to be entertained:

> The inundation of coastal cities; increasing risks for food production potentially leading to higher malnutrition rates; many dry regions becoming dryer, wet regions wetter, unprecedented heat waves in many regions, especially in the tropics; substantially exacerbated water scarcity in many

36 The anthropological and behaviouralist foundations of cooperation are discussed in Messner, Dirk, and Silke Weinlich, ed., *Global Cooperation and the Human Factor in International Relations* (Routledge: Oxford, 2016); see also the sociological contribution of the "theory of gift" introduced by Mauss, Marcel, "Essai sur le don: Formes et raisons de l'échange dans les sociétés primitives," in *Les classiques des sciences sociales* (Chicoutim: Université du Québec à Chicoutimi, 2002), accessed April 5, 2017. <http://classiques.uqac.ca/classiques/mauss_marcel/socio_et_anthropo/2_essai_sur_le_don/essai_sur_le_don.pdf>; cf. Moebius, Stephan, *Marcel Mauss* (Konstanz: UVK, 2006). It was elaborated upon in Ramel, Frédéric, "Marcel Mauss et l'étude des relations internationales: un héritage oublié," *Sociologie et sociétés* 36.2 (2004): 227-245; Adloff, Frank, and Steffen Mau, "Zur Theorie der Gabe und der Reziprozität," in *Vom Geben und Nehmen: Zur Soziologie der Reziprozität*, ed. Adloff, Frank, and Steffen Mau (Frankfurt a.M.: Campus, 2005), 9-57; Caillé, Alain, "Weder methodologischer Holismus noch methodologischer Individualismus: Marcel Mauss und das Paradigma der Gabe," in *Gift – Marcel Mauss' Kulturtheorie der Gabe*, ed. Stephan Moebius, and Christian Papilloud (Wiesbaden: Verlag für Sozialwissenschaften, 2006), 161-214; Adolff, Frank, and Christian Papilloud, "Einleitung," in *Anthropologie du don*, ed. Alain Caillé, (Frankfurt a.M.: Campus, 2008), 7-39.

regions; increased frequency of high-intensity tropical cyclones; and irreversible loss of biodiversity, including coral reef systems.[37]

Findings in the natural sciences lead almost exclusively to the conclusion that industrial development must be radically and swiftly turned around (in other words, developments shown in the ominous hockey-stick graphs have to be reversed)—meaning that the kind of development that took two centuries to spread across the globe step by step and from region to region as a result of the Industrial Revolution now has to proceed everywhere, within a period of around thirty to fifty years, on a completely different basis (therefore, a raft of authors speak of a "Great Transformation").[38]

By acknowledging this, the OECD and G7/8 and G20 states have by no means intended to introduce a planned economy. Consensus as to the basis for futurisation hinges on this: the desire to correct the maldevelopments of the past (specifically with regard to the labour market and to budgetary policy, but also generally with regard to industrialisation and to relations between "North" and "South") and—to use the standard figure of speech—, to safeguard the opportunities of future generations. Putting reductions front and centre in terms of policy goals should mean putting an end to the mentality of cutting costs, of passing on interest burdens, risks and the responsibility for technological innovation. Regarding the MDGs it should not mean reducing social services or putting an end to development. Rather, these should be decoupled from the received imperatives of economic growth and a system of production based on fossil fuels.

5 Saddle Period 2 (2000-2100)?

The concept of planetary boundaries emerged in 2009 from an international and interdisciplinary working group centred upon the Stockholm Environment Institute. The group identified nine areas of risk, seven of which were presented in numerical terms. In three areas, the limits had already been exceeded: climate change, the drop in biological diversity, and the release of nitrogen into the biosphere. Those areas that the research group identified as posing immanent risks are the stratospheric ozone layer, changes in land use, water

37 World Bank, *Turn Down the Heat* (New York: International Bank for Reconstruction and Development/The World Bank, 2012).

38 WBGU, *World in Transition*, 2011.

use, ocean acidification, the release of phosphorous into the biosphere and the sea, atmospheric aerosol loading, and chemical pollution.

The strain of human activities on the Earth system is now such that, in the eyes of natural scientists, sudden global changes in the environment can no longer be ruled out. In identifying tipping points in the Earth system, Tim Lenton, John Schellnhuber and others[39] are concerned not with an apocalyptic vision but scenarios based on measurable natural phenomena. These tipping points constitute a cornerstone of the Anthropocene thesis, according to which *human* intervention in nature has become a key geophysical determinant.[40] While it is possible that a cascade effect will ensue, it is neither inevitable nor unstoppable. But the prospect does demand—wherein lies the reflexive dimension of a geo-scientific anthropocentrism—robust human action of a corrective nature.

One way of interpreting the concept of planetary boundaries[41] is the following: in order to be sure of continuing to live with some degree of safety, humans must act within certain critical environmental boundaries and respect the Earth system's climatic, geophysical, atmospheric and ecological processes as well as their systemic interactions. Overstepping these boundaries may have devastating effects on humanity's chances of survival. Climate protection is the outstanding example of how natural crises jeopardise the future, curb a society's options, impose time pressures, and render political and economic decisions subject to—one might say—the reservations of nature itself. Until now, this has been scarcely acknowledged in political research and comes across among decision-makers as a narcissistic disease, if animate and inanimate nature forces them into a corner. They have to act with and behind guard rails.

The guard rails articulated as a result of the findings offer a clear indication of how futurisation was initially translated into operational policy. The underlying insight is that, for example, a warming of the Earth by two degrees Celsius

---

39 Lenton, Timothy M., and Hans Joachim Schellnhuber, "*Tipping the Scales,*" *Nature Reports Climate Change* 1.22 (2007). doi:10.1038/climate.2007.65; Lenton, Timothy M., et al., "*Tipping Elements in the Earth's Climate System,*" *PNAS* 105.6 (2008): 1786-1793. doi:10.1073/pnas.0705414105.

40 Leggewie, Claus, "Der Menschen entscheidet im Anthropozän. Konsequenzen aus der Entdeckung eines neuen Erdzeitalters," in *Sind Umweltkrisen Krisen der Natur oder der Kultur?*, ed. Bernd Herrmann (Berlin: Springer, 2015), 63-78 doi:10.1007/978-3-662-48139-4_5.

41 Rockström, Johan., et al., "Planetary Boundaries: Exploring the Safe Operating Space for Humanity," *Ecology and Society* 14.2 (2009): Art. 32, accessed November 9, 2015. <http://www.ecologyandsociety.org/vol14/iss2/art32/>.

could have destructive, even catastrophic effects; this knowledge has during recent years prompted political actors worldwide to highlight the guard rails in guidelines on global climate negotiations (G8/20 Summit 2009, Copenhagen 2009, G7 Summit 2015). All of the climate and energy policy targets established in the multi-tiered political system, from the World Bank through the European Union to local authorities, are no longer merely rhetorically oriented toward these guard rails and implementation has also begun in many newly industrialised countries. This is supposed to provide the basis for a binding contract under public international law in 2015. For, given the current trend in greenhouse gas emissions, we can expect the average temperature on Earth to rise by three or (considerably) more degrees by the end of the century.

Regarding the scientific evidence for the probability of planetary boundaries and the necessity of guard rails as well the political conclusions from this evidence, we may not yet talk about a new *Sattelzeit*. But the paradigm shift is supposed to be fundamental. In the rich OECD countries, as in the newly industrialised countries, the assumption persisted up until now that there must be a constant increase in wealth and security, only occasionally interrupted by cyclical crises and the odd catastrophe; along with the implicit rule that future options must not be robbed of opportunities by being overly rash when making provisions available in the here and now. However, the view is now widely shared that it is by no means certain that children will be better off than their parents and grandparents. Thus, what results from many different sources and motives is a politics in the mode of "Future II" (future perfect progressive)—one assesses from the perspective of the coming decades, what one will have had to have done in order to be able to achieve the goals declared today.[42]

6 Discussion: Possible Objections to Futurisation

Let me now address briefly possible objections to a paradigm shift and examine a range of sceptical objections, in particular for democratic governance.

A) *"Futurisation" vs. Presentism*

Democracy is government *pro tempore*, democratic politics deal with people 'here and now'. Under pressure from urgent ecosystemic needs, it is worth to remember:

42 Leggewie, Claus, and Harald Welzer, *Das Ende der Welt, wie wir sie kannten: Klima, Zukunft und die Chancen der Demokratie* (Frankfurt a.M.: S. Fischer, 2010).

> Compared to other forms of government, democracy is not disposed to sacrifice citizens or a whole generation for some distant future goal. It is less vulnerable to the claims of utopian idealists, religious zealots, or radical revolutionaries who call for great sacrifices from the present generation to bring about even greater good for the future of mankind.[43]

Yet, it is absolutely not unusual for actors within the political system to look to the future. As a rule, career politicians even define their nobler tasks as to 'introduce innovation' or to 'shape things', i.e. to deliver a design or a figuration of possible futures. Modern politics is conducted in line with the innovation paradigm, to name just two outstanding approaches from political theory: the systemic function of politics in structural functionalism was tagged "goal attainment",[44] whereas for Hannah Arendt, it was precisely the "capacity to begin"[45] that, in accordance with classical political ideas of antiquity and early Christianity, was the quintessence of the political. Thus the political sphere, often, alas, diverted by its presentism, is accustomed and devoted to dealing with "things future" Thus,

> this virtue of democracy becomes a vice when the good of future citizens is at stake. Presentism manifests itself in laws that neglect of long-term environmental risks, the consequences of genetic engineering, problems of populations growth, and development of the democratic process itself.[46]

B) *One Cannot Foresee the Future*

The methodology for forecasting future scenarios is soft, unreliable, biased toward the present and dependent upon existing ways of thinking. A whole branch of mostly interdisciplinary future studies (*Zukunftsforschung, futurología, futurologie*) surfaced during the Cold War only to disgrace itself by drawing on the wrong indicators and poorly conceived scenarios,[47] which often revealed more about the present and the idiosyncrasies of futures

43 Thompson, Dennis F., "Representing Future Generations: Political Presentism and Democratic Trusteeship," *Democracy, Equality, and Justice, Critical Review of International Social and Political Philosophy* 13.1 (2010): 17-37.

44 Parsons, Talcott, *The Structure of Social Action* (New York: MacGraw Hill, 1937).

45 Arendt, *The Human Condition.*

46 Thompson, "Representing Future Generations," 17-18.

47 Cf. Andersson, Jenny, "The Great Future Debate and the Struggle for the World," *The American Historical Review* 117.5 (2012): 1411-1430; *American History Review Forum: Histories of the Future: The American Historical Review* 117.5 (2012).

researchers than about the actual future of societies in the East and West that were locked into antagonistic relations.[48] The optimism of cybernetics as applied to political problems became overwhelmed by the dystopia of thermo-nuclear self-destruction; humanity tended "to imagine what it feared [with regard to certain risks], and to fear what it imagined".[49]

Nevertheless, government planners as well as corporate managers and military analysts left the world with a whole range of scenario techniques helping to face the structural condition of uncertainty. Scenarios became popular again with prognostics on climate change and other planetary boundaries and with social constructivism in general.[50] In these approaches, linear thinking was replaced by non-linear system thinking, and the projection of "one future" was substituted by the possibility of multiple futures. Scenario techniques focus more on the process perspective than on prediction, more on foresight than forecast. Probabilistic thinking is nowadays complemented by possibilistic views.[51] And it needs the imaginative power of "science fiction" and insights from storytelling in literature, film and games where dystopian narratives reign[52] and imaginations of a "better future" catch up.[53]

C) *The Democratic Deficit*

There is another worrying consequence of policy prescriptions deducted from the time pressure: related negotiations and democratic processes play for or 'buy' time in order to diffuse conflicts of interest and reach compromises.[54] A bypassing of established checks and balances, accelerated by a decisionism that rests on climate and energy politics, could threaten the poise of pluralistic

---

48 Gerhold, Lars et al., *Standards und Gütekriterien der Zukunftsforschung: Ein Handbuch für Wissenschaft und Praxis* (Wiesbaden: Springer, 2015).

49 The formula was uttered by the artist Adam Curtis in a talk with the author during "Tumbetalk 3" at the Ruhrtriennale- Festival on September 1, 2013.

50 Fuller, Ted, and Krista Loogma, "Constructing Futures: A Social Constructionist Perspective on Foresight Methodology," *Futures* 41.2 (2009): 71-79.

51 Clarke, Lee, "Possibilistic Thinking: A New Conceptual Tool for Thinking about Extreme Events," *Social Research* 75. 3 (2008): 669-691.

52 Horn, Eva, *Zukunft als Katastrophe* (Frankfurt a.M.: S. Fischer Wissenschaft, 2014); Kaplan, E. Ann, *Climate Trauma: Foreseeing the Future in Dystopian Film and Fiction* (New Brunswick: Rutgers University Press, 2015).

53 E.g. Ogilvy, James A., Creating Better Futures: Scenario Planning as a Tool for a Better Tomorrow (New York, NY: Oxford University Press, 2002); WBGU, World in Transition.

54 Günther, Klaus, *Politik des Kompromisses: Dissensmanagement in pluralistischen Demokratien* (Wiesbaden: VS Verlag, 2006).

societies; an aversion to loss and the defiance of morally justified limits to freedom is the result.

It is however only right that presentism, the orientation of political discourse toward the present, should be overcome; as should political actors who depend upon ever-shorter cycles of popularity as indicated by opinion polls, on electoral opportunism and media rankings—while ignoring many issues that require more working time than a single legislative term allows. Put as a norm, democracies should cope with ecological risks, yet many observers claim their inferiority, compared to autocratic regimes, in coming to terms with climate change and other planetary boundaries. The daunting hypothesis is rejected by a range of empirical studies[55] and there is some evidence that democracies can cope with these challenges. Nevertheless, they are stressed and in "things future" of well-equipped for public deliberation and consultation. Coming generations are systematically neglected in everyday bargaining and unattributed in conflict-resolution.

D) *Do Generations Matter?*

Any reference to 'future' in politics has to be firmed up in terms of intergenerational relations, an aspect that has become firmly embedded in the concept of sustainability. The undeservingly neglected concept of "temporality" (*Zeitlichkeit*) has only been weakly operationalised in social research, despite the pioneering works of Karl Mannheim[56] and Joachim Matthes[57]; which may come to no surprise as social and cultural sciences are predominantly concerned with spatially organised groups. Generations, or subjectively convergent peers, are not 'groups' in particular 'zones'; relations between generations hinge on ways of perceiving the world that are chronologically opposed to one another, with which people identify and, thus, render the synchronism (or, in this case: asynchronism) of their otherness as negotiable. It is within generations that the complex temporality of modern societies arranges itself. However, this can scarcely be illustrated by the way parents and children relate to one another within (enlarged) families; but the "cultural implications of people behaving in a way befitting of the stage of life in which they

55 Bättig, Michèle B., and Thomas Bernauer, "National Institutions and Global Public Goods: Are Democracies More Cooperative in Climate Change Policy?" *International Organization* 63.2 (2009): 281-308.

56 Mannheim, Karl, "The Problem of Generations," in *Essays on the Sociology of Knowledge: Collected Works,* vol. 5, ed. Paul Kecskemeti (New York: Routledge 1952): 276-322.

57 Matthes, Joachim, "Karl Mannheims 'Das Problem der Generationen', neu gelesen: Generationen-'Gruppen' oder 'gesellschaftliche Regelung von Zeitlichkeit'," *Zeitschrift für Soziologie,* 14.5 (1985): 363-372.

find themselves"[58] can no longer be resolved within the context of family relations.

This concept of generation, closely connected to a notion of social temporality, is underpinned by what may upon first sight seem like a simplistic notion of sustainability: always replant as many trees as you cut down. Such a notion serves one's own needs with regards to reproduction but also, above all—since trees require long stretches of time to grow and thrive—secures the basis for the subsistence of future generations.[59] This aspect of providing for the future is fundamental to rural communities, where its patriarchal flipside is no less apparent: the elderly expect from their heirs that established practices are continued and set them upon a path from which they are pressured hard to diverge from without paying the price in terms of generational conflict.

With the notion of generations as a central concern, another problem for democratic polities arises. Democracy can be considered as hinging on one central idea: decisions that affect citizens should be made or supervised by citizens themselves. This concord rule has already been undermined by economic, cultural and also political globalisation, the effects of which lead to the break-up of the nation-state as container. The sovereign state is increasingly exposed to decisions made externally that it cannot justify; at the same time, decisions made 'here' affect residents 'outside', in other regions of the world.

This does not only apply spatially but temporally, too: decisions that fall to the sovereign state 'today' affect people 'tomorrow'. This is not fundamentally new: infrastructure, conflicts, debts, etc. have always led to obligations that must be fulfilled by the next generation and their children and further generations to come. But anyone who, for example, committed themselves to the civil use of nuclear power in the 1950s, in accordance with a parliamentary majority but without providing a fail proof method for permanently disposing of

58 Matthes, "Karl Mannheims 'Das Problem der Generationen', neu gelesen: Generationen-'Gruppen' oder 'gesellschaftliche Regelung von Zeitlichkeit'," 369.

59 Instead of referring to a plethora of literature on the notion and praxis of sustainability, I quote from the UN-Document "Our Common Future", Chapter 2: "Sustainable development is development that meets the needs of the present without compromising the ability of future generations to meet their own needs. It contains within it two key concepts: the concept of 'needs'—in particular the essential needs of the world's poor, to which overriding priority should be given—, and the idea of limitations imposed by the state of technology and social organisation on the environment's ability to meet present and future needs. Cf. NGO Committee on Education, "Our Common Future, Chapter 2: Sustainable Development," *UN Documents: Gathering a Body of Global Agreements*, Report No. A/42/427, published 1987, accessed November 11, 2015. <http://www.un-documents.net/ocf-02.htm#III.3>.

nuclear waste, intrudes all the more significantly and lastingly upon the freedom of choice of future generations, and the options available to them.[60]

The Brundlandt Report of 1987 already stated: development can only be sustainable if "it meets the needs of the present without compromising the ability of future generations to meet their own needs."[61] Sustainability, when viewed from the perspective of democratic politics, is nothing other than (intergenerational) justice, and justice is the foundation of democracy. It is therefore not sufficient to concentrate exclusively on current constituencies within a given national territory (and to orient policy principally with these constituencies in mind); the scope must be widened to include 'others' elsewhere, both now and in the future.

Politics, however, is oriented toward a people in the 'here and now', not one that is tied to traditions or strives for utopia. We cannot know how future generations wish to live, neither, therefore, can we know whether we shall cause them harm. To pursue a politics in full consciousness of one's responsibilities, one must ensure the continuation of the democratic process and, from the perspective of today, that future generations are spared avoidable damages and impairments.

E) *Symbolic and Virtual Representation of Future People*

'Generation' has been identified above as the only temporal category around which societies are structured. Generations are biosocial hybrids: biological grandparents, parents and children constitute a familial triad, within which material and immaterial values are handed on. This happens during the lifetimes of the persons involved, or within their living memory. The public dimension of this private interaction consists in (re-)distributional measures implemented via social and pension policies, whereby there is always a risk of provoking conflicts between generations.

In view of the 'futurisation of politics', extended intergenerational relations now come into play that reach beyond directly related constellations and personal relationships to more abstract interactions—not between social classes or between regions in the present and in the foreseeable future but between 'today's' situation and that of the 'future'. In this respect, children and youths can only be considered symbolic representatives. Must this form of representation remain symbolic or can it—whether directly or indirectly—become a democratic one?

---

60 For the German case, cf. Radkau, and Hahn, *Aufstieg und Fall der deutschen Atomwirtschaft*.

61 NGO Committee on Education, "Our Common Future," 16.

The recognition of interests that have yet to be considered in the democratic process is known as surrogate representation.[62] This follows the inclusive dynamic of modern democracies that developed from the census suffrage that granted the propertied bourgeoisie certain voting rights through universal male suffrage in the twentieth century to women's suffrage and the lowering of the voting age to 18 (and lower), and to the partial introduction of voting rights for non-nationals in local elections. The notion of intergenerational justice seems to suggest extending the scope of representation temporally, too, and to establish associated institutions within representative democracy that can be developed and made operational.[63]

Social scientists are currently debating "proxy electorates" in connection with the incorporation of "the rights of future generations" into constitutions, with panels of experts devoted to sustainability and with local civil society initiatives that see to it—with more or less vigour—that issues affecting the future are introduced into current policy formulations. Such representation can be incorporated into existing parliaments or lead to the establishment of a new forum; some propose using various forms of children's voting rights to allow decisions that concern the hypothetical will of future generations to be passed on to the parents of the next generation. "Youth quotas" are also under consideration.[64]

---

62 Mansbridge, Jane, "Rethinking Representation," in *American Political Science Review* 97.4 (2003): 515-527.

63 Mayer, Lukas, "Intergenerational Justice," published in *Stanford Encyclopedia of Philosophy*, April 2, 2003, accessed October 13, 2015. <http://plato.stanford.edu/entries/justice-intergenerational/>.

64 Cf. Dobson, Andrew, "Representative Democracy and the Environment," in *Democracy and the Environment*, ed. William M. Laffetty, and James Meadowcroft (Cheltenham: Edward Elgar, 1996), 124-139; Ekeli, Kristian S. , "Giving a Voice to Posterity–Deliberative Democracy and Representation of Future People," *Journal of Agricultural and Environmental Ethics* 18.5 (2005): 429-450; Gosseries, Axel, and Inigo Gonzalez Ricoy, ed., *Institutions for Future Generations* (Oxford: Oxford University Press, 2015); Tännsjö, Torbjörn, "Future People, the All Affected Principle, and the Limits of the Aggregation: Model of Democracy," in *Hommage à Wlodek. 60 Philosophical Papers Dedicated to Wlodek Rabinowicz*, ed. Jonas Josefsson and Dan Egonsson, published by Lund University, 2007, accessed October 13, 2015. <http://www.fil.lu.se/hommageawlodek/site/papper/TannsjoTorbjorn.pdf>; Beckman, Ludvig, "Do Global Climate Change and the Interest of Future Generations Have Implications for Democracy?" *Environmental Politics* 4 (2008), 610-624; Beckman, Ludvig, "Democracy and Future Generations: Should the Unborn Have a Voice?" in *Spheres of Global Justice: Volume 2. Fair Distribution–Global Economic, Social and Intergenerational Justice*, ed. Jean-Christophe Merle (Dordrecht: Springer, 2013), 775-788. doi:10.1007/978-94-007-5998-5_62; Caney, Simon, "Political Institutions for the Future:

Attempts to root the needs of future generations in institutions (the creation of a virtual 'posterity') can be considered in three variants: a 'soft' constellation of trustees and tribunes dealing with issues that affect the future; the (democratically problematic) distribution of future rights among parents or ombudspersons; and the creation of advisory councils and parliamentary fora, which as consultative institutions would demand more consideration from other bodies within a democracy based on the principle of separation of powers. Whether a convincing level of institutionalisation succeeds can only ultimately be determined in practice. It will, however, certainly be necessary to overcome, on both an epistemic and moral level, the presentism that dominates today's democracies.

F) *Is Futurisation Just?*

Considerations of intergenerational and interregional justice are overlain by the intrasocietal debate on distribution, to which there are two sides: which social actors contribute most to climate change, and which must sacrifice the most in the name of climate protection? The effects of distribution become apparent both with regard to the strain placed on the environment and to the improvement of the quality of the environment, both of which are politically explosive issues. After all, the chances of the global energy transition succeeding and of the project being realised (and of sustainability goals succeeding in general) essentially hinge on questions of distribution. The politics of environment and climate is no island. It primarily concerns avoiding harmful emissions but, since the energy transition, it has increasingly become a matter of economic and industrial policy. The dominant paradigm is that of the green economy or green growth, which is supposed to prompt economic growth, marketable products, jobs and profits, but all based on a fundamentally different use of resources—and there is evidence to prove that it is succeeding (at least partially) in doing so.

The portfolio of an environmental and social politics oriented toward the future is not limited to indirect welfare effects such as the generation of incomes and jobs, the reduction of subventions and trade deficits, and eschewing both energy dependency and cumulative environmental degradation, as significant as these may be. The maintenance of ecosystem services or 'natural capital' is, for future generations, a basic requirement for life. A thoroughly

A Five-Fold Package," in *Institutions for Future Generations*, ed. Axel Gosseries, and Inigo Gonzalez Ricoy (Oxford: Oxford University Press, 2015), forthcoming; Ekeli, Kristian S., "Constitutional Experiments: Representing Future Generations Through Submajority Pules," *The Journal of Political Philosophy* 17.4 (2009): 440-461.

thought through precautionary principle can support a sustained level of development across generations and regions alike and counter the dominant mentality of cutting costs. This is only about new technologies but also about the form they take in society; it would be safe to assume that the post-fossil energy sector can be brought closer to people and accommodate a greater level of participation.[65]

According to the classic liberal social contract, those alive today cannot pass down orders to future generations because their needs are not known and it is therefore not permissible to deny them certain options. We cannot, today, force anyone to do anything in the future. While this remains true in principle, advocates of more robust generational justice say that today, given the limits of the Earth system, we must in a way 'voluntarily force' ourselves to take the likely needs and options of future generations into consideration and, in the process, to ensure that the positive aspects of such a transformation are not once again distributed unequally, along the same lines as the injustices that have increased throughout history.

7 Conclusion

Perhaps, then, not a new *Sattelzeit*, but likely a postmodern 'history writing of the future' is on the agenda. This does not rest upon a philosophy of history, neither does it have the utopia of "an alternative society" in its sights. It is rather as if mankind is on an expedition; the expedition will continue, based on what lies 'within sight'. No law of history pulls actors along with it; rather, they feel their way based on improvised experimentation. Thus, at best, a pluralistic narrative of possible developments emerges, and not a strict master narrative or even master plan. The future is not sketched out as a teleology, but rather made plausible through stories and scenarios.[66] In addition to which, one must familiarise oneself with the complex calculations based on modelling in the natural sciences and in engineering disciplines, and demand that more sociocultural and political reflexivity be factored into these. Predictions regarding climate have been firmed up since data collected during the previous two decades has been matching with previous predictions or exceeding them; however, the cultural foundations that would give meaning to these predictions are lacking, along with sociological fantasy and a system of economic

65 Leggewie, Claus, and Patrizia Nanz, *Die Konsultative* (Berlin: Wagenbach, 2015).

66 At the Centre for Global Cooperation Research (University du Duisburg, <http://www.gcr21.de>) a working group entitled "Models, Scenarios, Narratives" has been established.

calculations capable of connecting with life—in other words, everything that constitutes the signature of a great transformation. The social sciences and the humanities can content themselves in this situation with critiquing the worsening circumstances and devote themselves to a radical constructivism—or, together with other actors, exercise some influence over the narratives that are currently woven into designs for the future; even if these are becoming somewhat difficult to decipher as a result of a mixture of engagement and alienation.

This also means that, in a normative sense, sustainability policies demand more, not less democracy. As the French pioneer of political ecology, André Gorz, once said, the politics of environmental protection, which is not necessarily a politics of freedom, merely propagates

> the development of the industrial technosphere [...] without compromising the self-regenerating capacities of the ecosphere. This approach does not involve a fundamental break with industrialism and its hegemony of instrumental reason.[67]

And prohibitions, regulations, taxes, subsidies and fines only increase the extent to which society is subjected to external controls. Ecological politics don't work without the "need for change in the mentalities, values, motivations or economic interests of the social actor". If this change is to take place, it will not be as a result of some sort of authoritarian state measures or the rule of "green" experts. The defence of the lifeworld is defined instead as a set of activities that correspond "to the intentions that gave rise to them; in other words, that social individuals see, understand and accept the result as the outcome of their actions".[68] For Gorz, in democratic terms this constituted the chief attraction of ecology, which is intimately linked to

> the problem of retroactive coupling of necessity with normativeness or, if you prefer, the transformation of objective necessities into normative behaviour corresponding to needs that have been experienced, and that influence the form in which the objective necessities are perceived.[69]

---

67 André Gorz, "Die politische Ökologie zwischen Expertokratie und Selbstbegrenzung," in *Auswege aus dem Kapitalismus. Beiträge zur Politischen Ökologie,* ed. André Gorz (Zürich: Rotpunktverlag, 2009), 33 (my translation).

68 Ibid.

69 Ibid, 40.

Completely in line with the contractual theory of Hobbes', through Rousseau to Locke, and Kant, the great transformation is supposed to open up spaces of freedom, even if it definitely rules out certain options. The responsibility of the (*consumer*) *citizen* remains true to their ability to reflectively assess themselves, particularly where the relation between *second-order* and *first-order desires* is concerned.[70] A person's will is only free if they identify with their wishes at a higher level and *second-order* desires take effect.

Future, "that shadowy, inaccessible place" (James Gleick), can thus be anticipated by intelligent, careful, and imaginative people. Modernity has first invented past and future, whereas

> (i)n bygone times, people had no more than the barest glimmerings of visiting either the future or the past. It seldom occurred to anyone. It wasn't in their repertoire. Even travel through space was rare, by modern standards, and slow, before the railroads came.

Railroads and telecommunications marked the beginning of a new phase of globalization in space, contemporary futurism and futurology were the artistic and scientific symbols of that breakthrough discovery in the fin de siècle. A hundred years later, with the globalization of time, the creative mind "says we are here because things have yet to be".[71]

---

70 Frankfurt, Harry, *Freiheit und Selbstbestimmung: Ausgewählte Texte* (Berlin: Akademie-Verlag, 2001).

71 Wells, H.G., *The Time Machine* (1895), quoted in Gleick, James, (New York: Pantheon Books, 2016), 28.

Bibliography

50Hertz Transmission GmbH, Amprion GmbH, EnBW Transportnetze AG, and TenneT TSO GmbH. "Prognose der EEG-Umlage nach AusglMechV: Prognosekonzept und Berechnung der ÜNB, Oktober 14, 2011." Accessed July 14, 2015. <http://www.netz-transparenz.de/de/EEG-Umlage-2012.htm>.

"A 50th Anniversary Few Remeber: LBJ's Warning on Carbon Dioxode." Published by *The Daily Climate*. Accessed July 12, 2016. <http://www.dailyclimate.org/tdc-newsroom/2015/02/president-johnson-carbon-climate-warning>.

Abbot, George. *An Exposition Upon the Prophet Ionah*. London, 1600.

Abel, Wilhelm. *Agrarkrisen und Agrarkonjunktur*. Hamburg, Berlin: Parey, 1978.

Actas de las Cortes de Castilla. 17 Vols. Madrid, 1861-1891.

Adloff, Frank, and Christian Papilloud. "Einleitung." In *Anthropologie du don*, edited by Alain Caillé. Frankfurt a.M.: Campus, 2008.

Adloff, Frank, and Steffen Mau. "Zur Theorie der Gabe und der Reziprozität." In *Vom Geben und Nehmen: Zur Soziologie der Reziprozität*, edited by Frank Adloff, and Steffen Mau, 9-57. Frankfurt a.M.: Campus, 2005.

Adorno, Theordor W. "Was bedeutet Aufarbeitung der Vergangenheit? [1959]" In *Theodor W. Adorno: Gesammelte Schriften* 10.2, 555-572. Frankfurt a.M.: Suhrkamp 1977.

Adriana, Mica, Arkadiusz Peisert, Jan Winczorek, ed. *Sociology and the Unintended. Robert Merton Revisited*. Frankfurt a.M., 2011.

AG Energiebilanzen. "Bruttostromerzeugung in Deutschland von 1990 - 2014". Accessed July 14, 2015. <http://www.ag-energiebilanzen.de>.

Agentur für erneuerbare Energien. "Eigentumsverteilung erneuerbarer Energien – Anlagen 2012." Accessed July 14, 2015. <http://www.unendlich-viel-energie.de/media/image/1121.AEE_Erneuerbare_Energien_in_Buergerhand_2012_apr13.jpg>.

Alemán, Mateo. *Guzmán de Alfarache*. 2 Parts. Madrid 1599, Lisbon, 1604.

Alexandre, Pierre. *Le climat en Europe au Moyen Âge*. Paris: Éditions de l'Ecole des Hautes Études en Sciences Sociales, 1987.

Amann, Klaus, and Stefan Hirschauer. "Die Befremdung der eigenen Kultur: Ein Programm." In *Die Befremdung der eigenen Kultur: Zur ethnographischen Herausforderung soziologischer Empirie*, edited by Stefan Hirschauer, 7-52. Frankfurt a.M.: Suhrkamp, 1997.

American History Review Forum. "Histories of the Future." The American Historical Review 117.5 (2012).

Amery, Carl. *Das Ende der Vorsehung: Die gnadenlosen Folgen des Christentums*. Hamburg: Rowohlt, 1972.

Andersson, Jenny. "The Great Future Debate and the Struggle for the World." *American Historical Review* 117.5 (2012): 1411-1430.

Andresen, Steinar, and Agrawala Shardul. "Leaders, Pushers and Laggards in the Making of the Climate Regime." Global Environmental Change 12.1 (2002): 41-51.

Anonymous [Henry Arthington]. *Provision for the Poore, Now in Penurie, Out of the Store-House of God's Plenty*. London, 1597.

Anonymous. *Relazione delle feste fatte in Fiorenza sopra il ghiaccio del fivme d'Arno l'vltimo dì di dicembre MDCIV.* Florence, 1604.

Anonymus, ed. "[…] Die Anonyme Chronik bei Schnitt, sammt Fortsetzung: 1495-1541 […]." In *Basler Chroniken*, vol. 6, edited by Historische und Antiquarische Gesellschaft in Basel, 23-73. Leipzig: S. Hirzel, 1902.

Anonymus. MS 2543. Ghent: University Library.

Appadurai, Arjun. "Theory in Anthropology: Center and Periphery." *Comparative Studies in Society and History* 28.1 (1986): 356-361.

Appleby, Andrew B. *Famine in Tudor and Stuart England.* Liverpool: Liverpool University Press, 1978.

Archer, David, and Raymond T. Pierrehumbert, ed. *The Warming Papers: The Scientific Foundation for the Climate Change Forecast.* Hoboken: Wiley-Blackwell, 2011.

Arendt, Hannah. *The Human Condition.* Chicago: University of Chicago Press, 1958.

Arens, Bas, M.A.M. Löffler, and E.M. Nuijen. *Evaluatie Dynamisch Kustbeheer Friese Waddeneilanden.* Published 2007. Accessed January 25, 2014. <https://publicwiki.deltares.nl/download/attachments/17334280/Arens+ea+%282007%29+Dynamisch+Kustbeheer+Friese+Waddeneilanden.pdf>.

Assmann, Aleida. *Ist die Zeit aus den Fugen? Aufstieg und Fall des Zeitregimes der Moderne.* München: Hanser, 2013.

Atkinson, Ernest G., ed. *Calendar of State Papers Relating to Ireland of the Reign of Elizabeth, 1600-1601.* London: Public Record Office, 1905.

Atwell, William S. "Ming Observers of Ming Decline: Some Chinese Views on the '17th Century Crisis' in Comparative Perspective." *Journal of the Royal Asiatic Society of Great Britain and Ireland* 120 (1988): 316-348.

Atwell, William S. "Volcanism and Short-Term Climatic Change in East Asian and World History, c. 1200-1699." *Journal of World History* 12 (2001): 29-98.

Auer, Johann, and Joseph Ratzinger. *Kleine Katholische Dogmatik Bd. III: Die Welt – Gottes Schöpfung.* Regensburg: Pustet, 1975.

Bacon de Verulamio, Franciscus, *Novum Organum Scientiarum*, Editio Secunda (Amstelædami: Joannes Ravesteinius, 1660), 27 (Lib. 1, Aph. 1), 84-85 (Lib. 1, Aph. 82), 154-158 (Lib. II, Aph. 6-10). Accessed March 27, 2017. <http://www.mdz-nbn-resolving.de/urn/resolver.pl?urn=urn:nbn:de:bvb:12-bsb10042656-8>.

Baechtold, Jacob. *Hans Salat: Ein schweizerischer Chronist und Dichter aus der ersten Hälfte des XVI. Jahrhunderts: Sein Leben und seine Schriften.* Basel: Bahnmaier's Verlag, 1876.

Bähler, Arnold, ed. *Bendicht Rechbergers Bielerchronik 1524-1566*. Biel, 1902. Reprinted Biel: Werner Ruck-Beck, 1989.

Bals, Christoph, Charlotte Cuntz, Oldag Caspar, and Jan Burck. *The End of EU Climate Leadership*. Bonn: Germanwatch, 2013.

Barbazza, Marie-Catherine. *La société paysanne en Nouvelle Castille: Famille, mariage et transmission de biens à Pozuelo de Aravaca (1580-1640)*. Madrid: Casa de Velázquez, 2000.

Barclay, Alexander ed. [Sebastian Brant]. *The Ship of Fools*. 2 vols. Edinburgh, London: William Paterson/ Henry Sotheran 1874.

Barcley, John. *Spiegel Menschlicher Gemüths Neigungen*. Frankfurt a.M.: Erhard Berger, 1660. Accessed March 27, 2017. <http://www.mdz-nbn-resolving.de/urn/resolver.pl?urn=urn:nbn:de:bvb:12-bsb10913288-8>.

Bard, Edouard, Grant Raisbeck, Françoise Yiou, and Jean Jouzel, *Reconstructed Solar Irradiance Data. IGBP PAGES/World Data Center for Paleoclimatology Data Contribution Series #2003-006*. Boulder: National Climatic Data Center, 2003.

Bardi, Ugo. *Extracted: How the Quest for Mineral Wealth Is Plundering the Planet: A Report to the Club of Rome*. White River Junction. Vermont: Chelsea Green Publishing, 2014.

Barriendos, Mariano, and Fernando S. Rodrigo. "Study of Historical Flood Events of Spanish Rivers Using Documentary Data." *Hydrological Sciences* 51.5 (2006): 765-783.

Barroso, José Manuel Durão. "Statement of President Barroso at the UN High Level Segment COP 15 Copenhagen, 16 December 2009." Speech/09/583. Accessed June 15, 2011. <http://europa.eu/rapid/pressReleasesAction.do?reference=SPEECH/09/583>.

Barroso, José Manuel Durão, and Connie Hedegaard. "Statements by the President of the European Commission, José Manuel Barroso, and Connie Hedegaard, European Commissioner for Climate Action on the Cancún Agreement on Climate Change." MEMO/10/673. Published December 11, 2010. Accessed May 28, 2013. <http://europa.eu/rapid/pressReleasesAction.do?reference=MEMO/10/673&format=HTML&aged=0&language=EN&guiLanguage=fr>.

Bättig, Michèle B., and Thomas Bernauer. "National Institutions and Global Public Goods: Are Democracies More Cooperative in Climate Change Policy?" *International Organization* 63.2 (2009): 281-308.

Bauer, Steffen, and Clara Brandi. "Marrakesh Climate Conference: Time to Act!" *Diplomatisches Magazin* 10 (2016): 30-33.

Baulant, Michèle, and Jean Meuvret. *Prix des céréales de la mercuriale de Paris I: 1520-1620*. Paris: SEVPEN, 1960.

Baykal, Baykal. S., ed. *Peçevi Tarihi* [The Chroncle of Pecevi]. 2 Vols. Ankara: Kültür ve Turizm Bakanlığı, 1982.

Becker, Karin éd. *La pluie et le beau temps dans la littérature Française. Discours scientifique et transformations littéraires, du Moyen Âge à lépoque moderne*. Paris: Hermann, 2012.

Beckman, Ludvig. "Do Global Climate Change and the Interest of Future Generations Have Implications for Democracy?" *Environmental Politics* 4 (2008), 610-624.

Beckman, Ludvig. "Democracy and Future Generations: Should the Unborn Have a Voice?" In *Spheres of Global Justice: Volume 2. Fair Distribution–Global Economic, Social and Intergenerational Justice*, edited by Jean-Christophe Merle, 775-788. Dordrecht: Springer, 2013. doi:10.1007/978-94-007-5998-5_62.

Begert, Michael, Thomas Schlegel, and Walter Kirchhofer. "Homogeneous Temperature and Precipitation Series of Switzerland from 1864 to 2000." *International Journal of Climatology* 25.1 (2005): 65-80. doi:10.1002/joc.1118.

Behringer, Wolfgang. *Witches and Witch-Hunts: A Global History*. Cambridge: Polity Press, 2004.

Behringer, Wolfgang. *Kulturgeschichte des Klimas: Von der Eiszeit zur globalen Erwärmung*. München: Beck, 2009.

Behringer, Wolfgang, Hartmut Lehmann, and Christian Pfister, ed. *Kulturelle Konsequenzen der Kleinen Eiszeit—Cultural Consequences of the Little Ice Age*. Göttingen: Vandenhoeck & Ruprecht, 2005.

Behringer, Wolfgang. *Kulturgeschichte des Klimas. Von der Eiszeit zur globalen Erwärmung*. München: Beck, 2009.

Benedict, Philip. "Civil War and Natural Disaster in Northern France." In *The European Crisis of the 1590s: Essays in Comparative History*, edited by Peter Clark, 84-105. London: George Allen & Unwin, 1985.

Bennassar, Bartolomé. *Recherches sur les grandes épidémies dans le nord de l'Espagne à la fin du XVIe siècle: Problèmes de documentation et de méthode*. Paris: SEVPEN, 1969.

Bennassar, Bartolomé. *Valladolid au siècle d'Or: Une ville de Castille et sa campagne au XVIe siècle: Problèmes de documentation et de méthode*. Paris, The Hague: Mouton, 1967.

Bennett, Jane. *Vibrant Matter: A Political Ecology of Things. Durham*. N.C.: Duke University Press, 2010.

Benz, Arthur, and Yannis Papadopoulos. "Actors, Institutions and Democratic Governance: Comparing Across Levels." In *Governance and Democracy: Comparing National, European and International Experiences*, edited by Arthur Benz und Yannis Papadopoulos, 273-295. London: Routledge, 2006.

Benz, Arthur, Susanne Lütz, Uwe Schimank, and Georg Simonis. "Einleitung." In *Handbuch Governance: Theoretische Grundlagen und empirische Anwendungsfelder*, edited by Arthur Benz, Susanne Lütz, Uwe Schimank, and Georg Simonis, 9-26. Wiesbaden: VS Verlag für Sozialwissenschaften, 2007.

Bercé, Yves. *Histoire des croquants*. Paris: PUF, 1986.

Berger, Peter L., and Thomas Luckmann. *Die gesellschaftliche Konstruktion der Wirklichkeit: Eine Theorie der Wissenssoziologie*. Frankfurt a. M.: Fischer, 2009.

Berry, Mary Elizabeth. *Hideyoshi*. Cambridge: Harvard University Press, 1982.

Biesbroeck, Robbert G., Rob J. Swart, Timothy R. Carter, Caroline Cowan, Thomas Henrichs, and Hanna Mela."Europe Adapts to Climate Change: Comparing National Adaptation Strategies." *Global Environmental Change* 20.3 (2010): 440-450.

Blazejczak, Jürgen, Jochen Diekmann, Dietmar Edler, Claudia Kemfert, Karsten Neuhoff, and Wolf-Peter Schill. "Energiewende erfordert hohe Investitionen." *DIW Wochenbericht* 26 (2013): 19-30.

Blumenberg, Hans. *Paradigms for a Metaphorology*. Ithaca: Cornell University Press, 2010.

Böhme, Gernot, and Hartmut Böhme. *Feuer, Wasser, Erde, Luft. Eine Kulturgeschichte der Elemente*. München: Beck, 1996.

Böhme, Hartmut. "'Was birgt die Wolke?' Zur Kultur- und Kunstgeschichte von Wolken und Wetter." In *Wolkenbilder: Die Erfindung des Himmels*, edited by Stephan Kunz, Johannes Stückelberger, and Beat Wismer, 11-21. München: Hirmer, 2005.

Boff, Leonardo. *Ethik für eine neue Welt*. Düsseldorf: Patmos, 2000.

Bohle, Hans-Georg. "Geographische Entwicklungsforschung." In *Geographie*, edited by Hans Gebhardt, Rüdiger Glaser, Ulrich Radtke, and Paul Reuber, 797-815. Heidelberg: Spektrum, 2007.

Bojanowski, Axel. "Hitze-Jahr 1540: Wetterdaten enthüllen Europas größte Naturkatastrophe." Published by Spiegel Online Wissenschaft, July 2, 2014. Accessed October 14, 2014. <http://www.spiegel.de/wissenschaft/natur/hitze-und-duerre-1540-katastrophe-in-europa-im-mittelalter-a-978654.html>.

Bolz, Norbert, and David Bosshart. *Kult-Marketing: Die neuen Götter des Marktes*. Düsseldorf: ECON, 1995.

Bond, Edward, ed. *Russia at the Close of the Sixteenth Century*. London: Hakluyt Society, 1856.

Borghi, Carlo, ed. *Tomasino de Bianchi detto de Lancellotti: Cronica Modenese, 1538-1540. Monumenti di Storia Patria delle province modenesi, Serie delle Cronache*, vol. 7. Parma: Fiaccadori, 1868.

Borst, Arno. *Das Buch der Naturgeschichte. Plinius und seine Leser im Zeitalter des Pergaments. Abhandlungen der Heidelberger Akademie der Wissenschaften, Philosophisch-historische Klasse, Jg. 1994*, 2. Heidelberg: Winter, 1994.

Botzem, Sebastian, Jeanette Hofmann, Sigrid Quack, Gunnar Folke Schuppert, and Holger Straßheim. "Die Dynamik des Governance-Ansatzes: Vier Dimensionen von Wandel." In *Governance von und durch Wissen: Interdisziplinäre Tagung* "Governance von und durch Wissen" (Berlin), edited by Gunnar Folke Schuppert and Andreas Voßkuhle, 11-28. Baden-Baden: Nomos, 2008.

Bouteiller, Ernest de, ed. *Journal de Jean le Coullon, 1537-1587*. Paris, Nancy: D. Dumoulin, 1881.

Bradley, Raymond S. *Paleoclimatology: Reconstructing Climates of the Quaternary*. San Diego: Academic Press, 1999.

Bräuninger, Michael, and Sven Schulze. "Konsequenzen der Energiewende." Accessed August 17, 2010. <http://www.hwwi.org/fileadmin/hwwi/Publikationen/Studien/HWWI-Studie-Energiewende-2012.pdf>.

Brague, Remi. *Europa, seine Kultur, seine Barbarei. Exzentrische Identität und römische Sekundarität*. Wiesbaden: Springer, 2012.

Brandi, Clara. "EU Climate Leadership in Copenhagen and Beyond." In *The European Union and Global Development: An "Enlightened Superpower" in the Making?*, edited by Stefan Gänzle, Sven Grimm, and Davina Makhan, 175-189. Houndmills: Palgrave Macmillan, 2012.

Brandi, Clara, Steffen Bauer, Pieter Pauw, Sander Chan, and Okka Lou Mathis. *EU Climate Leadership: Five Building Blocks for Ambitious Action*. Bonn: German Development Institute/Deutsches Institut für Entwicklungspolitik (DIE), 2015.

Brant, Sebastian. *Das Narrenschyff*. Basel: Jo[hann] B[ergmann], 1494. Accessed March 27, 2017. <http://nbn-resolving.de/urn:nbn:de:bvb:12-bsb00036978-3>.

Brázdil, Rudolf. "Fir Tree-Ring Reconstruction of March–July Precipitation in Southern Moravia (Czech Republic), 1376-1996." *Climate Research* 20.3 (2002): 223-239. doi:10.3354/cr020223.

Brázdil, Rudolf, and Christian Pfister. "Historical Climatology in Europe—the State of the Art." *Climatic Change* 70 (2005): 363-430.

Brázdil, Rudolf, Petr Dobrovolný, Libor Elleder, Vilibald Kakos, Oldřich Kotyza, Vít Květoň, Jarmila Macková, Miloslav Müller, Josef Štekl, Radim Tolasz, and Hubert Valášek. *Historické a současné povodně v České republice* [Historical and Recent Floods in the Czech Republic]. *Historie počasí a podnebí v českých zemích* [History of Weather and Climate in the Czech Lands], vol. 7. Brno, Prague: Masaryk University and Czech Hydrometeorological Institute, 2005.

Brázdil, Rudolf, Zbigniew W. Kundzewicz, and Gerardo Benito. "Historical Hydrology for Studying Flood Risk in Europe." *Hydrological Sciences Journal* 51.5. *Special Issue Historical Hydrology* (2006): 739-64.

Brázdil, Ruldolf, and Karel Kirchner. *Vybrané přírodní extrémy a jelich dopady na Moravě a ve Sleszku* [Selected Natural Extremes and their Impacts in Moravia and Silesia]. Brno, Praha, Ostrava, 2007.

Brázdil, Rudolf, Petr Dobrovolný, Jürg Luterbacher, Anders Moberg, Christian Pfister, Dennis Wheeler, and Eduardo Zorita. "European Climate of the Past 500 Years: New Challenges for Historical Climatology." *Climatic Change* 101.1-2 (2010): 7-40. doi:10.1007/s10584-009-9783-z.

Brázdil, Rudolf, Oldřich Kotyza, Petr Dobrovolný, Ladislava Řezníčková, and Hubert Valášek, *Climate of the Sixteenth Century in the Czech Lands. History of Weather and Climate in the Czech Lands*, vol. 10. Brno: Masaryk University, 2013.

Bretherton, Charlotte, and John Vogler. *The European Union as a Global Actor*. London: Routledge, 1999.

Briffa, Keith R., P.D. Jones, R.B. Vogel, F.H. Schweingruber, M.G.L. Baillie, S.G. Shiyatov, and E.A. Vaganov."European Tree-rings and Climate in the 16th Century." *Climatic Change* 43 (1999): 151-168.

Briffa, Keith R., and Timothy J. Osborn. "Blowing Hot and Cold." *Science* 295 (2002): 2227-2228.

Bruch, Michael, Volker Münch, Markus Aichinger, Michael, Kuhn, Martin Weymann, and Gerhard Schmid. *Power Blackout Risks: Risk Management Options*. CRO Forum Position Paper. Published 2011. Accessed August 26, 2014. <https://www.allianz.com/v_1339677769000/media/responsibility/documents/position_paper_power_blackout_risks.pdf>.

Brückner, Eduard. "Der Einfluß der Klimaschwankungen auf die Ernteerträge und Getreidepreise in Europa." *Geographische Zeitschrift* 1.1/2 (1895): 39-51.

Brüning, Rainer, and Peter Exner, ed. *Wege aus der Armut: Baden in der ersten Hälfte des 19. Jahrhunderts*. Karlsruhe: Generallandesarchiv: 2007.

Brugge, Rutger van der, Jan Rotmans, and Derk Loorbach. "The Transition in Dutch Water Management." *Regional Environmental Change* 5.4 (2005): 164-176.

Brugger, Silvia. "European Climate Leadership under Siege—the 2030 Climate and Energy Debate." Published by Heinrich-Böll-Stiftung, October 7, 2013. Accessed January 7, 2014. <http://www.boell.de/en/2013/10/07/european-climate-leadership-under-siege-2030-climate-and-energy-debate>.

Brumont, François. *Paysans de Vieille Castille aux XVIe et XVIIe siècles*. Madrid: Casa de Velázquez, 1993.

Brunekreeft, Gert, and Roland Meyer. "Netzinvestitionen im Strommarkt: Anreiz- oder Hemmniswirkungen der deutschen Anreizregulierung?" *Energiewirtschaftliche Tagesfragen* 1-2 (2011): 2-5.

Brunner, Kurt. "Klimaentwicklungen in alten Karten und Bildern." In *Natur und Mensch in Mitteleuropa im letzten Jahrtausend*, edited by Bayerische Akademie der Wissenschaften, 65-80. *Rundgespräche der Kommission für Ökologie* 32. München: Friedrich Pfeil, 2007.

Bucher, Klaus, and Eva R. Wanka. "Das Problem der Wetterfühligkeit." *Promet* 33 (2007): 133-139.

Buisman, Jan, and Aryan F.V. van Engelen. *Duizend jaar weer, wind en water in de lage landen,* vol. 3 (1450-1575). Franeker: Van Wjnen, 1998.

Buisman, Jan. *Duizend jaar weer, wind en water in de Lage Landen IV: 1575-1675*. Franeker: Van Wijnen, 2000.

Bullinger, Heinrich. "Diarium (Annales Vitae) der Jahre 1504-1574." In *Quellen zur Schweizer Reformationsgeschichte*, vol. 2, edited by Emil Egli, 1-124. Basel: Basler Buch- und Antiquariatshandlung, 1904.

Bundesamt für Umwelt, Wald und Landschaft (BUWAL). "Hitzesommer 2003: Synthesebericht 2005." Accessed March 28, 2017. <https://web-beta.archive.org/web/20160802130808>/<http://www.occc.ch/products/heatwave03/PDF_D/Hitzesommer03_d.pdf>. *Auswirkungen des Hitzesommers 2003 auf die Gewässer. Schriftenreihe Umwelt* 369. Bern, 2004.

Bundesanstalt für Gewässerkunde (BfG), ed. *Niedrigwasserperiode 2003 in Deutschland: Ursachen–Wirkungen–Folgen.* Mitteilungen, Nr. 27. Koblenz: 2006.

Bundesblatt, 40, vol. IV, No. 53, 8 December 1888.

Bundesministerium für Umwelt, Naturschutz und Reaktorsicherheit. *Ausbau erneuerbarer Energien im Strombereich: EEG-Vergütungen, -Differenzkosten und -Umlage sowie ausgewählte Nutzeneffekte bis zum Jahr 2030.* Published Berlin: BUNR, 2008.

Bundesministerium für Umwelt, Naturschutz und Reaktorsicherheit. *Leitstudie 2010, Langfristszenarien und Strategien für den Ausbau der erneuerbaren Energien in Deutschland bei Berücksichtigung der Entwicklung in Europa und global.* Berlin: BUNR, 2010.

Bundesministerium für Wirtschaft und Energie (BMWi), ed. *Expertenkommission zum Monitoring-Prozess „Energie der Zukunft". Stellungnahme zum ersten Fortschrittsbericht der Bundesregierung für das Bereichtsjahr 2013.* BMWi: Berlin, 2014. Accessed March 30, 2017. <https://www.bmwi.de/Redaktion/DE/Publikationen/Energie/zweiter-monitoring-bericht-energie-der-zukunft.pdf?__blob=publicationFile&v=8>.

Bundesnetzagentur. "Monitoringbericht." Accessed, July 14, 2015. <http://www.bundesnetzagentur.de/DE/Sachgebiete/ElektrizitaetundGas/Unternehmen_Institutionen/DatenaustauschundMonitoring/Monitoring/Monitoringberichte/Monitoring_Berichte_node.html>.

Bundesverband Solarwirtschaft, and Prognos. *Kosten der Solarstromförderung.* Berlin, Basel: BS, 2011.

Büntgen, Ulf, Jan Esper, David C. Frank, Kurt Nicolussi, and Martin Schmidhalter. "A 1052-Year Tree-Ring Proxy for Alpine Summer Temperatures." *Climate Dynamics* 25 (2005): 141-53.

Büntgen, Ulf, David C. Frank, Daniel Nievergelt, and Jan Esper. "Summer Temperature Variation in the European Alps: AD 755-2004." *Journal of Climate* 19 (2006): 5606-5623.

Büntgen, Ulf, Rudolf Brázdil, Karl-Uwe Heussner, Jutta Hofmann, Raymond Kontic, Tomás Kyncl, Christian Pfister, Katerina Chromá, and Willy Tegel. "Combined Dendro-Documentary Evidence of Central European Hydroclimatic Springtime Extremes over the Last Millennium." *Quaternary Science Reviews* 30 (2011): 3947-3959. doi:10.1016/j.quascirev.2011.10.010.

Büntgen, Ulf, Willi Tegel, Kurt Nicolussi, Michael McCormick, David C. Frank, Valerie Trouet, Jed O. Kaplan, Franz Herzig, Karl-Uwe Heussner, Heinz Wanner, Jürg Luterbacher, and Jan Esper. "2500 Years of European Climate Variability and Human Susceptibility." *Science* 331.6017 (2011): 578-82.

Büntgen, Ulf, Willy Tegel, Marco Carrer, Paul J. Krusic, Michael Hayes, and Jan Esper. "Limited Tree-Ring Evidence for a 1540 European 'Megadrought': Commentary to Wetter et al. (2014)." *Climatic Change Letters* (2015).

Burck, Jan, Franziska Marten, Christoph Bals. "Climate Change Performance Index 2016." Published by Climate Action Network. Accessed July 17, 2016. <https://germanwatch.org/en/download/13626.pdf>.

Burigozzo, Giovanni M. "Cronaca di Milano dall'anno 1500 sino al 1544." In *Archivio Storico Italiano*, vol. 3, edited by G.P. Viesseux, 419-552. Florence: Leo S. Olschki, 1842.

Burke, Edmund. "The Big Story: Human History, Energy Regimes and the Environment." In *The Environment and World History*, edited by Edmund Burke, and Kenneth Pomeranz, 33-53. Berkeley: 2009.

Burmeister, Karl Heinz. "'Der heiße Sommer' 1540 in der Bodenseeregion." In *Schriften des Vereins für Geschichte des Bodensees und seiner Umgebung*, vol. 126, edited by Jürgen Klöckler, 59-87. Ostfildern: Jan Thorbecke, 2008.

Buszello, Horst. "Teuerung und Hungersnot am Ober- und Hochrhein im Spätmittelalter und in der frühen Neuzeit (circa 1300-1800)." In *Kriege, Krisen und Katastrophen am Oberrhein vom Mittelalter bis zur Frühen Neuzeit*, 32-71. Schopfheim: Geschichtsverein Markgräflerland e.V, 2007.

Cabourdin, Guy. Terre et hommes en Lorraine, 1550-1635: Toulois et comté de Vaudémont. 2 Vols. Nancy: Annales de l'Est, 1977.

Caillé, Alain. "Weder methodologischer Holismus noch methodologischer Individualismus. Marcel Mauss und das Paradigma der Gabe." In *Gift – Marcel Mauss' Kulturtheorie der Gabe*, edited by Stephan Moebius and Christian Papilloud, 161-214. Wiesbaden: Verlag für Sozialwissenschaften, 2006.

Callendar, Guy S. "The Artificial Production of Carbon Dioxide and Its Influence on Temperature." *Quarterly Journal of the Royal Meteorological Society* 64.275 (1938): 223-240.

Calmet, Augustin. "Chronique de la noble cité de Metz: Preuves servant à l'histoire de Lorraine." In *Histoire de Lorraine*, vol. 3, edited by Augustin Calmet, François Jean-Baptiste Noël and Auguste Prost, col. Cccxxviij. Paris: Éditions du Palais Royal, 1973.

Camenisch, Chantal, Kathrin M. Keller, Melanie Salvisberg, Benjamin Amann, Martin Bauch, Sandro Blumer, Rudolf Brázdil, Stefan Brönnimann, Ulf Büntgen, Bruce M.S. Campbell, Laura Fernández-Donado, Dominik Fleitmann, Rüdiger Glaser, Fidel González-Rouco, Martin Grosjean, Richard C. Hoffmann, Heli Huhtamaa, Fortunat Joos, Andrea Kiss, Oldřich Kotyza, Flavio Lehner, Jürg Luterbacher, Nicolas Maughan, Raphael Neukom, Theresa Novy, Kathleen Pribyl, Christoph C. Raible, Dirk Riemann,

Maximilian Schuh, Philip Slavin, Johannes P. Werner, Oliver Wetter. "The Early Spörer Minimum. A Period of Extraordinary Climate and Socio-Economic Changes in Western and Central Europe." *Climate of the Past Discussions* 12 (2016): 1-33. Accessed April 4, 2017. <http://www.clim-past-discuss.net/cp-2016-7/cp-2016-7.pdf>.

Campe, Rüdiger. *Spiel der Wahrscheinlichkeit. Literatur und Berechnung zwischen Pascal und Kleist*. Göttingen: Wallstein, 2002. [*The Game of Probability. Literature and Calculation from Pascal to Kleist*. Stanford: University Press, 2012.]

Camuffo, Dario. "History of the Long Series of Daily Air Temperature in Padova (1725-1998)." *Climatic Change* 53 (2004): 7-75.

Camuffo, Dario, and Chiara Bertolin. "The Earliest Temperature Observations in the World: The Medici Network (1654-1670)." *Climatic Change* 111.2 (2012): 335-63.

Caney, Simon. "Political Institutions for the Future: A Five-Fold Package." In *Institutions for Future Generations*, edited by Axel Gosseries, and Inigo Gonzalez Ricoy. Oxford: Oxford University Press, 2015.

Carnap, Rudolf. *Der logische Aufbau der Welt*. Berlin: Weltkreis, 1928.

Carnap, Rudolf. *The Logical Structure of the World. Pseudoproblems in Philosophy*. London: Routledge & Kegan, 1967.

Carroll, Lewis. *Through the Looking-Glass, and What Alice Found There* (1871).

Cartwright, Nancy. *The Dappled World. A Study of the Bounderies of Science*. Cambridge: University Press 1999.

Casey, James G. "Spain: A Failed Transition." In *The European Crisis of the 1590s: Essays in Comparative History*, edited by Peter Clark, 209-228. London: George Allen & Unwin, 1985.

Castillo, Alvaro. "El 'Servicio de Millones' y la población del reino de Granada en 1591." *Saitabi* 11 (1961): 61-91.

Casty, Carlo, Heinz Wanner, Jürg Luterbacher, Jan Esper, and Reinhard Böhm. "Temperature and Precipitation Variability in the European Alps since 1500." *International Journal of Climatology* 25 (2005): 1825-1880.

Caviedes, Cesar. *El Niño in History: Storming across the Centuries*. Gainesville: University Press of Florida, 2001.

Chan, Sander, Clara Brandi, and Steffen Bauer. "Aligning Transnational Climate Action with International Climate Governance: The Road from Paris." *Review of European Community & International Environmental Law* (RECIEL) 25.2 (2016): 238-247. Accessed March 29, 2017. <https://www.die-gdi.de/externe-publikationen/article/aligning-transnational-climate-action-with-international-climate-governance-the-road-from-paris/>.

Charter, David, and Rory Watson. "EU Climate Change Push in Disarray as Italy Joins Iron Curtain Revolt." *The Times*, October 17, 2008.

Chassany, Jean-Philippe. *Les couleurs du vent. La météo d'hier et d'aujourd'hui*. Paris: Maisonneuve et Larose, 1986.

Chaunu, Huguette, and Pierre Chaunu. Séville et l'Atlantique (1594-1650). VI.1: Tables statistiques. Paris: SEVPEN, 1956.

Christmann, Gabriela B. "Kommunikative Raumkonstruktionen als (Proto-) Governance." In *Governance und Raum*, edited by Heiderose Kilper, 27-48. Baden-Baden: Nomos, 2010.

Christmann, Gabriela B., and Kerstin Büttner. "Raumpioniere, Raumwissen, Kommunikation—zum Konzept kommunikativer Raumkonstruktion." *Berichte zur deutschen Landeskunde* 85 (2011): 361-378.

Christmann, Gabriela B., Oliver Ibert, Heiderose Kilper, Timothy Moss. *Vulnerabilität und Resilienz in sozio-räumlicher Perspektive: Begriffliche Klärungen und theoretischer Rahmen*. Working Paper, 44. Erkner: Leibniz-Institut für Regionalentwicklung und Strukturplanung, 2011. Accessed January 25, 2014. <http://www.irs-net.de/download/wp_vulnerabilitaet.pdf>.

Christmann, Gabriela B., and Oliver Ibert. "Vulnerability and Resilience in a Socio-spatial Perspective." *Raumforschung und Raumordnung* 70 (2012): 259-272.

Chroma, Kateřina, Ruldolf Brázdil, and Radim Tolasz. "Spatio-temporal Variability of Hailstorms for Moravia and Silesia in the Summer Half-Year of the Period 1961-2000." *Meteorologický časopis* 8 (2005).

Chronik der Stadt Kitzingen. Stadtarchiv Kitzingen. Manuscript, Msc. 314.

Chuine, Isabelle, Pascal Yiou, Nicolas Viovy, Bernard Seguin, Valérie Daux, and Emmanuel Le Roy Ladurie. "Historical Phenology: Grape Ripening as a Past Climate Indicator: Summer Temperature Variations are Reconstructed from Harvest Dates since 1370." *Nature* 432 (2004): 289-290.

Chuine, Isabelle. *Burgundy Grape Harvest Dates and Spring–Summer Temperature Reconstruction*. Published by IGBP PAGES/World Data Center for Paleoclimatology, Data Contribution Series #2005-007. Boulder, 2005.

Clarke, Lee. "Possibilistic Thinking. A New Conceptual Tool for Thinking about Extreme Events." *Social Research* 75.3 (2008): 669-691.

Climatological Database for the World's Oceans (CLIWOC). "Climatological Database for the World's Oceans 1750-1850." Accessed September 13, 2014. <http://pendientedemigracion.ucm.es/info/cliwoc/>.

Clinton, Bill. "How We Ended Welfare, Together" *The New York Times*, August 22, 2006. Accessed November 18, 2015. <http://www.nytimes.com/2006/08/22/opinion/22clinton.html?_r=0>.

Coats, Sloan, Jason E.Smerdon, Richard Seager, Benjamin Cook, and Jesús Fidel González-Ruoco. "Megadroughts in Southwestern North America in ECHO-G Millennial Simulations and their Comparison to Proxy Drought Reconstructions." *Journal of Climate* 26 (2013): 7635-7649.

Coler, Johann. *Calendarium perpetuum et libri oeconomici:* [...]. Wittemberg: Paul Helwig, 1603. Accessed March 27, 2017. <http://www.mdz-nbn-resolving.de/urn/resolver.pl?urn=urn:nbn:de:bvb:12-bsb10228551-6>.

Coler, Johann. *Calendarium perpetuum* [...]. *Das ist: Ein stetswehrender Calender/ Auch Zwantzig nothwendige vnd gantz nützliche Haußbücher:* [...]. Wittemberg: Johann Wilhelm Fincelius, Paul Helwig, 1632.

Collier, J.P., ed. *The Egerton Papers.* London: Camden Society, 1840.

Collinson, Patrick. "Ecclesiastical Vitriol: Religious Satire in the 1590s and the Invention of Puritanism." In *The Reign of Elizabeth I: Court and Culture in the Last Decade*, edited by John A. Guy, 150-170. Cambridge: Cambridge University Press 1995.

Columella, Iunius Moderatus. *De re rustica libri XII.* Lugduni: Seb. Gryphius, 1548. Accessed March 27, 2017. <http://www. mdz-nbn-resolving.de/urn/resolver.pl?urn=urn:nbn:de: bvb:12-bsb10171885-3>.

Complutense University of Madrid. "Andalusian Seasonal and Annual Rainfall Index, 1501-1997." Accessed November 28, 2014. <http://pendientedemigracion.ucm.es/info/reclido/es/basesdatos/Rainfallindex.txt>.

Condorcet, Marquis de. Esquisse d'un tableau historique des progrès de l'esprit humain. Paris: Agasse, 1795.

Cook, Edward R. "Multi-Proxy Reconstructions of the North Atlantic Oscillation (NAO) Index: A Critical Review and a New Well-Verified Winter NAO Index Reconstruction Back to AD 1400." In *The North Atlantic Oscillation: Climatic Significance and Environmental Impact*, edited by Hurell, James W., Yochanan Kushnir, Geir Ottersen, and Martin Visbeck, 63-79. *Geophysical Monograph Series*, vol. 134. Washington D.C.: American Geophysical Union, 2013.

Council of the European Union. *Energy and Climate Change—Elements of the Final Compromise*, 17215/08. Brussels: Council of the European Union, 2008.

Council of the European Union. *Environment Council of Ministers Conclusions*, 14790/09. Brussels: Council of the European Union, 2009.

Croix, Alain. *Bretagne aux XVIe et XVIIe siècles: La vie, la mort, la foi.* 2 Vols. Paris: Maloine 1981.

Croix, Alain. Nantes et le pays nantais au XVIe siècle: Étude démographique. Paris: SEVPEN, 1974.

Crosby, Alfred W. *The Columbian Exchange: Biological and Cultural Consequences of 1492.* Westport, Conn.: Greenwood Pub. Co., 1972.

Crouch, Colin. *Post-Democracy.* Cambridge: Polity, 2004.

Crowley, Thomas J., and Thomas S. Lowery. "How Warm Was the Medieval Warm Period?" *Ambio* 29.1 (2000): 51-54.

Crutzen, Paul J. and Eugene F. Stoermer. "The 'Anthropocene'," *Global Change Newsletter 41* (2000): 17-18.

Curtin, Joseph. *The Copenhagen Conference: How Should the EU Respond?* Dublin: Institute of International and European Affairs, 2010.

Cyriandrus [Greek for Herrmann], Uranophilus [Greek for: Friend of the Heaven], *Meteorologica. Das ist: Vier vnd zwantzig Jährige eigentliche und trewfleissige Observation vnd tägliche verzeichnüß des Gewitters/ vom 1. Januarii 1623 an/ biß zum letzten Decembris 1646 in dreyen membris verfasset.* [...] *Alles Zu Rettung der bißher sehr beschimpfften Meteorologiae/ den Kunstliebenden vnd Prognosticanten aber zu gefallen/ vnd mächtiger vorarbeit sich künfftig besser in der Natur vmbzusehen.* Cassel: Salomon Schadewitz, Sebald Köhler, 1551 [1651] . Accessed March 27, 2017. <http://www.mdz-nbn-resolving.de/urn/resolver.pl?urn=urn:nbn:de:bvb:12-bsb10058302-0>.

Cysat, Renward. "Stationes Annorum: Witterung, Missjahre, Teuerung." In *Collectanea Chronica und denkwürdige Sachen pro Chronica Lucernensi et Helvetiae*, vol. 4.2 edited by Josef Schmid, 898-962. Luzern: Diebold Schilling Verlag, 1969,.

Cysat, Renward. *Collectanea Chronica und denkwürdige Sachen pro Chronica Lucernensi et Helvetiae*, vol. 4.2, edited by Josef Schmid. Lucerne: Diebold-Schilling-Verlag, 1969.

Dalton, Russell J. *Democratic Challenges, Democratic Choices: The Erosion of Political Support in Advanced Industrial Democracies*. Oxford: Oxford University Press, 2004.

Damro, Chad. "EU–UN Environmental Relations: Shared Competence and Effective Multilateralism." In *The European Union at the United Nations: Intersecting Multilateralism*, edited by Katie Verlin Laatikainen, and Karen E. Smith, 175-192. Hampshire: Palgrave Macmillan, 2006.

Das Regierungs-Blatt für das Königreich Württemberg im Auszuge. 2 Vols (1806-1816, and 1817-1824). Stuttgart: 1840 and 1841.

Dasent, John R. *Acts of the Privy Council of England: New Series XXVII, 1597*. London: HMSO, 1908.

Daston, Lorraine. *Classical Probability in the Enlightenment*. Princeton, N.J.: Princeton University Press, 1988.

Davidson, Nicholas, S. "Northern Italy in the 1590s." In *The European Crisis of the 1590s: Essays in Comparative History*, edited by Peter Clark, 157-176. London: George Allen & Unwin, 1985.

Davies, Timothy B. "Sicily." In *The European Crisis of the 1590s: Essays in Comparative History*, edited by Peter Clark, 191-208. London: George Allen & Unwin, 1985.

de Vries, Jan, "Population." In *Handbook of European History 1400-1600. Late Middle Ages, Renaissance and Reformation,* vol. 1, edited by Thomas A. Brady, Heiko A. Oberman, and James D. Tracy, 13. Leiden: Brill, 1995.

Dean, David. *Law-making and Society in Late Elizabethan England: The Parliament of England, 1584-1601*. Cambridge: Cambridge University Press, 1996.

Debelke, Jos, and Peter Vis, ed. "EU Climate Policy Explained." Accessed May 21, 2017. <https://ec.europa.eu/clima/sites/clima/files/eu_climate_policy_explained_en.pdf>.

Dee, Megan. "Evaluating European Union Leadership in Multilateral Negotiations: A Framework for Analysis." Paper presented at the "European Union Studies Association (EUSA) 12th Biennial International Conference," Regency Hyatt Hotel Boston, March 3-5, 2011.

Degroot, Dagomar. "Testing the Limits of Climate History: The Quest for a Northeast Passage during the Little Ice Age, 1594-1597." *Journal of Interdisciplinary History* 45 (2015): 459-84.

Dekker, Paul, and Josje den Ridder. "Politisches Unbehagen in den Niederlanden: Wie sich das kleine liberale und tolerante Land verändert hat." *INDES – Zeitschrift für Politik und Gesellschaft* (2011): 46-55.

Delgado, Mariano. "Verliebter Blick: Die Natur der Neuen Welt in der Apologética historia sumaria des Bartolomé de las Casas." In *Begriff und Darstellung der Natur in der spanischen Literatur der frühen Neuzeit, Hispanistisches Kolloquium*, vol. 4, edited by Wolfgang Matzat, and Gerhard Poppenberg, 291-304. München: Fink, 2012.

Delsalle, Paul. *La Franche Comté au temps de Charles Quint*.Besançon: Presses Universitaires de Franche Comté, 2004.

Deltacommissie. *Samen werken met Water: Een Land dat Leeft, Bouwt aan zijn Toekomst. Bevindingen van de Deltacommissie 2008*. Accessed January 25, 2014. <http://www.deltacommissie.com/doc/2008-09-03%20Advies%20Deltacommissie.pdf>.

Des Forges, Roger V. Cultural Centrality and Political Change in Chinese History: Northeast Henan in the Fall of the Ming. Stanford: Stanford University Press, 2003.

Descartes, René. *Discours de la méthode pour bien conduire sa raison, & chercher la verité dans les sciences*. Leyde: Ian Maire, 1637. Accessed March 27, 2017. <http://reader.digitale-sammlungen.de/resolve/display/bsb10860336.html>.

Descola, Philippe. *Par-delà nature et culture*. Paris: Gallimard, 2005/ 2015.

Deutsche Bischofskonferenz. "Climate Change: A Focal Point of Global, Intergenerational and Ecological Justice." *Commission for Society and Social Affairs and Commission for International Church Affairs*, vol. 29. Bonn: DBK, 2007.

Deutsche Energy-Agentur. "Integration erneuerbarer Energien in die deutsche Stromversorgung im Zeitraum 2015-2020 mit Ausblick 2025." Accessed July 14, 2015. <http://www.dena.de/publikationen/energiesysteme/dena-netzstudie-ii.html>.

Deutsche Energie-Agentur. "Der Dena-Gebäudereport 2012: Statistiken und Analysen zur Energieeffizienz im Gebäudebestand." Accessed October 01, 2013. <http://www.dena.de/projekte/gebaeude/dena-gebaeudereport.html>.

Deutsche Welle. "A Quick Guide to 'Agenda 2010'." Published October 17, 2003. Accessed November 9, 2015. <http://www.dw.com/en/a-quick-guide-to-agenda-2010/a-988374-1>.

Deutscher Bundestag. *Ausschreibungsbericht nach § 99 des Erneuerbare-Energien-Gesetzes, Drucksache 18/7287*. Published 2016. Accessed July 17, 2016. <http://dip21.bundestag.de/dip21/btd/18/072/1807287.pdf>.

Deutscher Wetterdienst – Wetter und Klima aus erster Hand. "Deutschlandwetter im Jahr 2015." Last modified December 30, 2015. Accessed November 8, 2016. <https://www.dwd.de/DE/presse/pressemitteilungen/DE/2015/20151230_deutschlandwetter_jahr2015_news.html>.

Diamond, Jared M., *Collapse: How Societies Choose to Fail or Succeed.* New York: Viking Press, 2005.

"Die Aufzeichnungen Adelberg Meyers 374-1542." In *Basler Chroniken*, vol. 8, edited by Historische und Antiquarische Gesellschaft in Basel, 345-380 Leipzig: S. Hirzel, 1902.

Die deutschen Bischöfe. "Handeln für die Zukunft der Schöpfung." In *Erklärungen der Kommissionen*, vol. 19, edited by Sekretariat der Deutschen Bischofskonferenz. Bonn: Sekretariat der Deutschen Bischofskonferenz, 1998.

Dijk, Judith M. van. "Water Assessment in the Netherlands." *Impact Assessment and Project Appraisal* 24.3 (2006): 199-209.

Diller, Hans ed. *Hippokrates, Über die Umwelt* (Corpus medicorum Graecorum 1, 1, 2). Berlin: Akademie, 1999.

Dillinger, Johannes, "Organized Arson as a Political Crime: The Construction of a 'Terrorist' Menace in the Early Modern Period." *Crime, History & Societies* 10.2 (2006): 102-119.

Dimitrov, Radoslav S. "Inside Copenhagen: The State of Climate Governance." *Global Environmental Politics* 10.2 (2010): 18-24.

Dimitrov, Radoslav S. "The Paris Agreement on Climate Change: Behind Closed Doors." *Global Environmental Politics* 16.3 (2016): 1-11.

Dobrovolný, Petr, Anders Moberg, Rudolf Brázdil, Christian Pfister, Rüdiger Glaser, Rob Wilson, Aryan van Engelen, Danuta Limanówka, Andrea Kiss, Monika Halíčková, Jarmila Macková, Dirk Riemann, Jürg Luterbacher, and Reinhard Böhm. "Monthly, Seasonal, and Annual Temperature Reconstructions for Central Europe Derived from Documentary Evidence and Instrumental Records since AD 1500." *Climatic Change* 101.1-2 (2010): 69-107. doi:10.1007/s10584-009-9724-x.

Dobrovolný, Petr, Rudolf Brázdil, Mirek Trnka, Oldrych Kotyza, and Hubert Valasek Hubert. "Precipitation Reconstruction for the Czech Lands, AD 1501-2010." *International Journal of Climatology* 2 (2014): 1-14. doi: 10.1002/joc.3957.

Dobson, Andrew. "Representative Democracy and the Environment." In *Democracy and the Environment*, edited by William M. Laffetty, and James Meadowcroft, 124-139. Cheltenham: Edward Elgar, 1996.

Douglass, Andrew E. "Some Topographic and Climatic Characters in the Annual Rings of the Yellow Pines and Sequoias of the Southwest." *Proceedings of the American Philosophical Society* 61 (1922): 117-22.

Drelichman, Mauricio, and Hans-Joachim Voth. *Lending to the Borrower from Hell: Debt, Taxes, and Default in the Age of Philip II*. Princeton: Princeton University Press, 2014.

Dröge, Susanne, and Oliver Geden, "The EU and the Paris Climate Agreement." *SWP Comments*. Berlin: Stiftung Wissenschaft und Politik (SWP), 2015.

Dröge, Susanne, and Oliver Geden, "After the Paris Agreement New Challenges for the EU's Leadership in Climate Policy." *SWP Comments*. Berlin: Stiftung Wissenschaft und Politik (SWP), 2016.

Dubos, Jean Baptiste. *Réflexions critiques sur la poésie et sur la peinture. Deuxième partie.* Paris: Jean Mariette, 1719. Accessed March 27, 2017. <http://www.mdz-nbn-resolving.de/urn/resolver.pl?urn=urn:nbn:de:bvb:12-bsb10573734-5>.

Ducot, Joëlle, and Claude Thomasset, ed. *Le temps qu'il fait au Moyen-Âge: Phénomènes atmosphériques dans la littérature, la pensée scientifique et religieuse*. Paris: Presses de l'Université, 1998.

Dürr, Friedrich. *Heilbronner Chronik*. Heilbronn: Verlag Eugen Salzer, 1895.

Dunning, Chester S.L. *Russia's First Civil War: The Time of Troubles and the Founding of the Romanov Dynasty*. University Park: Pennsylvania State Press, 2001.

Dupâquier, Jacques, ed. *Histoire de la population française I: Des origines à la Renaissance*. Paris: PUF, 1988.

Dupont, Claire. "Leading or Muddling Through? The EU and Durban." *Ideas on Europe: European Council and Foreign Affairs* (2012). Accessed April 6, 2017. <http://www.ies.be/node/1297>.

Edwards, Paul N. "History of Climate Modeling." *Wiley Interdisciplinary Reviews: Climate Change* 2.1 (2011): 128-139.

Egenhofer, Christian, and Anton Georgiev. *Why the Transatlantic Alliance Climate Change Partnership Matters More Than Ever.* Brussels: Centre for European Policy Studies (CEPS), 2010.

Ekeli, Kristian S. "Giving a Voice to Posterity—Deliberate Democracy and Representation of Future People." *Journal of Agricultural and Environmental Ethics* 18.5 (2005): 429-450.

Ekeli, Kristian S. "Constitutional Experiments: Representing Future Generations Through Submajority Pules." *The Journal of Political Philosophy* 17.4 (2009): 440-461.

Elgar, Andrew. *Habermas: The Key Concepts.* Oxford, New York: Routledge, 2006.

Elgström, Ole. "Outsiders' Perceptions of the EU in International Trade Negotiations." *Journal of Common Market Studies* 45.4 (2007): 949-967.

Elgström, Ole. "EU Leadership in an Emerging New World Order." Paper presented at the 5th Pan-European Conference on EU Politics, Porto, June 24-26, 2010.

Elison, George. "The Cross and the Sword: Patterns of Momoyama History." In *Warlords, Artists and Commoners: Japan in the Sixteenth Century*, edited by George Elison and Bardwell L. Smith, 55-85. Honolulu: University of Hawai'i Press, 1981.

Elliott, John H. "Yet another Crisis?" In *The European Crisis of the 1590s: Essays in Comparative History*, edited by Peter Clark, 301-311. London: George Allen & Unwin, 1985.

Elliott, John H. *Imperial Spain, 1469-1716*. London: Arnold, 1963.

"Energy Boss Warns of Blackouts in Europe." Published by The Local. Accessed November 7, 2013. <http://www.thelocal.de/20131029/energy-boss-warns-of-blackouts-in-europe>.

Engelen, Aryan F. van, Jan Buisman, and Frank Ijnsen. "Reconstruction of the Low Countries Temperature Series AD 764-1998." In *International Conference on Climate Change and Variability—Past, Present and Future*, edited by Takehiko Mikami, 151-157. Tokyo: Tokyo Metropolitan University, 2000.

Engelen, Aryan F. van, Jan Buisman, and Frank Ijnsen. "A Millennium of Weather, Winds and Water in the Low Countries." In *History and Climate: Memories of the Future?*, edited by Philip D. Jones, Astrid E.J. Ogilvie; Trevor D. Davies; and Keith R. Briffa, 101-124. New York: Kluwer Academic/Plenum Publishers, 2001.

Erdmann, Georg. "Kosten des Ausbaus erneuerbarer Energien." Study conducted by Technischen Universität Berlin." Accessed April 23, 2013. <http://www.vbw-bayern.de/Redaktion-%28importiert-aus-CS%29/04_Downloads/Downloads_2011/3_Wirtschaftspolitik/3.5_Energie/Publikationen/studie_eeg_energie_kosten_110907_final_NGa_2.pdf>.

Erlebach, Peter. *Natur und Spiritualität in der englischen Literatur- und Geistesgeschichte*. Heidelberg: Winter, 2011.

Esper, Jan, Rob Wilson, David C. Frank, Anders Moberg, Heinz Wanner, and Jürg Luterbacher. "Climate: Past Ranges and Future Changes." *Quaternary Science Reviews* 24.20-21 (2005): 2164-2166. doi:10.1016/j.quascirev.2005.07.001.

Etheridge, David M., L. Paul Steele, Ray Langenfelds, Roger J. Francey, Jean-Marc Barnola, and Vin I. Morgan. "Historical CO_2 Records from the Law Dome DE08, DE08-2, and DSS Ice Cores." In *Trends: A Compendium of Data on Global Change*, edited by Carbon Dioxide Information Analysis Center, Oak Ridge National Laboratory, U.S. Department of Energy. Oak Ridge: Oak Ridge National Laboratory, 1998. Accessed September 13, 2014. <http://cdiac.ornl.gov/trends/co2/lawdome>.

Etheridge, David M., L. Paul Steele, Roger J. Francey, and Ray Langenfelds. *Ice Core, Firn Air and Archived Air Atmospheric Methane Concentration Data*. IGBP PAGES/World Data Center for Paleoclimatology Data Contribution Series #2002-039. Boulder: National Climatic Data Center, 2002.

EU Climate Action. "How the EU helped build the ambition coalition. Tweet by Tweet." Accessed October 18, 2016. <https://storify.com/EUClimateAction/how-the-eu-helped-build-the-coalition-ambition>.

European Commission. *Limiting Global Climate Change to 2 Degrees Celsius: The Way Ahead for 2020 and Beyond: Communication to the European Parliament, the Council, the European Economic and Social Committee and the Committee of the Regions*. COM 2. Luxembourg: Office for Official Publications of the European Communities, 2007.

European Commission. *Brussels European Council, Presidency Conclusions*, 17271/08. Brussels: European Commission, December 11-12, 2008.

European Commission. “Copenhagen Conference Must Produce Global, Ambitious and Comprehensive Agreement to Avert Dangerous Climate Change.” IP/09/1867. Published December 2, 2009. Accessed June 15, 2011. <http://europa.eu/rapid/pressReleasesAction.do?reference=IP/09/1867>.

European Commission. “2020 Climate & Energy Package.” Last modified July 8, 2011. Accessed July 11, 2016. <http://ec.europa.eu/clima/policies/strategies/2020/index_en.htm>.

European Commission. “2030 Climate & Energy Framework.” Last modified July 8, 2011. Accessed July 11, 2016. <http://ec.europa.eu/clima/policies/strategies/2030/index_en.htm>.

European Commission. “2050 Low-Carbon Economy.” Last modified July 8, 2011. Accessed July 11, 2016. <http://ec.europa.eu/clima/policies/strategies/2050/index_en.htm>.

European Commission. “COM (2011) 885/2: EU Roadmap 2050.” Accessed July 14, 2015. <http://eur-lex.europa.eu/legal-content/EN/TXT/PDF/?uri=CELEX:52011DC0885&from=EN>.

European Commission. “Report from the Commission to the European Parliament and the Council (COM 2015),” page 4. Accessed May 21, 2017. <http://eur-lex.europa.eu/resource.html?uri=cellar:adf850de-8de4-11e5-b8b7-01aa75ed71a1.0007.02/DOC_1&format=PDF>.

European Commission. “Historic Climate Deal in Paris: Speech by Commissioner Miguel Arias Cañete at the Press Conference on the Results of COP21 Climate Conference in Paris.” *SPEECH/15/6320*. Published December 14, 2015. Accessed October 18, 2016 <http://europa.eu/rapid/press-release_SPEECH-15-6320_de.htm>.

European Commission. “2050 Low-Carbon Economy.” Last modified March 24, 2017. Accessed March 24, 2017. <http://ec.europa.eu/clima/policies/strategies/2050/index_en.htm>.

European Council. Presidency Conclusions 8-9 March 2007, 7224/1/07. Brussels: European Council. 2007.

European Council. *Conclusions 23-24 October 2014. 2030 Climate and Energy Framework*. Brussels: European Council. 2014.

European Council. “A European Strategy for Low-Emmision Mobility.” Last modified May 25, 2017. Accessed May 29, 2017. <https://ec.europa.eu/clima/policies/transport_en>.

Eybl, Jutta, Reinhold Godina, Petra Lalk, Peter Lorenz, Gabriele Müller, Viktor Weilguni. *Trockenheit in Österreich im Jahr 2003: Ein hydrologischer Situationsbericht*. Wien: Bundesministerium für Land und Forstwirtschaft, Umwelt und Wasserwirtschaft Abteilung VII/3, 2004. Accessed November 20, 1014. <https://www.google.de/search?q=Trockenheit+in+%C3%96sterreich+im+Jahr+2003%3A+Ein+hydrologischer+Si

tuationsbericht+&ie=utf-8&oe=utf-8&client=firefox-b-ab&gfe_rd=cr&ei=cfvlWKCFPLLVXqW6tPAB>.

Fäh, Emil. *Die Hagelversicherung in der Schweiz in Vergangenheit und Gegenwart. Zum 75jährigen Bestehen der Schweizerischen Hagel-Versicherungs-Gesellschaft 1880-1954*. Zürich: Schweizerische Hagel-Versicherungs-Gesellschaft, 1954.

Falkner, Robert. "The Political Economy of 'Normative Power' Europe: EU Environmental Leadership in International Biotechnology Regulation." *Journal of European Public Policy* 14.4 (2007): 507-526.

Fels, Edwin. *Der wirtschaftende Mensch als Gestalter der Erde*. Erde und Weltwirtschaft. Ein Handbuch der Allgemeinen Wirtschaftsgeographie, vol. 5. Stuttgart: Franckh, 1967 [1935].

Fleischmann, Sigmund. *Die Agrarkrisis von 1845-1855 mit besonderer Berücksichtigung von Baden*. Heidelberg: Karl Rössler, 1902.

Fleming, James R. *The Callendar Effect. The Life and Work of Guy Stewart Callendar (1898-1964), the Scientist Who Established the Carbon Dioxide Theory of Climate Change*. Boston, Mass.: American Meteorological Society, 2007.

Field, Christopher B., Vicente Barros, Thomas F. Stocker, and Qin Dahe, ed. *Managing the Risks of Extreme Events and Disasters to Advance Climate Change Adaptation: Special Report of the Intergovernmental Panel on Climate Change*. Cambridge: Cambridge University Press, 2012.

Fierro, Alfred. *Histoire de la météorologie*. Paris: Denoël, 1991.

Film on the "Sand Motor" in Ter Heijde. Uploaded by "Rijksoverheid", 2011. Accessed January 27, 2014. <http://www.youtube.com/watch?v=hsA4FebE6_8>.

Fischer, Erich M., Sonia. I. Seneviratne, Pier. L. Vidale, Daniel Lüthi, Cristoph Schär. "Soil Moisture–Atmosphere Interactions during the 2003 European Summer Heat Wave." *Journal of Climate* 20 (2007): 5081-5099. doi:10.1175/JCLI4288.1.

Fischer, Torben, and Matthias N. Lorenz. *Lexikon der "Vergangenheitsbewältigung" in Deutschland: Debatten- und Diskursgeschichte des Nationalsozialismus nach 1945*. Bielefeld: Transcript, 2007.

Fitzherber[t], [Anthony]. *The Booke of Husbandry. Very Profitable and Necessarye for All Maner of Persons* [...]. London: Iohn Awdely, 1573.

Fludd, Robert. *Philosophia sacra & vere Christiana Seu Meteorologia Cosmica*. Francofurti: Officina Bryana, 1626. Accessed March 27, 2017. digital.slub-dresden.de/werkansicht/dlf/16704/1/.

Food and Agriculture Organization of the United Nations. "Climate Change and Food Security: Risks and Responses." Accessed November 10, 2016. <http://www.fao.org/3/a-i5188e.pdf>.

Forum Netzintegration Erneuerbare Energien. "Das Stromnetz der Zukunft." Accessed April 23, 2013. <http://www.duh.de/uploads/media/DUH_Broschuere_NetzintEE_2010.pdf>.

Fossati, Francesco, ed. *Opere scelte di Benedetto Giovio.* Como: Fostinelli, 1887, reprint 1982.

Foucault, Michel. *Discipline and Punish: The Birth of the Prison.* Translated by Alan Sheridan. London: Vintage Books, 1995.

Fox, Matthew. *A New Reformation: Creation Spirituality and the Transformation of Christianity.* Rochester: Inner Tradition, 2006.

Fox, Matthew. *The Coming of the Cosmic Christ.* San Francisco: Herper, 1988.

Fox, Matthew. *Vision vom Kosmischen Christus: Aufbruch ins dritte Jahrtausend.* Stuttgart: Kreuz, 1991.

Fox, Nicols. *Against the Machine: The Hidden Luddite History in Literature, Art, and Individual Lives.* Washington, D.C.: Island Press, 2003.

Frankenberg, Peter. *Ähnlichkeitsstrukturen von Ernteertrag und Witterung in der Bundesrepublik Deutschland.* Wiesbaden: Steiner, 1984.

Frankfurt, Harry. *Freiheit und Selbstbestimmung: Ausgewählte Texte.* Berlin: Akademie-Verlag, 2001.

Franziskus. *Laudato Si'. Encyclica on Care for Our Common Home.* Vatican: Vatican Press, 2015.

Fraunhofer-Institut für System- und Innovationsforschung (ISI), Deutsches Institut für Wirtschaftsforschung (DIW), Gesellschaft für Wirtschaftliche Strukturforschung mbH (GWS), and Institut für ZukunftsEnergieSysteme (IZES). "Einzel- und gesamtwirtschaftliche Analyse von Kosten- und Nutzenwirkungen des Ausbaus Erneuerbarer Energien im deutschen Strom- und Wärmemarkt: Update der quantifizierten Kosten- und Nutzenwirkungen für 2010."

Fraunhofer-Institut für solare Energiesysteme (ISE). "Energy Transformation Index 2013". Accessed July 17, 2016. <https://www.ise.fraunhofer.de/de/downloads/pdf-files/aktuelles/ise-ises-eti.pdf>.

Freiburger Zeitung. May 2, 1832 (Tagesausgabe).

Freiburger Zeitung. May 7, 14, 18, 1847.

Friedrich, Michael, Sabine Remmele, Bernd Kromer, Jutta Hofmann, Marco Spurk, Klaus F. Kaiser, Christian Orcel, and Manfred Küppers. "The 12,460-Year Hohenheim Oak and Pine Tree-Ring Chronology from Central Europe: A Unique Annual Record for Radiocarbon Calibration and Paleoenvironment Reconstructions." *Radiocarbon* 46.3 (2004): 1111-1122.

Fritts, Harold C. *Tree Rings and Climate.* London, New York: Academic Press, 1976.

Frossard, Anatoile. "Livre de raison de la famille de Froissard-Broissia de 1532 à 1701." In *Mémoires de la Société d'émulation du Jura*, 27-105. Published 1886.

Fuhr, Lili, Liane Schalatek, Tigere Chagutah, and Kulthoum Omari. *COP 17 in Durban: A Largely Empty Package.* Berlin: Heinrich-Boll-Stiftung, 2011.

Fuhr, Lili, and Liane Schalatek. "The 'Doha Climate Gateway': Will the Camel Go Through the Eye of the Needle?" Published by Heinrich-Böll-Stiftung, December 12, 2012. Accessed October 21, 2013. <http://www.boell.de/de/node/276935>.

Fuller, Ted, and Krista Loogma. "Constructing Futures; a Social Constructionist Perspective on Foresight Methodology." *Futures* 41.2 (2009): 71-79.

Gabriel, Karl. *Christentum zwischen Tradition und Postmoderne. Quaestiones Disputatae*, vol. 141. Freiburg: Herder, 1994.

Galilei, Galileo. *Dialogo sopra i due massimi sistemi des mondo Tolemaico e Copernicano* [..]. Fiorenza: Giovanni Batista Landini, 1632.

Ganoczy, Alexandre. *Theologie der Natur*. Zürich: Benziger, 1982.

García España, Emilio, and Annie Molinié-Bertrand, ed. *Censo de Castilla, 1591: Vecindario y estudio analítico.* 2 Vols. Madrid: Instituto Nacional de Estadística, 1980-1986.

García-Herrera, Ricardo, Julio Díaz, Ricardo M. Trigo, Jürg Luterbacher, and Erich M. Fischer. "A Review of the European Summer Heat Wave of 2003." *Critical Reviews in Environmental Science and Technology* 40.4 (2010): 267-306. doi:10.1080/10643380802238137.

Gardner, Gary. "Engaging Religion in the Quest for a Sustainable World." In *State of the World 2003*, edited by The Worldwatch Institute, 152-176. New York: Norton & Company, 2003.

Garnier, Emmanuel, V. Daux, P. Yiou, and I. García de Cortazár-Atauri. "Grapevine Harvest Dates in Besançon (France) between 1525 and 1847: Social Outcomes or Climatic Evidence?" *Climatic Change* 104.3-4 (2011): 703-727. doi:10.1007/s10584-010-9810-0.

Gehlinger, G. "Überblick über die Entwicklung der Landwirtschaft in Württemberg seit der Mitte des 18. Jahrhunderts." *Württembergische Jahrbücher* 1897. I.49-76.

Gerard, John. *The Herball or General Historie of Plantes*. London, 1597.

Gergis, Joëlle L., and Anthony M. Flower. "A History of ENSO Events since A.D. 1525: Implications for Future Climate Change." *Climatic Change* 92 (2009): 343-387.

Gerhold, Lars, Dirk Holtmannspötter, Christian Neuhaus, Elmar Schüll, and Axel Zweck. *Standards und Gütekriterien der Zukunftsforschung: Ein Handbuch für Wissenschaft und Praxis.* Wiesbaden: Springer, 2015.

German Advisory Council on Global Change (WBGU). *World in Transition: Ways Towards Global Environmental Solutions*. Berlin: Springer Verlag, 1996.

German Advisory Council on Global Change (WBGU). *World in Transition—A Social Contract for Sustainability*. Berlin: WBGU, 2011.

German Advisory Council on Global Change (WBGU). *Climate Protection as a World Citizen Movement*. Berlin: WBGU, 2014.

German Advisory Council on Global Change (WBGU), ed. *Humanity on the Move: Unlocking the Transformative Power of cities.* Flagship Report. Berlin 2016: WBGU.

Ghosh, Amitav. *Incendiary Circumstances: A Chronicle of the Turmoil of Our Times*. Boston: Houghton Mifflin, 2005.

Ghosh, Amitav. *The Great Derangement: Fiction, History and Politics in the Age of Global Warming*. Chicago, IL: Chicago University Press, 2016.

Giaiotti, Dario, Sergio Nordio, and Fulvio Stel. "The Climatology of Hail in the Plain of Friuli Venezia Giulia." *Atmospheric Research* 67-68 (2003): 247-259.

Giddens, Anthony. *The Politics of Climate Change*. Cambridge: Polity Press, 2009.

Gille, Zsuzsa, and Seán Ó Riain. "Global Ethnography." *Annual Review of Sociology* 28 (2002): 271-295.

GIZ and Ecofys. *Lessons for the Tendering System for Renewable Electricity in South Africa from International Experience in Brazil, Morocco and Peru*. Pretoria, 2013. Accessed March 302017. <http://www.ecofys.com/files/files/ecofys-giz-2013-international-experience-res-tendering.pdf>.

Glacken, Clarence J. *Traces on the Rhodian Shore. Nature and Culture in Western Thought from Ancient Times to the End of the Eighteenth Century*. Berkeley: University of California, 1967.

Gladstones, John. *Wine, Terroir and Climate Change*. Kent Town: Wakefield Press, 2011.

Glaser, Rüdiger. *Klimarekonstruktion für Mainfranken, Bauland und Odenwald*. Stuttgart: Fischer, 1991.

Glaser, Rüdiger. *Klimarekonstruktion für Mainfranken, Bauland und Odenwald anhand direkter und indirekter Witterungsdaten seit 1500. Paläoklimaforschung*, vol. 5. Stuttgart, New York: Fischer, 1991.

Glaser, Rüdiger. *Klimageschichte Mitteleuropas. Tausend Jahre Wetter, Klima, Katastrophen*. Darmstadt: Wissenschaftliche Buchgesellschaft, 2001.

Glaser, Rüdiger, Rudolf Brázdil, Christian Pfister, Petr Dobrovolný, Mariano Barriendos Vallvé, Anita Bokwa, Dario Camuffo, Oldrich Kotyza, Danuta Limanówka, Lajos Rácz, and Fernando S. Rodrigo. "Seasonal Temperature and Precipitation Fluctuations in Selected Parts of Europe During the Sixteenth Century." *Climatic Change* 43.1 (1999): 169-200. doi:10.1023/A:1005542200040.

Glaser, Rüdiger, Christoph Beck, and Heiko Stangl. "Zur Temperatur- und Hochwasserentwicklung der letzten 1000 Jahre in Deutschland." In *DWD Klimastatusbericht 2003*, edited by Deutscher Wetterdienst, 55-67. Offenbach: Deustcher Wetterdienst, 2004.

Glaser, Rüdiger, Heiko Stangl, and Michel Lang. "Crues en Europe Centrale depuis l'an 1300 et dans leur contexte régional." *La Houille Blanche* 5 (2004): 43-49.

Glaser, Rüdiger, and Dirk Riemann. "A Thousand-Year Record of Climate Variations for Germany and Central Europe Based on Documentary Data." *Journal Of Quaternary Science* 24.5 (2009): 437-49.

Glaser, Rüdiger, Dirk Riemann, Rudolf Brázdil, Petr Dobrovolný, Aryan van Engelen, and Danuta Limanowka. "The Variability of European Floods since AD 1500." *Climatic Change* 101 (2010): 235-256.

Gloy, Karen. *Das Verständnis der Natur. Zweiter Band: Die Geschichte des ganzheitlichen Denkens*. München: Beck, 1996.

Goethe, Johann Wolfgang von. *Poetische und prosaische Werke in zwei Bänden*, vol. 1. Stuttgart: Cotta, 1836.

Göttmann, Frank. "Die Versorgungslage in Überlingen zur Zeit der Hungerkrise 1770/1771." In *Vermischtes zur neueren Sozial-, Bevölkerungs- und Wirtschaftsgeschichte des Bodenseeraumes*, edited by Frank Göttmann, 75-134. Konstanz: Gorre, 1990.

Goldschmid, Johann Jakob. *Erzellung seltsammer Natur-Geschichten* [...] *so sich bey uns in der Statt Wintertur zugetragen haben* (*bis 1543*), vol. 3. Stadtbibliothek Winterthur Ms Fol. 3.

González Muñoz, María del Carmen. *La población de Talavera, siglos XVI-XX. Estudio socio-geográfico*. Toledo: Diputación Provincial, 1974.

Goody, Jack, and Ian Watt. "The Consequences of Literacy." *In Comparitive Studies in Society and History* 5 (1962/1963), 304-345.

Gorz, André. *Écologie et Liberté*. Paris: Galilée, 1977.

Gorz, André. "Die politische Ökologie zwischen Expertokratie und Selbstbegrenzung." In *Auswege aus dem Kapitalismus. Beiträge zur Politischen Ökologie*, edited by André Gorz. Zürich: Rotpunktverlag, 2009.

Gosseries, Axel, and Inigo Gonzalez Ricoy, ed. *Institutions for Future Generations*. Oxford: Oxford University Press, 2015.

Goudsblom, Johan. *Fire and Civilisation.* London: Penguin Books, 1992.

Gracía Rivas, Manuel. *La 'invasión' de Aragón en 1591: Una solución militar a las alteraciones del reino*. Zaragoza: Diputación General de Aragón, 1992.

Grafton, Richard. *Grafton's Chronicle, or History of England* [...] *from the Year 1189 to 1558 Inclusive,* vol. 2. London: J. Johnsen, 1809.

Grau, Thilo. "Comparison of Feed-in Tariffs and Tenders to Remunerate Solar Power Generation." *DIW Discussion Paper* 1363. Accessed July 7, 2016. <https://www.diw.de/documents/publikationen/73/diw_01.c.437464.de/dp1363.pdf>.

Gray, Robert. *A Good Speed to Virginia*. London, 1609.

Greenberg, Arthur. *From Alchemy to Chemistry in Picture and Story*. Hoboken: Wiley, 2007.

Greengrass, Mark. "The Later Wars of Religion in the French Midi." In *The European Crisis of the 1590s: Essays in Comparative History*, edited by Peter Clark, 105-134. London: George Allen & Unwin, 1985.

Griffin, David R. *Religion and Scientific Naturalism*. Albany, N.Y.: Suny, 2000.

Griswold, William J. *The Great Anatolian Rebellion, 1000-1020/1591-1611*. Berlin: K. Schwarz Verlag, 1983.

Grigg, David B. *Population Growth and Agrarian Change. A Historical Perspective.* Cambridge: University Press, 1980.

Groen, Lisanne, and Arne Niemann. "Challenges in EU External Climate Change Policy-making in the Early Post-Lisbon Era: the UNFCCC Copenhagen Negotiations." In *EU External Relations Law and Policy in the Post-Lisbon Era,* edited by Paul James Cardwell, 315-334. The Hague: T.M.C. Asser Press, 2012.

Groen, Lisanne, Arne Niemann, and Sebastian Oberthür. "The EU as a Global Leader? The Copenhagen and Cancun UN Climate Change Negotiations." *Journal of Contemporary European Research* 8.2 (2012): 173-191.

Groen, Lisanne, and Arne Niemann. "The European Union at the Copenhagen Climate Negotiations: A Case of Contested EU Actorness and Effectiveness." *International Relations* 27.3 (2013): 308-324.

Groen, Lisanne, Arne Niemann, and Sebastian Oberthür. "The EU's Role in International Climate Change Policy-Making: A Global Leader in Decline?" In *Global Power Europe,*vol. 2. *Policies, Actions and Influence of the EU's External Relations*, edited by Astrid Boening, Jan-Frederik Kremer, and Aukje Van Loon, 37-54. Heidelberg: Springer, 2013.

Groen, Lisanne, "European Foreign Policy on the Environment and Climate Change." In *The SAGE Handbook of European Foreign Policy*, edited by K.E. Jorgensen et al., 884-897. London: SAGE, 2015.

Groenleer, Martijn L.P., and Louise G. van Schaik. "United We Stand? The European Union's International Actorness in the Cases of the International Criminal Court and the Kyoto Protocol." *Journal of Common Market Studies* 45.5 (2007): 969-998.

Großherzoglich-Badisches Landwirtschaftliches Wochenblatt. July/August 1844.

Großherzoglich-Badisches Landwirtschaftliches Wochenblatt. July/August 1845.

Großherzoglich-Badisches Regierungsblatt, vol. 14, and 15. Karlsruhe, 1809-1868.

Grove, Jean M. *The Little Ice Age*. London, New York: Routledge, 1988.

Grove, Richard H., and John Chappell, ed. *El Niño: History and Crisis*. Cambridge: The White Horse Press, 2000.

Guaman Poma de Ayala, Felipe. *El primer nueva corónica, y buen gobierno por D. Filippe Guaman Poma de Aiala*. 1615/1616.

Güll, R. "Auswanderung aus Württemberg und Baden." *Statistisches Monatsheft Baden-Württemberg* 9 (2013): 41-48.

Guha, Sumit. *Health and Population in South Asia from Earliest Times to the Present*. London, Hurst, 2001.

Guldin, Rainer. "Windschiffe. Die Wolke als Fahrzeug und Bühnenrequisit." In *Wolkenbilder: Die Erfindung des Himmels*, edited by Stephan Kunz, Johannes Stückelberger, and Beat Wismer, 39-44. München: Hirmer 2005.

Guldin, Rainer. *Die Sprache des Himmels. Eine Geschichte der Wolken*. Berlin: Kadmos, 2006.

Gunderson, Lance H., and Crawford S. Holling, ed. *Panarchy: Understanding Transformations in Human and Natural Systems*. Washington, D.C.: Island Press, 2002.

Günther, Klaus. *Politik des Kompromisses: Dissensmanagement in pluralistischen Demokratien.* Wiesbaden: VS Verlag, 2006.

Gupta, Akhil, and James Ferguson. "Beyond 'Culture': Space, Identity, and the Politics of Difference." *Cultural Anthropology* 7.1 (1992): 6-23.

Gutmann, Myron P. *War and Rural Life in the Early Modern Low Countries.* Princeton: Princeton University Press, 1980.

Habermas, Jürgen. *Zwischen Naturalismus und Religion.* Frankfurt a.M.: Suhrkamp, 2005 (English: *Between Naturalism and Religion*, 2005).

Habermas, Jürgen. *The Philosophical Discourse of Modernity.* Oxford: Polity, 1987.

Habermas, Jürgen. *Zwischen Naturalismus und Religion.* Frankfurt: Suhrkamp, 2005.

Hacking, Ian. *The Taming of Chance.* Cambridge: University Press, 1990.

Hald, Anders. *A History of Probability and Statistics and Their Applications before 1750.* New York: Wiley, 1990.

Hald, Anders. *A History of Mathematical Statistics from 1750 to 1930.* New York: Wiley, 1998.

Hamblin, Jacob D. *Arming Mother Nature: The Birth of Catastrophic Environmentalism.* New York: Oxford University Press, 2013.

Harley, Trevor A. "Nice Weather for the Time of the Year: the British Obsession with the Weather." In *Weather, Climate, Culture*, edited by Sarah, Strauss, and Benjamin S. Orlove, 103-118. Oxford: Berg, 2003.Haverkamp, Anselm. Metapher. Die Ästhetik in der Rhetorik. Bilanz eines exemplarischen Begriffes. (München: Fink, 2007).

Harvey, Fiona. "Doha Climate Talks: EU Weakened Over New Emissions Targets." *The Guardian*, November 23, 2012.

Haverkamp, Anselm. *Metapher. Die Ästhetik in der Rhetorik. Bilanz eines exemplarischen Begriffes.* München: Fink, 2007.

Hawkins, Ed, and Philip D. Jones. "On Increasing Global Temperatures: 75 Years after Callendar." *Quarterly Journal of the Royal Meteorological Society* 139.677 (2013): 1961-1963.

Hedegaard, Connie. "Europe Has Brought About a New Phase in Global Climate Policy." *The Guardian*, December 13, 2011.

Hellmann, Gustav. *Repertorium der deutschen Meteorologie: Leistungen der Deutschen in Schriften, Erfindungen und Beobachtungen auf dem Gebiete der Meteorologie und des Erdmagnetismus von den ältesten Zeiten bis zum Schlusse des Jahres 1881.* Leipzig: Engelmann, 1883.

Hellmann, Gustav, ed. *Meteorologische Beobachtungen vom XIV. bis XVII. Jahrhundert. Neudrucke von Schriften und Karten über Meteorologie* 13. Berlin: A. Asher & Co, 1901.

Hellmann, Gustav, ed. *E. Halley, W. Whiston, J.C. Wilcke, A. von Humboldt, C. Hansteen. Die ältesten Karten der Isogonen, Isoklinen, Isodynamen 1701, 1721, 1768, 1804, 1825, 1826. Neudrucke von Schriften und Karten über Meteorologie und Erdmagnetismus*, Vol4 Berlin: Asher, 1894.

Hellmann, Gustav. *Meteorologische Volksbücher. Ein Beitrag zur Geschichte der Meteorologie und zur Kulturgeschichte.* Berlin: Paetel, 1895.

Hellmann, Gustav, ed. *Evangelista Torricelli, Esperienza dell'Argento Vivo. Accademia del Cimento, Istrumenti per conoscer l'Alterazioni dell'Aria. Neudrucke von Schriften und Karten über Meteorologie und Erdmagnetismus*, vol.7. Berlin: Asher, 1897.

Hellmann, Gustav, ed. *E. Halley, A. von Humboldt, E. Loomis, U.J. Le Verrier, E. Renou. Meteorologische Karten 1688, 1817, 1846, 1863, 1864. Neudrucke von Schriften und Karten über Meteorologie und Erdmagnetismus*, vol.8. Berlin: Asher, 1897.

Hellmann, Gustav, ed. *Meteorologische Beobachtungen vom 14. bis 17. Jahrhundert.* (*Neudrucke von Schriften und Karten über Meteorologie und Erdmagnetismus*, vol. 13. Berlin: Asher 1901.

Hellmann, Gustav. *Beiträge zur Geschichte der Meteorologie*, vol. 1. Berlin: Preußisches Meteorologisches Institut 1914.

Hellmann, Gustav. *Die Meteorologie in den deutschen Flugschriften und Flugblättern des 16. Jahrhunderts. Ein Beitrag zur Geschichte der Meteorologie. Abhandlungen der Preußischen Akademie der Wissenschaften Jg. 1921, Phys.-Math. Klasse 1.* Berlin: Akademie der Wissenschaften, 1921.

Hellmann, Gustav. *Versuch einer Geschichte der Wettervorhersage im 16. Jahrhundert.* (*Abhandlungen der Preussischen Akademie der Wissenschaften Jg. 1924 Phys.-Math. Klasse 1.*Berlin: Akademie der Wissenschaften, 1924.

Helm, Dieter. "EU Climate-Change Policy—A Critique." In *The Economics and Politics of Climate Change*, edited by Dieter Helm, and Cameron Hepburn, 222-244. Oxford: Oxford University Press, 2009.

Henning, Richard. "Katalog bemerkenswerter Witterungsereignisse von den ältesten Zeiten bis zum Jahre 1800." *Abhandlungen des Preußischen Meteorologischen Instituts* 2.4 (1904): 93.

Henrici, Peter. "Il senso cristiano del tempo: Prospettive per un nuovo ordinamento del lavoro e del riposo in una società." *Notiziario* 4 (2002): 130-150.

Henshaw, Anne. "Climate and Culture in the North: The Interface of Archaeology, Paleoenvironmental Science, and Oral History." In *Weather, Climate, Culture*, edited by Sarah Strauss, and Benjamin S. Orlove, 217-231. Oxford: Berg 2003.

Herder, Johann Gottfried. *Outlines of a Philosophy of the History of Man.* New York: Bergman, 1966.

Herrmann, Bernd. "Kartoffel, Tod und Teufel," in *"... mein Acker ist die Zeit": Aufsätze zur Umweltgeschichte* ed. by Bernd Hermann (Göttingen: Universitätsverlag, 2011), 293-347.

Herrmann, Bernd. *Umweltgeschichte. Eine Einführung in Grundbegriffe.* Berlin, Heidelberg: Springer, 2013.

Heunisch, A.J. von. *Das Grossherzogthum Baden, historisch-geographisch-statistisch-topographisch beschrieben. Mit Beigaben von J. Bader*. Heidelberg: Verlag der Julius Groos'schen Universitätsbuchhandlung, 1857.

Heymann, Matthias, and Dania Achermann. "From Climatology to Climate Science in the 20th Century." In *Palgrave Handbook of Climate History*, edited by White, Sam, Christian Pfister and Franz Mauelshagen. Basingstoke: 2017.

Hildebrandt, Helmut, Martin Gudd. "Getreidebau, Missernten und Witterung im südwestlichen unteren Vogelsberg und dem angrenzenden Vorland während des 16. und frühen 17. Jahrhunderts." In *Archiv für hessische Geschichte*, vol. 49, 85-147. Darmstadt: Historischer Verein für Hessen e. V., 1991.

Hille, Martin. "Mensch und Klima in der frühen Neuzeit: Die Anfänge regelmässiger Wetterbeobachtung, 'Kleine Eiszeit', und ihre Wahrnehmung bei Renward Cysat (1545-1613)." *Archiv für Kulturgeschichte* 83 (2001): 63-91.

Himmelsbach, Iso, Rüdiger Glaser, Johannes Schönbein, Dirk Riemann, and Brice Martin. "Flood Risk along the Upper Rhine since AD 1480." *Hydrology and Earth System Sciences Discussions* 12.1 (2015): 177-211.

Hindle, Steve. "Dearth, Fasting and Alms: The Campaign for General Hospitality in late Elizabethan England." *Past and Present* 172 (2001): 44-86.

Hippel, Wolfgang von. *Auswanderung aus Südwestdeutschland: Studien zur württembergischen Auswanderung und Auswanderungspolitik im 18. und 19. Jahrhundert*. Stuttgart: Klett-Cotta, 1984.

Hippocrates on Airs, Waters, and Places. (London: Wyman, 1881).

Historical Manuscripts Commission. *The Manuscripts of the Duke of Somerset, Fifteenth Report*. Appendix VII. London: HMSO, 1897.

Höhlbaum, Konstantin, ed. *Das Buch Weinsberg: Kölner Denkwürdigkeiten aus dem 16. Jahrhundert*, vol. 1. Leipzig: Dürr, 1886.

Hölscher, Lucian. *Die Entdeckung der Zukunft*. Frankfurt a.M.: Fischer-Taschenbuch-Verlag, 1999.

Hofmeister, Georg. *Ethikrelevantes Natur- und Schöpfungsverständnis: Umweltpolitische Herausforderungen, naturwissenschaftlich-philosophische Grundlagen, schöpfungstheologische Perspektiven, Fallbeispiel: grüne Gentechnik. Darmstädter Theologische Beiträge zu Gegenwartsfragen*, vol. 4. Frankfurt a. M.: Peter Lang, 2000.

Hohberg, Wolf Helmhardt von. *Georgica Curiosa oder Adeliches Land=Leben* (...), 2 vols. Nürnberg: Endter, 1695. Accessed March 27, 2017. <http://digital.slub-dresden.de/werkansicht/dlf/8112/1/>; <http://digital.slub-dresden.de/werkansicht/dlf/8113/1/>.

Holinshed, Raphaell. *The Description and Histories of England*. London: Henry Denham, 1587.

Holzinger, Marcus. "Welcher Realismus? Welcher Sozialkonstruktivismus? Ein Kommentar zu Georg Kneers Verteidigung des Sozialkonstruktivismus und zu Bruno Latours Akteur-Netzwerk-Theorie." *Zeitschrift für Soziologie* 38.6 (2009): 521-534.

Horn, Eva. *Zukunft als Katastrophe*. Frankfurt a.M.: S. Fischer, 2014.

Horstmanshoff, Herman F.J., Antonie M. Luyendijk-Elshout, and F.G.Schlesinger. ed. *The four Seasons of Human Life. Four Anonymous Engravings from the Trent Collection. Nieuwe Nederlandse bijdragen tot de geschiedenis der geneeskunde en der natuurwetenschappen,* vol. 60; *Pantaleon Reeks*, vol. 32. Rotterdem, Durham: Erasmus Publishing, Trent Collection Duke University, 2002.

Hovi, Jon, Tora Skodvin, and Steinar Andresen. "The Persistence of the Kyoto Protocol: Why Other Annex I Countries Move on Without the United States." *Global Environmental Politics* 3.4 (2003): 1-23.

Howard, Luke. *On the Modification of Clouds, and on the Principles of Their Production, Suspension, and Destruction:* [...]. London: J. Taylor, 1803.

Howe, Joshua P. *Behind the Curve: Science and the Politics of Global Warming*. Seattle: University of Washington Press, 2014.

Huarte de San Juan, Juan. *Johann Huarts Prüfung der Köpfe zu den Wissenschaften: Worinne er die Verschiedenen Fähigkeiten die in den Menschen liegen zeigt; Einer jeden den Theil der Gelehrsamkeit bestimmt der für sie eigentlich gehöret; Und endlich den Aeltern Anschläge ertheilt wie sie fähige und zu den Wissenschaften aufgelegte Söhne erhalten können.* Zerbst: Zimmermann, 1752. Accessed March 27, 2017. <http://www.mdz-nbn-resolving.de/urn/resolver.pl?urn=urn:nbn:de:bvb:12-bsb10401189-3>.

Hubbert, Marion K. "Energy from Fossil Fuels." *Science* 109.2823 (1949): 103-109.

Huber, Bruno, and Veronika Giertz-Siebenlist. "Unsere tausendjährige Eichen-Jahrringchronologie durchschnittlich 57 (10-150)-fach belegt." In *Sitzungsberichte der Akademie der Wissenschaften, Mathematisch-Naturwissenschaftliche Klasse Abteilung 1*, 178: 37-42 Wien: Verlag der Österreichischen Akademie der Wissenschaften, 1970.

Huber, Oswald. *Schaffhauser Chronik 1521-1582*, edited by Carl August Bächtold. Schaffhausen: Buchdruckerei H. Meier und Cie, 1906.

Hünermann, Peter ed. *Das Zweite Vatikanische Konzil und die Zeichen der Zeit heute. Anstöße zur weiteren Rezeption*. Freiburg: Herder, 2006.

Hünermann, Peter. "Gottes Handeln in der Geschichte: Theologie als Interpretatio temporis." In *Freiheit Gottes und der Menschen: Festschrift für Thomas Pröpper*, edited by Michael Böhnke, 109-135. Regensburg: Pustet, 2006.

Humboldt, Alexander von, *Cosmos: A Sketch of the Physical Description of the Universe*, vol. 1. Translated by E.C. Otte. Baltimore: Johns Hopkins University Press 1997. First published as *Kosmos: Entwurf einer physischen Weltbeschreibung*, 5 Vols. Tübingen: Cotta, 1845-1862.

Humpert, Magdalene. *Bibliographie der Kameralwissenschaften*. Köln: Schroeder, 1937.

Illich, Ivan. *Tools for Convivability*. New York: Harper & Row, 1973.

Im-Thurn, Eduard. *Chronik der Stadt Schaffhausen, Volume 2: Von Wiedererlangung der Reichsfreiheit bis zum Eintritt in den Bund der Eidgenossen 1415-1501*. Schaffhausen: Brodtmann, 1844.

Íñiguez de Lequerica, Juan, ed. *Sermones funerales, en las honras del rey nuestro señor don Felipe II*. Madrid, 1599.

International Institute for Sustainable Development (IISD). "Summary of the Copenhagen Climate Change Conference." *Earth Negotiations Bulletin* 12.459 (2009). Accessed April 19, 2011. <http://www.iisd.ca/vol12/enb12459e.html>.

International Institute for Sustainable Development (IISD). "Summary of the Cancun Climate Change Conference." *Earth Negotiations Bulletin* COP16 (2010). Accessed October 2, 2013. <http://www.iisd.ca/vol12/enb12498e.html>.

International Institute for Sustainable Development (IISD). "Summary of the Doha Climate Change Conference." *Earth Negotiations Bulletin* COP18 (2012). Accessed October 21, 2013. <http://www.iisd.ca/vol12/enb12567e.html>.

International Institute for Sustainable Development (IISD). "Summary of the Warsaw Climate Change Conference." *Earth Negotiations Bulletin* COP19 (2013). Accessed November 27, 2013. <http://www.iisd.ca/vol12/enb12594e.html>.

International Institute for Sustainable Development (IISD). "Summary of the Paris Climate Change Conference." *Earth Negotiations Bulletin* COP21 (2015). Accessed October 18, 2016. <http://www.iisd.ca/vol12/enb12663e.html>.

Intergovernmental Panel on Climate Change (IPCC), ed. *Climate Change 2007: Synthesis Report*. Published 2007. Accessed January 25, 2014. <http://www.ipcc.ch/pdf/assessment-report/ar4/syr/ar4_syr.pdf>.

Intergovernmental Panel on Climate Change (IPCC), ed. *Climate Change 2013: The Physical Science Basis. Fifth Assessment Report*. Cambridge, UK: Cambridge University Press, 2013.

Intergovernmental Panel on Climate Change (IPCC), ed. *Climate Change 2014: Impacts, Adaption, and Vulnerability*. Cambridge, New York: Cambridge University Press, 2014.

İpşirli, Mehmet., ed. *Tarih-i Selâniki* [The Chronicle of Selâniki]. 2 Vols. Istanbul: İstanbul Üniversitesi Edebiyat Fakültesi Yayınları, 1989.

Issar, Arie S., and Mattanyah Zohar. *Climate Change—Environment and History of the Near East*. Berlin, Heidelberg, New York: Springer, 2007.

Jacobeit, Jucundus, Heinz Wanner, Gerhard Koslowski, and Martin Gudd. "European Surface Pressure Patterns for Months with Outstanding Climatic Anomalies during the Sixteenth Century." *Climatic Change* 43.1 (1999): 201-221. doi:10.1023/A:1005594216878.

Jacobs, Michael, "The Doha Climate Talks Were a Start, But 2015 Will be the Moment of Truth," *The Guardian*, December 20, 2012.

Jaeger, Friedrich. "Kultur." *In Enzyklopädie der Neuzeit*, vol. 7, edited by Friedrich Jaeger, 253-281. Stuttgart: Metzler, 2008.

Jago, Charles. "Taxation and Political Culture in Castile 1590-1640." In *Spain, Europe and the Atlantic World: Essays in Honour of John H. Elliott*, edited by Richard L. Kagan and Geoffrey Parker, 48-72. Cambridge: Cambridge University Press, 1995.

James, Jamie. *The Music of the Spheres. Music, Science, and the Natural Order of the Universe*. London: Little & Brown, 1994.

Jeismann, Michael. "Grüne Gegenwart. Die Grenze der Ökologie: Reinhart Kosellecks Begriffsgeschichten." *Frankfurter Allgemeine Zeitung*, November 12, 2006.

Jentsch, Mareike, Tobias Trost, Lukas Emele, and Michael Sterner. "Power to Gas als Langzeitspeicher." *Energy* 2.0 (2011): 46.

Johannes Paul II. *Evangelium vitae. Encylica about the Gospel of Life*. Vatican: Vatican Press 1995.

Johannes Paul II. "Friede mit Gott dem Schöpfer, Friede mit der ganzen Schöpfung: Botschaft von Papst Johannes Paul II. zur Feier des Weltfriedenstages am 1. Januar 1990." *L'Osservatore Romano* 50 (1989): 1; 7f.

Jolly, William M., Matthias Dobbertin, Niklaus E. Zimmermann, and Markus Reichstein. "Divergent Vegetation Growth Responses to the 2003 Heat Wave in the Swiss Alps." *Geophysical Research Letters* 32 (2005): L18409. doi:10.1029/2005GL023252.

Jones, Eric Lionel. *The European Miracle: Environments, Economies, and Geopolitics in the History of Europe and Asia*. Cambridge: Cambridge University Press, 1987.

Jordan, Andrew, Dave Huitema, and Harro van Asselt. "Climate Change Policy in the European Union: An Introduction." In *Climate Change Policy in the European Union: Confronting the Dilemmas of Mitigation and Adaptation?*, edited by Andrew Jordan, Dave Huitema, Harro van Asselt, Tim Rayner and Frans Berkhout, 3-26. Cambridge: Cambridge University Press, 2010.

Jouanna, Jacques. *Hippocrate*. Paris: Fayard, 1992. [am.: *Hippocrates*. Baltimore: Johns Hopkins University Press, 1999].

Jouanna, Jacques. *Hippocrate. Airs, eaux, lieux*. Paris: Les Belles Lettres, 1996.

Jouanna, Jacques. *Hippocrate. Pour une archéologie de l'École de Cnide*, vol. 141. Paris: Belles Lettres, 2009.

Jouanna, Jacques. "Rhetoric and medicine in the Hippocratic corpus". In *Rhetoric and medicine in the Hippocratic Corpus. Greek Medicine from Hippocrates to Galen. Selected Papers*, edited by Jacques Jouanna, 39-53. Leiden: Brill, 2012.

Kabat, Pavel, Louise O. Fresco, Marcel J.F. Stive, Cees P. Veerman, van Alphen, P.L.J. Jos, Parmet, W.A.H. Bart et al. "Dutch Coasts in Transition." *Nature Geoscience* 2.7 (2009): 450-452.

Kaijser, Arne. "System Building from Below: Institutional Change in Dutch Water Control Systems." *Technology and Culture* 43.3 (2002): 521-548.

Kambartel, Friedrich. *Erfahrung und Striktur. Bausteine zu einer Kritik des Empirismus und Formalismus*. Frankfurt a.M.: Suhrkamp, 1968.

Kaniewski, David, Elise Van Campo, and Harvey Weiss. "Drought Is a Recurring Challenge in the Middle East." *Proceedings of the National Academy of Sciences* 109 (2012): 3862-3867.

Kant, Immanuel. "Kritik der reinen Vernunft." In *Werke in sechs Bänden*, vol. 2, edited by Wilhelm Weischedel. Darmstadt: Wissenschaftliche Buchgesellschaft, 1966.

Kaplan, E. Ann. *Climate Trauma: Foreseeing the Future in Dystopian Film and Fiction.* New Brunswick: Rutgers University Press, 2015.

Karlsson, Christer, Charles Parker, Mattias Hjerpe, and Björn-Ola Linnér. "Looking for Leaders: Perceptions of Climate Change Leadership among Climate Change Negotiation Participants." *Global Environmental Politics* 11.1 (2011): 89-107.

Kartha, Sivan, and Peter Erickson. *Comparison of Annex 1 non-Annex 1 Pledges under the Cancun Agreements. Working Paper* 6. Stockholm: Stockholm Environment Institute, 2011.

Kates, Robert W. "The Interaction of Climate and Society." In *Climate Impact Assessment. Studies of the Interaction of Climate and Society*, edited by Robert W. Kates, Jesse H. Ausubel and Mimi Berberian, 3-36. *ICSU/SCOPE Report* 27. Chichester: Wiley, 1985.

Kates, Robert W., Jesse H. Ausubel, and Mimi Berberian ed. *Climate Impact Assessment. Studies of the Interaction of Climate and Society.* (*International Council of Scientific Unions*), vol. 27. Chichester: Wiley, 1985.

Kaulbach, Hans-Martin, Reinhart Schleier. *"Der Welt Lauf". Allegorische Graphikserien des Manierismus*. Ostfildern: Hatje, 1997.

Keating, Dave. "Shaky Leadership on the World Stage." *European Voice*, May 1, 2013.

Keller, Reiner. "Wissenssoziologische Diskursanalyse als interpretative Analytik." In *Die diskursive Konstruktion von Wirklichkeit: Zum Verhältnis von Wissenssoziologie und Diskursforschung*, edited by Reiner Keller, Andreas Hirseland, Werner Schneider and Willy Viehöver, 49-75. Konstanz: UVK, 2005.

Kemfert, Claudia, and Dorothea Schäfer. "Finanzierung der Energiewende in Zeiten großer Finanzmarktinstabilität." *DIW Wochenbericht* 31 (2012): 3-14.

Kemfert, Claudia. *Battle about Electricity, Myths, Power and Monopolies*. Hamburg: Murmann, 2013.

Kessler, Josua. *Chronologie Santgallischer Begebenheiten vom Jahr 1540 bis Ende des Jahres 1645. Handschriften* Nr. 74, Manuscript. St. Gallen: Stiftsarchiv St. Gallen, n. d.

Kiesraad. "Databank Verkiezingsuitslagen." Accessed May 21, 2017. <http://www.verkiezingsuitslagen.nl/>.

Kilar, Piotr. *500 Year Poland Temperature and Precipitation Indices. IGBP PAGES/World Data Center for Paleoclimatology Data Contribution Series* 2010-135. Boulder: National Climatic Data Center, 2010 Accessed November 7, 2013. <https://www.ncdc.noaa.gov/paleo/study/10050>.

Kilian, Bertil, and Ole Elgström. "Still a Green Leader? The European Union's Role in International Climate Negotiations." *Cooperation and Conflict* 45.3 (2010): 255-273.

Kilper, Heiderose. "Governance und die soziale Konstruktion von Räumen: Eine Einführung." In *Governance und Raum*, edited by Heiderose Kilper, 9-24. Baden-Baden: Nomos, 2010.

Kippenberg, Hans G., and Tilman Seidensticker, ed. *The 9/11 Handbook: Annotated Translation and Interpretation of the Attackers' Spiritual Manual*. London: Equinox, 2006.

Kiss, Andrea, Rob Wilson, and Istvan Bariska. "An Experimental 392-Year Documentary-Based Multi-Proxy (Vine and Grain) Reconstruction of May–July Temperatures for Kőszeg, West-Hungary." *International Journal of Biometeorology* 55.4 (2011): 595-611.

Klessmann, Corinna, Silvana Tiedemann, Fabian Wigand, Tobias Kelm, Jenny Winkler, Mario Ragwitz, Marie-Christin Haufe, and Karl-Martin Ehrhart. "Ausschreibungen für PV-Freiflächenanlagen Auswertung der ersten zwei Runden." Accessed July, 7, 2016. <https://www.erneuerbare-energien.de/EE/Redaktion/DE/Downloads/ausschreibungen-pv-freiflaechanlagen-auswertung-erste-zwei-runden.pdf?__blob=publicationFile&v=3>.

Knauer, Mauritius. *Calendarium Oeconomicum Practicum Perpetuum. Das ist: Immerwährender Curieuser Hauß=Calender* [...]. Nürnberg: Johann Georg Lochner, 1741. Accessed March 27, 2017. <http://www.mdz-nbn-resolving.de/urn/resolver.pl?urn=urn:nbn:de:bvb:12-bsb10300775-3>.

Knoblauch, Hubert. *Kommunikationskultur: Die kommunikative Konstruktion kultureller Kontexte*. Berlin: de Gruyter, 1995.

Köpke, Peter, Marianne Placzek, Henning Staiger, and Peter Winkler: "Solare UV-Strahlung und ihre Wirkung auf den Menschen." In *Promet* 33 (2007): 95-108.

Körber, Hans-Günther. *Vom Wetteraberglauben zur Wetterforschung*. Leipzig: Edition, 1989.

Kohl, Karl-Heinz. *Die Macht der Dinge. Geschichte und Theorie sakraler Objekte*. München: Beck, 2003.

Komlos, John. *Nutrition and Economic Development in the Eighteenth-Century Habsburg Monarchy: an Anthropometric History*. Princeton: University Press, 1989.

Koningsveld, Mark van, J.P.M. Mulder, M.J: F. Stive, L. van der Valk, and A.W. van der Weck. "Living with Sea-Level Rise and Climate Change: A Case Study of the Netherlands." *Journal of Coastal Research 24.2* (2008): 367-379.

Kopfmüller, Jürgen. "Einführung." In *Risiko Klima: Der Treibhauseffekt als Herausforderung für Wissenschaft und Politik*, edited by Jürgen Kopfmüller und Reinhard Coenen, 9-26. Frankfurt a.M.: Campus, 1997.

Kopperschmidt, Josef. "Das Ende der Verleumdung. Einleitende Anmerkungen zur Wirkungsgeschichte der Rhetorik." In *Rhetorik Band II: Wirkungsgeschichte der Rhetorik*, edited by Josef Kopperschmidt, 1-33. Darmstadt: Wissenschaftliche Buchgesellschaft 1991.

Koselleck, Reinhart. Kritik und Krise: Eine Studie zur Pathogenese der bürgerlichen Welt. Frankfurt a.M.: Suhrkamp, 1973.

Koselleck, Reinhart. Vergangene Zukunft: Zur Semantik geschichtlicher Zeiten. Frankfurt a.M.: Suhrkamp, 1979.

Koselleck, Reinhart. *Futures Past: On the Semantics of Historical Time*. Cambridge: MIT Press, 1985.

Koselleck, Reinhart. *Critique and Crisis: Enlightenment and the Pathogenesis of Modern Society*. Cambridge: MIT Press, 1988.

Koselleck, Reinhart. *Begriffsgeschichten*. Frankfurt a.M.: Suhrkamp, 2006.

Koselleck, Reinhart. *Vom Sinn und Unsinn der Geschichte: Aufsätze und Vorträge aus vier Jahrzehnten*. Berlin: Suhrkamp, 2010.

Krämer, Daniel. "Vulnerabilität und die konzeptionellen Strukturen des Hungers. Eine methodische Annäherung." In *Handeln in Hungerkrisen. Neue Perspektiven auf soziale und klimatische Vulnerabilität*, edited by Collet, Dominik, Thore Lassen, and Ansgar Schanbacher, 45-65. Göttingen: Universitätsverlag, 2012.

Kreditanstalt für Wiederaufbau (KfW). "Der energetische Sanierungsbedarf und der Neubaubedarf von Gebäuden der kommunalen und sozialen Infrastruktur." Study conducted by Bremer Energieinstitut for KfW. Published 2011. Accessed July 14, 2015. <http://www.bremer-energie-institut.de/download/publications/BEI500_024_0431_Gutachten.pdf>.

Kreditanstalt für Wiederaufbau (KfW). "Energieeffizient sanieren—Soziale Organisationen." Accessed July 14, 2015. <https://www.kfw.de/inlandsfoerderung/%C3%96ffentliche-Einrichtungen/Energetische-Stadtsanierung/Finanzierungsangebote/Energieeffizient-Sanieren-kommunale-Unternehmen-%28219%29>.

Kunz, Stephan, Johannes Stückelberger, and Beat Wismer ed. *Wolkenbilder. Die Erfindung des Himmels*. München: Hirmer, 2005.

Kurz, Heinrich ed. *Johann Fischart's sämmtliche Dichtungen. 3 Theile*. Leipzig: Weber, 1867. Accessed March 27, 2017. <http://www.mdz-nbn-resolving.de/urn/resolver.pl?urn=urn:nbn:de:bvb:12-bsb10105532-1>.

L'Honoré Naber, Samuel P., ed. *Reizen van Jan Huyghen van Linschoten naar het Noorden 1594-1595*. The Hague: Linschoten Vereeniging, 1914.

Labbé, Thomas, and Fabien Gaveau. "Les dates de bans de vendange à Dijon: Etablissement critique et révision archivistique d'une série ancienne." *Revue Historique* 657.1 (2011): 19-51. doi:10.3917/rhis.111.0019.

Lachiver, Marcel. *Les années de misère: La famine au temps du Grand Roi, 1680-1720*. Paris: Fayard, 414-417.

Laconicum Europae Speculum quo IX praecipuae nationes et populi, absque omni praeiudicio, ludibrio et opprobrio, saltemcuriositatis et oblectationis gratia, certis characteribus et symbolis discernuntur [...]. Augsburg: Martin Engelbrecht, 1736/1737.

Lamb, Hubert H. *Climate: Present, Past and Future: Climatic History and the Future. Climate: Present, Past and Future*, vol. 2. London: Methuen, 1977.

Landes, Richard. *Heaven on Earth: The Varieties of the Millennial Experience*. Oxford: Oxford University Press, 2011.

Landsteiner, Erich. "The Crisis of Wine Production in Late 16th Century Central Europe: Climatic Causes and Economic Consequences." *Climatic Change* 43 (1999): 323-334.

Läpple, Dieter. "Essay über den Raum: Für ein gesellschaftswissenschaftliches Raumkonzept." In *Stadt und Raum: Soziologische Analysen*, edited by Hartmut Häußermann, and Walter Siebel, 157-207. Pfaffenweiler: Centaurus, 1991.

Latour, Bruno. *We Have Never Been Modern*. Translated by Catherine Porter. Cambridge, MA: Harvard University Press, 1993.

Latour, Bruno. *Politics of Nature. How to Bring the Sciences into Democracy.* Translated by Catherine Porter. Cambridge, London: Harvard University Press, 2004.

Latour, Bruno. "From Realpolitik to Dingpolitik or How to Make Things Public." In *Making Things Public: Atmospheres of Democracy. Exhibition at ZKM, Center for Art and Media Karlsruhe 20.03.–3.10.2005*, edited by Bruno Latour and Peter Weibel, 14-43. Cambridge: MIT Press, 2005.

Lavoisier, Antoine Laurent de. *Traité élémentaire de chimie : présenté dans un ordre nouveau, et d'après les découvertes modernes*, 2 Vols. Paris: Deterville, IX/1801. Accessed March27,2017.<http://www.mdz-nbn-resolving.de/urn/resolver.pl?urn=urn:nbn:de:bvb:12-bsb10073155-7>.

Lawson, Peter. "Property Crime and Hard Times in England, 1559-1624." *Law & History Review* 4.1 (1986): 95-127.

Le Roy Ladurie, Emmanuel. *Histoire du climat depuis l'an mil.* Paris: Flammarion, 1967.

Le Roy Ladurie, Emmanuel. *Historie humaine et comparée du climat I: Canicules et glaciers (XIIIe–XVIIIe siècle)*. Paris: Fayard, 2004.

Le Roy Ladurie, Emmanuel. *Historie humaine et comparée du climat II: Disettes et révolutions (1740-1860)*. Paris: Fayard, 2006.

Le Roy Ladurie, Emmanuel, and Daniel Rousseau. *Naissance De L'*histoire *Du Climat*. Paris: Hermann, 2013.

Leggewie, Claus, and Harald Welzer. *Das Ende der Welt, wie wir sie kannten: Klima, Zukunft und die Chancen der Demokratie*. Frankfurt a.M.: S. Fischer, 2010.

Leggewie, Claus, and Patrizia Nanz. *Die Konsultative.* Berlin: Wagenbach, 2015.

Leggewie, Claus. "Der Menschen entscheidet im Anthropozän. Konsequenzen aus der Entdeckung eines neuen Erdzeitalters." In *Sind Umweltkrisen Krisen der Natur oder der Kultur?*, edited by Bernd Herrmann, 63-78. Berlin: Springer, 2015. doi:10.1007/978-3-662-48139-4_5.

Leggewie, Claus, et al. *WGBU 2016, Policy Paper G20*, forthcoming.

Lehman, Christian. [...] *Historischer Schauplatz derer natürlichen Merckwürdigkeiten in dem Meißnischen Ober=Erzgebirge*. Meissen: Lankisch, 1699.

Lehmann, Hartmut. "Endzeiterwartung im Luthertum im späten 16. und im frühen 17. Jahrhundert." In *Die lutherische Konfessionalisierung in Deutschland: Wissenschaftliches Symposion des Vereins für Reformationsgeschichte 1988*, edited by Hans-Christof Rublack, 545-558. Gütersloh: Gerd Mohn, 1992.

Lehmann, Hartmut. "Frömmigkeitsgeschichtliche Auswirkungen der 'Kleinen Eiszeit'." In *Volksreligiosität in der modernen Sozialgeschichte*, edited by Wolfgang Schieder, 31-50. Göttingen: Vandenhoeck & Ruprecht, 1986.

Leijonhufvud, Lotta, Rob Wilson, Anders Moberg, Johan Söderberg, Dag Retsö, and Ulrica Söderlind. "Five Centuries of Winter/Spring Temperatures in Stockholm Reconstructed from Documentary Evidence and Instrumental Observations." *Climatic Change* 101.1-2 (2010): 109-141.

Leinkauf, Thomas. *Mundus combinatus. Studien zur Struktur der barocken Universalwissenschaft am Beispiel Athanasius Kircher SJ (1602-1680)*. Berlin: Akademie, 2009.

Lengwiler, Martin. *Risikopolitik im Sozialstaat. Die schweizerische Unfallversicherung 1870-1970*. Köln: Böhlau, 2006.

Lenschow, Andrea, and Carina Sprungk. "The Myth of a Green Europe." *Journal of Common Market Studies* 48.1 (2010): 133-154.

Lenton, Timothy M., and Hans Joachim Schellnhuber. "Tipping the Scales." *Nature Reports Climate Change* 1.22 (2007). doi:10.1038/climate.2007.65.

Lenton, Timothy M., Hermann Held, Elmar Kriegler, Jim W. Hall, Wolfgang Lucht, Stefan Rahmstorf, and Hans Joachim Schellnhuber. "Tipping Elements in the Earth's Climate System," *PNAS* 105.6 (2008): 1786-1793. doi:10.1073/pnas.0705414105.

Lepenies, Wolf. *Das Ende der Naturgeschichte: Wandel kultureller Selbstverständlichkeiten in den Wissenschaften des 18. und 19. Jahrhunderts*. München: Hanser, 1976.

Leppin, Volker. *Antichrist und Jüngster Tag: Das Profil apokalyptischer Flugschriftenpublizistik im deutschen Luthertum 1548-1618.* Heidelberg: Gütersloher Verlagshaus, 1999.

Leroux, Alfred, and Auguste Bosvieux, ed. "Extraits du Journal de Me Pierre de Teysseulh: Chanoine de l'église de Limoges, 1533-1568." In *Chartes, chroniques et mémoriaux pour servir à l'histoire de la Marche et du Limousin*, edited by Alfred Leroux, and Auguste Bosvieux, 253-261. Tulle: Crauffon, 1886.

Leuschner, Hans H., and Fritz H. Schweingruber. "Dendroökologische Klassifizierung und Auswertung häufig auftretender intraannueller holzanatomischer Merkmale bei Eichen und Kiefern." *Dendrochronologia* 14 (1996): 273-286.

Li, Tana. Nguyên Cochinchina: Southern Vietnam in the Seventeenth and Eighteenth Centuries. Ithaca: Cornell University Press, 1998.

Link, Christian. *Schöpfung: Handbuch Systematische Theologie*, vol. 7.1, and 7.2. Gütersloh: Gütersloher Verlagshaus, 1991.

Linse, Ulrich. "Das 'natürliche' Leben: Die Lebensreform." In *Die Erfindung des Menschen: Schöpfungsträume und Körperbilder 1500-2000*, edited by Richard van Dülmen, 435-458. Böhlau: Wien, 1998.

Lintsen, Harry. "Two Centuries of Central Water Management in the Netherlands." *Technology and Culture* 43.3 (2002): 549-568.

Löning, Karl, and Erich Zenger. *Als Anfang schuf Gott: Biblische Schöpfungstheologien*. Düsseldorf: Patmos, 1997.

Löschel, Andreas, Georg Erdmann, Frithjof Staiß, and Hans-Joachim Ziesing. "Expertenkommission zum Monitoring-Prozess „Energie der Zukunft"." (Stellungnahme zum ersten Fortschrittsbericht der Bundesregierung für das Bereichtsjahr 2013.) Accessed November 23rd, 2016. <http://www.bmwi.de/DE/Themen/Energie/Energiewende/monitoring-prozess.html>.

Löw, Martina. *Raumsoziologie*. Frankfurt a.M.: Suhrkamp, 2001.

Lonergan, Bernard J.F. *Method in Theology*. London: Darton, Longman and Todd, 1972.

Lorenzo Cadarso, Pedro L. *Los conflictos sociales en Castilla (siglos XVI–XVII)*. Madrid: Siglo XXI, 1996.

Lübbe, Hermann. "Kontingenzerfahrung und Kontingenzbewältigung." In *Kontingenz: Poetik und Hermeneutik*, vol. 17, edited by Gerhart von Graevenitz and Odo Marquard, 35-47. München: Wilhelm Fink, 1998.

Lübken, Uwe. *Die Natur der Gefahr. Überschwemmungen am Ohio River im 19. und 20. Jahrhundert*. Göttingen: Vandenhoeck & Ruprecht, 2014.

Luc, Jean André de. *Neue Ideen über die Meteorologie*, 2 vols. Berlin, Stettin: Nicolai 1797. [Idées sur la météorologie, 2 vols. Paris: Duchesne, 1787.] Accessed March 27, 2017. <http://www.mdz-nbn-resolving.de/urn/resolver.pl?urn=urn:nbn:de:bvb:12-bsb10134206-1>;<http://www.mdz-nbn-resolving.de/urn/resolver.pl?urn=urn:nbn:de:bvb:12-bsb10134207-6>.

Luhmann, Niklas. *Die Religion der Gesellschaft*. Frankfurt a.M.: Suhrkamp, 2000.

Luterbacher, Jürg, Elena Xoplaki, Daniel Dietrich, Ralph Rickli, Jucundus Jacobeit, Christoph Beck, Dimitrios Gyalistras, Christoph Schmutz, and Heinz Wanner. "Reconstruction of Sea Level Pressure Fields Over the Eastern North Atlantic and Europe Back to 1500." *Climate Dynamics* 18.7 (2002): 545-561. doi:10.1007/s00382-001-0196-6.

Luterbacher, Jurg, and Elena Xoplaki. "500-Year Winter Temperature and Precipitation Variability over the Mediterranean Area and its Connection to the Large-Scale Atmospheric Circulation." In *Mediterranean Climate: Variability and Trends*, edited by Hans-Jürgen Boehle, 133-153. Heidelberg: Springer Verlag, 2003.

Luterbacher, Jürg, Mark A. Liniger, Annette Menzel, Nicole Estrella, Paul M. Della-Mara, Christian Pfister, This Rutishauser, and Elena Xoplaki. "The Exceptional European Warmth of Autumn 2006 and Winter 2007: Historical Context, the Unterlying Dynamics and its Phenological Impacts." *Geophysical Research Letters* 34.12 (2007): 1-6. doi:10.1029/2007GL029951.

Luther, Martin. „Luther an seine Ehefrau" [Eisenach,] 16/26. Juli 1540. In *D. Martin Luthers Werke: Kritische Gesamtausgabe*, edited by Otto Clemen. Abt. Briefwechsel, vol. 9 (1540-1528). Februar 1542, Nr 3519, 204-205. Weimar: Hermann Böhlaus Nachfolger, 1941.

Lynch, Peter. *The Emergence of Numerical Weather Prediction: Richardson's Dream*. Cambridge: University Press, 2006.

Mabey, Nick. "Understanding Europe's Unexpected Durban Success." Published January 2012. Accessed October 21, 2013. <http://esharp.eu/big-debates/the-green-economy/16-understanding-europe-s-unexpected-durban-success/>.

Malberg, Horst. *Bauernregeln. Aus meteorologischer Sicht*. Berlin: Springer, 2003.

Malla, Katak. *The EU and Climate Treaty Negotiations after the Durban Conference*. *European Policy Analysis* March 2012. Stockholm: Swedish Institute for European Policy Studies, 2012.

Manley, Gordon. "Central England Temperatures: Monthly Means 1659 to 1973." *Quarterly Journal of the Royal Meteorological Society* 100 (1974): 389-405.

Mann, Michael E., Mark A. Cane, Stephen E. Zebiak, and Amy Clement. "Volcanic and Solar Forcing of the Tropical Pacific over the Past 1000 Years." *Journal of Climate* 18.3 (2005): 447-456.

Mannheim, Karl. "The Problem of Generations." In *Essays on the Sociology of Knowledge*, edited by Karl Mannheim, 276-322. London: Routledge, 1952.

Manrique, Emilio, and Ángel Fernández-Cancio. "Extreme Climatic Events in Dendroclimatic Reconstructions from Spain." *Climatic Change* 44 (2000): 123-138.

Mansbridge, Jane. "Rethinking Representation." *The American Political Science Review* 97.4 (2003): 515-528.

Manuel, Frank E. *The Prophets of Paris*. Cambridge, MA: Harvard University Press 1962.

Marcos Martín, Alberto. *Economía, sociedad, pobreza en Castilla: Palencia 1500-1814*. Palencia: Diputación Municipal, 1985.

Marcos Martín, Alberto. *España en los siglos XVI, XVII y XVIII: Economía y sociedad*. Barcelona: Crítica, 2000.

Marcus, George E. "Ethnography in/of the World System: The Emergence of Multi-Sited Ethnography." *Annual Review of Anthropology* 24 (1995): 95-117.

Mathieu, Jon. "Klimawandel und Wirtschaftsgeschichte der Vormoderne: Zur Methodendiskussion." *Schweizerisches Jahrbuch für Wirtschafts- und Sozialgeschichte* 30 (2015): 21-33.

Matthes, Joachim. "Karl Mannheims 'Das Problem der Generationen', neu gelesen: Generationen-'Gruppen' oder 'gesellschaftliche Regelung von Zeitlichkeit'." *Zeitschrift für Soziologie* 14.5 (1985): 363-372.

Matzarakis, Andreas, and Helmut Mayer ed. *Proceedings zur sechsten Fachtagung BIOMET des Fachausschusses Biometeorologie der Deutschen Meteorologischen Gesellschaft*. Freiburg: Meteorologisches Institut der Universität, 2007.

Matzat, Wolfgang, and Gerhard Poppenberg ed. *Begriff und Darstellung der Natur in der spanischen Literatur der frühen Neuzeit. Hispanistisches Kolloquium*, vol. 4. München: Fink, 2012.

Mauelshagen, Franz. *Klimageschichte der Neuzeit*. Darmstadt: Wissenschaftiche Buchgesellschaft, 2010.

Mauelshagen, Franz. "Sharing the Risk of Hail: Insurance, Reinsurance and the Variability of Hailstorms in Switzerland, 1880-1932." *Environment & History* 17.1 (2011): 171-191.

Mauelshagen, Franz. "Climate Catastrophism: The History of the Future of Climate Change." In *Historical Disasters in Context: Science, Religion, and Politics*, edited by Andrea Janku, Franz Mauelshagen and Gerrit Schenk, 261-282. London, New York: 2012.

Mauelshagen, Franz. "Die Vergesellschaftung des Hagelrisikos: Zur Geschichte der landwirtschaftlichen Hagelversicherung in der Schweiz, 1818-1950." *Traverse* 2014.3 (2014): 60-71.

Mauelshagen, Franz. "Das Zeitalter der Ungewissheit: Zukunftsszenarien und globale Bedrohung nach dem Zweiten Weltkrieg." In *Vorsorgen in der Moderne,* edited by Nicolai Hannig. München: 2016.

Mauelshagen, Franz. "Bridging the Great Divide—the Anthropocene as a Challenge to the Social Sciences and Humanities." In *Religion and the Anthropocene*, edited by Celia Deane-Drummond, Markus Vogt and Sigurd Bergmann, 87-102. Eugene, OR: 2017.

Mauelshagen, Franz. "Modernity's Decline? Climate Change, the Anthropocene and the Return of Catastrophism." In *Decline,* edited by William O'Reilly and Richard Miles. Budapest, New York NJ: forthcoming 2017.

Mauss, Marcel. "Essai sur le don: Formes et raisons de l'échange dans les sociétés primitives." In *Les classiques des sciences sociales*. Chicoutim: Université du Québec à Chicoutimi, 2002. Accessed October 13, 2015. <http://classiques.uqac.ca/classiques/mauss_marcel/socio_et_anthropo/2_essai_sur_le_don/essai_sur_le_don.pdf>.

Mayer, Lukas. "Intergenerational Justice." Published in *Stanford Encyclopedia of Philosophy*, April 2, 2003. Accessed October 13, 2015. <http://plato.stanford.edu/entries/justice-intergenerational/>.

Mayewski, Paul A., Loren D. Meeker, Meeker S. Twickler, Sallie I. Whitlow, Qinzhao Yang, W.B. Lyons, and Michael Prentice. "Major Features and Forcing of High-Latitude Northern Hemisphere Atmospheric Circulation Using a 110,000-Year-Long Glaciochemical Series." *Journal of Geophysical Research* 102 (1997): 26345-26366.

Mayntz, Renate. "Von der Steuerungstheorie zu Global Governance." In *Governance in einer sich wandelnden Welt,* edited by Gunnar Folke Schuppert and Michael Zürn, 43-60. Wiesbaden: VS Verlag für Sozialwissenschaften, 2008.

Meadows, Dennis, Donella H. Meadows, Erich Zahn, and Peter Milling, ed. *Die Grenzen des Wachstums – Bericht des Club of Rome zur Lage der Menschheit*. Stuttgart: DVA, Stuttgart 1972.

Meetz, Karen S. "Tempora triumphant. Ikonographische Studien zur Rezeption des antiken Themas der Jahreszeitenprozession im 16. und 17. Jahrhundert und zu seinen naturphilosophischen, astronomischen und bildlichen Voraussetzungen." PhD diss., University of Bonn, 2003. Accessed March 27, 2017. <http://hss.ulb.uni-bonn.de/2003/0126/0126.pdf>.

Megerle, Klaus. *Württemberg im Industrialisierungsprozeß Deutschlands: Ein Beitrag zur regionalen Differenzierung der Industrialisierung.* Stuttgart: Klett-Cotta, 1982).

Meier, Nicole, This Rutishauser, Christian Pfister, Heinz Wanner, and Jürg Luterbacher. "Grape Harvest Dates as A Proxy for Swiss April to August Temperature Reconstructions Back to AD 1480." *Geophysical Research Letters* 34.20 (2007): L20705.

Meinel, Christoph. "Natur als moralische Anstalt. Die Meteorologia philosophico-politica des Franz Reinzer, S.J, ein naturwissenschaftliches Emblembuch aus dem Jahre 1698." *Nuncius: Annali di Storia della Scienza* 2 (1987): 37-94.

Meinel, Christoph. "Les météores de Froidmont et les météores de Descartes." In, *Libert Froidmont et les résistances scientifiques*, edited by Anne-Catherine Bernès, 105-129. Haccourt: Association, 1988.

Melzer, Christian. *Bergkläufftige Beschreibung der Churfürstl. Sächß. freyen und im Meißnischen Ober-Erz-Gebirge löbl. Bergkstadt Schneebergk.* Schneeberg: publisher not indicated 1684.

Mertens, Gerhard. *Umwelten: Eine humanökologische Pädagogik.* Paderborn: Schöningh, 1998.

Merton, Robert. "The Sociology of Knowledge." *Isis* 27 (1937): 493-503.

Messner, Dirk, and Silke Weinlich, ed. *Global Cooperation and the Human Factor in International Relations.* Routledge: Oxford, 2016.

Metz, Bert. "The Failure of Copenhagen: What Now for the EU?" *European Voice*, January 22, 2010. Accessed April 20, 2011. <http://www.europeanvoice.com/article/2010/01/the-failure-of-copenhagen-what-now-for-the-eu-/66967.aspx>.

Metz, Bert. "International Climate Change Agreements." In *Controlling Climate Change*, edited by Bert Metz, 318-350. Cambridge: Cambridge University Press, 2011.

Middleton, William E.K. *Invention of Meteorological Instruments.* Baltimore: Johns Hopkins, 1969.

Miller, Peter N. *Cultural Histories of the Material World.* Ann Arbor: The University of Michigan Press, 2013.

Ministerie van Infrastructuur en Milieu, ed. *Structuurvisie Infrastructuur en Ruimte: Nederland Concurrerend, Bereikbaar, Leefbaar en Veilig.* Published 2012. Accessed January 25, 2014. <http://www.rijksoverheid.nl/documenten-en-publicaties/rapporten/2012/03/13/structuurvisie-infrastructuur-en-ruimte.html>.

Ministerie van Verkeer en Waterstaat. *Kustverdediging na 1990: Beleidskeuze voor de Kustlijnzorg. Tweede Kamer, Vergaderjaar 1989-1990, 21 136, nrs. 5-6.* Published 1990. Accessed January 25, 2014. <http://repository.tudelft.nl/view/hydro/uuid:a4c7d579-d528-4f26-9364-1346817230ba/>.

Ministerie van Verkeer en Waterstaat, ed. *Tweede Kustnota: Kustbalans 1995*. Published 1996. Accessed January 25, 2014. <http://de.scribd.com/doc/76339527/Kustbalans-1995-de-tweede-kustnota>.

Ministerie van Verkeer en Waterstaat, ed. *Derde Kustnota: Traditie, Trends en Toekomst*. Published 2000. Accessed January 25, 2014. <http://stive51.home.xs4all.nl/delta links/3ekustnota.pdf>.

Ministerie van Verkeer en Waterstaat, ed. *Het Natuurfenomeen van Ballum*. Published 2006a.

Ministerie van Verkeer en Waterstaat, Ministerie van Volkshuisvesting, Ruimtelijke Ordening en Milieubeheer, Ministerie van LNV , ed. *Nationaal Water Plan 2009-2015*. Published 2009. <http://www.rijksoverheid.nl/documenten-en-publicaties/rap porten/2009/12/01/nationaal-waterplan-2009-2015.html>. Accessed January 25, 2014.

Ministerie van Volkshuisvesting, Ruimtelijke Ordening en Milieubeheer, Ministerie van Verkeer en Waterstaat, Ministerie van LNV, Ministerie van EZ, , ed. *Nationaal Programme Adaptatie Ruimte voor Klimaat*. Published 2006. Accessed January 25, 2014. <http://www.klimaatonderzoeknederland.nl/gfx_content/documents/ark/ARK_-_MR_-_9_maart_06.pdf>.

Mittelstraß, Jürgen. *Neuzeit und Aufklärung : Studien zur Entstehung der neuzeitlichen Wissenschaft und Philosophie*. Berlin: de Gruyter, 1970.

Moebius, Stephan. *Marcel Mauss*. Konstanz: UVK, 2006.

Molinié-Bertrand, Annie. *Au Siècle d'Or: L'Espagne et ses hommes: La population du royaume de Castille au XVIe siècle*. Paris: Economica, 1985.

Montaigne, Michel de. *Essais*. Frankfurt a.M.: Eichborn, 1999.

Monumenta Germaniae Historica: CD-ROM 3. München, Turnhout: Monumenta Germaniae Historica, Brepols Publishers, 2002.

Moryson, Fynes. *An Itinerary Containing His Ten Yeeres Travell: Through the Twelve Dominions of Germany, Bohmerland, Sweitzerland, Netherland, Denmarke, Poland, Italy, Turky, France, England, Scotland & Ireland*. 4 vols. Glasgow: John MacLehose and Sons, 1907.

Münk, Hans. "Die Würde des Menschen und die Würde der Natur: Theologisch-ethische Überlegungen zur Grundkonzeption einer ökologischen Ethik." *Stimmen der Zeit* 1 (1997): 17-29.

Mukheibir, Pierre. "Potential Consequences of Projected Climate Change Impacts on Hydroelectricity Generation." *Climatic Change* 121.1 (2013): 67-78. doi:10.1007/s10584-013-0890-5.

Munzar, Jan. "Weather Patterns in Czechoslovakia during the Years 1588-1598." In *European Climate Reconstructed from Documentary Data: Methods and Results*, edited by Burkhard Frenzel, Christian Pfister and Birgit Gläser, 51-56. *Paläoklimaforschung*, vol. 7. Stuttgart, New York: Fischer, 1992.

Munzar, Jan. "Early 17th Century Weather in Prague after J. Kepler's Observations." In *Proceedings of the Regional Workshop on Climate Variability and Climatic Change Vulnerability and Adaptation*, edited by Ivana Nemešová, 69-72. Prague: Institute of Atmospheric Physics Czech Academy of Sciences, 1996.

Musil, Robert. *The Man Without Qualities*. London: Picador, 1997,

Myllyntanus, Timo. "Summer Frost as a Natural Hazard with Fatal Consequences in Preindustrial Finland." In *Natural Disasters, Cultural Responses: Case Studies towards a Global Environmental History*, edited by Christoph Mauch, and Christian Pfister, 77-102. Lanham: Rowman & Littlefield, 2009.

Netzentwicklungsplan Strom. "Netzentwicklungsplan Strom 2012." Accessed August 17, 2012. <http://www.netzentwicklungsplan.de/content/netzentwicklungsplan-2012-2-entwurf>.

Neuhoff, Karsten, Hermann Amecke, Aleksandra Novikova, and Kateryna Stelmakh. "Energetische Sanierung: Handlungsbedarf auf vielen Ebenen." *DIW Wochenbericht* 34 (2011): 3-13.

Neukom, Raphael, David J. Karoly, Heinz Wanner, Mark Curran, Julie Elbert, Fidel González-Rouco, Braddock K. Linsley, Andrew D. Moy, Ignacio Mundo, Christoph C. Raible, Eric J. Steig, Tas van Ommen, Tessa Vance, Ricardo Villalba, Jens Zinke, and David Frank. "Inter-Hemispheric Temperature Variability over the Past Millennium." *Nature Climate Change* 4 (May 2014): 362-367.

Newton, Kenneth, and Pippa Norris. "Confidence in Public Institutions: Faith Culture or Performance?" In *Disaffected Democracies: What's Troubling the Trilateral Countries?*, edited by Susan J. Pharr and Robert D. Putnam. Princeton: Princeton University Press, 2000.

NGO Committee on Education. "Our Common Future, Chapter 2: Sustainable Development." *UN Documents: Gathering a Body of Global Agreements*, Report No. A/42/427. Published 1987. Accessed November 11, 2015. <http://www.un-documents.net/ocf-02.htm#III.3>.

Nicault, Antoine et al. "Mediterranean Drought Fluctuation During the last 500 Years Based on Tree-Ring Data." *Climate Dynamics* 31 (2008): 227-245. doi: 10.1007/s00382-007-0349-3.

Noordegraaf, Leo. "Dearth, Famine and Social Policy in the Dutch Republic at the End of the Sixteenth Century." In *The European Crisis of the 1590s: Essays in Comparative History*, edited by Peter Clark, 67-83. London: George Allen & Unwin, 1985.

Nordhaus, William D. "Can We Control Carbon Dioxide?" *IIASA Working Paper*. Accessed July 12, 2016. <http://pure.iiasa.ac.at/365/1/WP-75-063.pdf>.

Novgorodskija i pskovskija lětopisi [The Chronics of Novgorod and Pskov]. *Polnoe sobranie russkich letopisej* [Complete Edition of the Russion Chronics], vol. 4. Translated by Ursula Bieber. Saint Peterburg: v tipografii Ėduarda Praca, 1848. Reprinted Düsseldorf: Brückner 1973.

Nunn, Patrick D. *Climate, Environment and Society in the Pacific during the Last Millenium. Developments in Earth and Environmental Sciences*, vol. 6. London: Elsevier, 2007.

Oberthür, Sebastian, and Claire Roche Kelly. "EU Leadership in International Climate Policy: Achievements and Challenges." *The International Spectator* 43.3 (2008): 35-50.

Oberthür, Sebastian. "The Role of the EU in Global Environmental and Climate Governance." In *The European Union and Global Governance*, edited by Mario Telò, 192-209. London: Routledge, 2009.

Oberthür, Sebastian. "Global Climate Governance after Cancun: Options for EU leadership." *The International Spectator* 46.1 (2011): 5-13.

Oberthür, Sebastian. "Where to go from Paris? The European Union in Climate Geopolitic." *Global Affairs* 2.2 (2016):119-130.

Oberthür, Sebastian. "The Paris Agreement: Rebooting Climate Cooperation. Perspectives on EU Implementation of the Paris Outcome." *Carbon & Climate Law Review* 10.1 (2016): 34-45.

OcCC et al. *Klimaänderung und Die Schweiz 2050. Erwartete Auswirkungen auf Umwelt, Gesellschaft und Wirtschaft*. Bern, 2007.Rockström, Johan. Will Steffen, Kevin Noone, et al. "Planetary Boundaries: Exploring the Safe Operating Space for Humanity," *Ecology and Society* 14.2 (2009).

OECD. "Transition to a Low-Carbon Economy: Public Goals and Corporate Practices." Published by OECD, 2010. <http://dx.doi.org/10.1787/9789264090231-en>.

Oeschger, Hans, Bruno Messerli, and Maja Svilar ed. *Das Klima. Analysen und Modelle. Geschichte und Zukunft*. Berlin: Springer, 1980.

Ogilvie, Astrid E.J., and Gísli Pálsson. "Mood, Magic, and Metaphor: Allusions to Weather and Climate in the Sagas of Icelanders." In *Weather, Climate, Culture*, edited by Sarah Strauss, and Benjamin S. Orlove, 251-274. Oxford: Berg, 2003.

Ogilvy, James A. *Creating Better Futures: Scenario Planning as a Toolfor a Better Tomorrow*. New York: Oxford University Press, 2002.

Olivari, Micheli. *Entre el trono y la opinió: La vida política castellana en los siglos XVI y XVII*. Valladolid: Junta de Castilla y León, 2004.

Oosterveld, Hendrik R. "Een Wereld van Verschill. Vertig Jaar Beleid in het Waddengebied." Paper presented at the fourth symposium "Waddenacademie KNAW," Leeuwarden, June 6, 2010. Accessed April 5, 2017. <http://www.waddenacademie.nl/fileadmin/inhoud/pdf/04-bibliotheek/Een_wereld_van_verschil_web.pdf>.

Oppenheimer, Clive. "Climatic, Environmental and Human Consequences of the Largest Known Historic Eruption: Tambora Volcano (Indonesia) 1815." *Progress in Physical Geography* 27.2 (2013): 230-259.

Oreskes, Naomi, David A. Stainforth, and Leonard A. Smith. "Adaptation to Global Warming: Do Climate Models Tell Us What We Need to Know?" *Philosophy of Science* 77.5 (2010): 1012-1028.

Orth, René, Jakob Zscheischler, and Sionia I. Seneviratne, "Record Dry Summer in 2015 Challenges Precipitation Projections in Central Europe." *Scientific Reports* 6 (2016). doi: 10.1038/srep28334.

Orth, René, Martha Vogel, Jürg Luterbacher, Christian Pfister, and Sonia Seneviratne. "Did European Temperatures in 1540 Exceed Present-Day Records?," *Environmental Research Letters* 11.11 (2016). doi: 10.1088/1748-9326/11/11/114021.

Ostler, Nicholas. *Empires of the Word. A Language History of the World.* London: HarperCollins, 2005.

Outhwaite, Brian. "Dearth, the English Crown, and the 'Crisis of the 1590s'." In *The European Crisis of the 1590s: Essays in Comparative History*, edited by Peter Clark, 23-43. London: George Allen & Unwin, 1985.

Ozel, Oktay. "Population Changes in Ottoman Anatolia during the Sixteenth and Seventeenth Centuries: The 'Demographic Crisis' reconsidered." *International Journal of Middle Eastern Studies* 36 (2004): 183-204.

Palladius, Rutilius Taurus Aemilianus. *De re rustica Libri XIIII*. Parisiis: Robertus Stephanus, 1543. Accessed March 27, 2017. <http://www.mdz-nbn-resolving.de/urn/resolver.pl?urn=urn:nbn:de:bvb:12-bsb10172095-8>.

Pannenberg, Wolfhart. *Systematische Theologie*, vol. 2. Göttingen: Vandenhoeck & Ruprecht, 1991.

Papadopoulos, Yannis. "Governance und Demokratie." In *Governance – Regieren in komplexen Regelsystemen: Eine Einführung*, edited by Arthur Benz, 215-238. Wiesbaden: VS Verlag für Sozialwissenschaften, 2004.

Parker, Charles F., and Christer Karlsson. "Climate Change and the European Union's Leadership Moment: An Inconvenient Truth?" *Journal of Common Market Studies* 48.4 (2010): 923-943.

Parker, Charles, Christer Karlsson, Mattias Hjerpe, and Björn-Ola Linnér. "Fragmented Climate Change Leadership and the Ambiguous COP-15 Outcome: From Hopenhagen to Nopenhagen." Paper presented at the 3rd Uppsala Forum Workshop on the Politics, Policy, and Ethics of Global Climate Change, University of Uppsala, September 28, 2010.

Parker, Geoffrey, *Global Crisis. War, Climate Change and Catastrophe in the Seventeenth Century*. London, New Haven: Yale University Press, 2013.

Parker, Geoffrey. "The *Dreadnought* Revolution of Tudor England." *Mariner's Mirror* 82 (1996): 269-300.

Parker, Geoffrey. *Felipe II: La biografía definitiva*. Barcelona: Planeta, 2010.

Parsons, Talcott. *The Structure of Social Action.* New York: MacGraw Hill, 1937.

Patt, Anthony, Stefan Pfenninger, and Johan Lilliestam. "Vulnerability of Solar Energy Infrastructure and Output to Climate Change." *Climatic Change* 121.1 (2010): 93-102. doi:10.1007/s10584-013-0887-0.

Paul VI. *Populorum progressio. Encyclical on the Development of People.* Vatican: Vatican Press, 1967.

Pauling, Andreas, Jürg Luterbacher, Carlo Casty, and Heinz Wanner. "Five Hundred Years of Gridded High-Resolution Precipitation Reconstructions over Europe and the Connection to Large Scale Circulation." *Climate Dynamics* 26.4 (2006): 387-405. doi:10.1007/s00382-005-0090-8.

Pauling, Andreas, Jürg Luterbacher, Carlo Casty, and Heinz Wanner. "Five Hundred Years of Gridded High-resolution Precipitation Reconstructions over Europe and the Connection to Large-Scale Circulation." *Climate Dynamics* 26 (2006): 387-405.

Pérez Moreda, Vicente. *Las crisis de mortalidad en la España interior: Siglos XVI–XIX.* Madrid: Siglo XXI, 1980.

Pfaff, Karl. *Geschichte der Stadt Stuttgart nach Archival-Urkunden und anderen bewährten Quellen. 2. Teil.* Stuttgart: Sonnewald, 1846.

Pfister, Christian. *Das Klima der Schweiz von 1525 bis 1860 und seine Bedeutung in der Geschichte von Bevölkerung und Landwirtschaft.* 2 Vols. Bern: Haupt, 1984.

Pfister, Christian. *Klimageschichte der Schweiz 1525-1860: Das Klima der Schweiz von 1525-1860 und seine Bedeutung in der Geschichte von Bevölkerung und Landwirtschaft.* Bern, Stuttgart: Haupt, 1988.

Pfister, Christian. "Food Supply in the Swiss Canton of Bern, 1850." In *Hunger in History. Food Shortage, Poverty, and Deprivation*, edited by Lucile F. Newman. 281-303. Cambridge: Blackwell, 1990.

Pfister, Christian. *Wetternachhersage: 500 Jahre Klimavariationen und Naturkatastrophen 1496-1995.* Bern: Haupt, 1999.

Pfister, Christian, Rudolf Brázdil, Rüdiger Glaser, Anita Bokwa, Franz Holawe, Danuta Limanowka, Oldřich Kotyza, Jan Munzar, Lajos Rácz, Elisabeth Strömmer, and Gabriela Schwarz-Zanetti. "Daily Weather Observations in Sixteenth-Century Europe." *Climatic Change* 43 (1999): 111-150.

Pfister, Christian. "The '1950s Syndrome' and the Transition from a Slow-Going to a Rapid Loss of Global Sustainability." In *The Turning Points in Environmental History*, edited by Frank Ükötter, 90-118. Pittsburgh: 2010.

Pfister, Christian, Rudolf Brázdil, and Rüdiger Glaser. *Climatic Variability in Sixteenth-Century Europe and its Social Dimension.* Dordrecht: Kluwer Academic Publishers, 1999.

Pfister, Christian, Rudolf Brázdil, and Manuel Barriendos. "Reconstructing Past Climate and Natural Disasters in Europe Using Documentary Evidence." *Pages News* 10.3 (2002), 6-8.

Pfister, Christian. "Weeping in the Snow: The Second Period of Little Ice Age-Type Impacts, 1570-1630." In *Kulturelle Konsequenzen der "Kleinen Eiszeit"*, edited by Wolfgang Behringer, Harmut Lehmann, and Christian Pfister, 31-86. Göttingen: Vandenhoeck & Ruprecht, 2005.

Pfister, Christian, Rolf Weingartner, and Jürg Luterbacher. "Hydrological Winter Droughts over the Last 450 Years in the Upper Rhine Basin: A Methodological Approach." *Hydrological Sciences—Journal des Sciences Hydrologiques* 51.5 (2006): 966-985. doi:10.1623/hysj.51.5.966.

Pfister, Christian. "Climatic Extremes, Recurrent Crises and Witch Hunts: Strategies of European Societies in Coping with Exogenous Shocks in the Late Sixteenth and Early Seventeenth Centuries." *The Medieval History Journal* 10.1-2 (2007): 33-73.

Pfister, Christian, Jürg Luterbacher, Heinz Wanner, Dennis Wheeler, Rudolf Brázdil, Quansheng Ge, Zhixin Hao, Anders Moberg, Stefan Grab, and Maria del Prieto. "Documentary Evidence as Climate Proxies." "White Paper" written for the Proxy Uncertainty Workshop in Trieste, 9-11 June, 2008. Accessed November 6, 2013. <http://www.pages.unibe.ch/index.php/products/2029-proxy-data-uncertainties-white-paper-documentary-evidence-as-climate-proxies>.

Pfister, Christian, and Daniel Krämer. "The Relaunch of Historical Climate Impact Research—A Timely Challenge for Historical Climatology." Paper presented at the conference "Historical Climatology—Past and Future," German Historical Institute, Paris. Paris, September 5-6, 2011.

Pfister, Christian. *"The Monster Swallows You". Disaster Memory and Risk Culture in Western Europe, 1500-2000.* Rachel Carson Center Perspectives, 2011. Accessed May 5, 2017. <http://www.environmentandsociety.org/sites/default/files/web_edition_color.pdf>.

Pfister, Christian. "Renward Cysat – ein 'interdisziplinärer' Pionier der Klimaforschung im Alpenraum.» *Geschichtsfreund* 166 (2014): 188-208.

Pfister, Christian, and Oliver Wetter. "Tree-Rings and People—Different Views on the 1540 Megadrought: Reply to Büntgen et al. (2015)." *Climatic Change Letters* (2015).

Pfister, Christian, and Urs Dietrich-Felber. "Euro-Climhist: A Data-Base on Past Weather and Climate in Europe and its Human Dimension." Accessed November 7, 2013. <http://www.euroclimhist.com>.

Pfister, Christian. "When Europe Was Burning: The Multi-Seasonal Megadrought of 1540 and the Arsonist Paranoia." In *Disasters, Risks and Cultures: A Comparative and Transcultural Survey of Historical Disaster Experiences between Asia and Europe*, edited by Gerrit Jasper Schenk, in press. Heidelberg: Springer, 2016.

Pichard, George, and Emeline Roucaute. "Sécheresses du Bas Rhone et invasions de sauterelles." Accessed August 25, 2014. <http://histrhone.cerege.fr/squelettes/pdf/Chronologie_Secheresses_Bas_Rhone.pdf>.

Plinius Secundus, Gaius. *Naturalis historiae libri XXXVII*. Lutetia: Chaudière, 1516. Accessed March 27, 2017. <http://www. mdz-nbn-resolving.de/urn/resolver.pl?urn=urn:nbn:de:bvb:12-bsb10140829-8>.

Pohlmann, Angela. "Local Climate Change Governance." In Anita Engels (Hg.). Global Transformations Towards a Low Carbon Society. Hamburg University of Hamburg / KlimaCampus (Working Paper, 4). (2011).

Porter, Martin. *Windows of the Soul. Physiognomy in European Culture 1470–1780*. Oxford: Clarendon, 2005.

Poumadère, Marc, Claire Mays, Sophie Le Mer, and Russell Blong. "The 2003 Heat Wave in France: Dangerous Climate Change Here and Now." *Risk Analysis* 25.6 (2005): 1483-1494. doi:10.1111/j.1539-6924.2005.00694.x.

Power, Mary E., Rex L. Lowe, Paula C. Furey, Jill R. Welter, Michael P. Limm, Jacques C. Finlay, Collin A. Bode, S. Chang, Maria Goodrich, and John B. Sculley. "Algal Mats and Insect Emergence in Rivers under Mediterranean Climates: Towards Photogrammetric Surveillance." *Freshwater Biology* 54.10 (2009): 2101-2115. doi:10.1111/j.1365-2427.2008.02163.x.

Power, Michael J. "A 'Crisis' Reconsidered: Social and Demographic Dislocation in London in the 1590s." *London Journal* 12 (1986): 134-145.

Power, Michael J. "London and the Control of the 'Crisis' of the 1590s." *History* 70 (1985): 371-385.

ProClim Forum for Climate and Global Change, and Plattform of the Swiss Academy of Sciences. "Hitzesommer 2003: Synthesebericht 2005." Published Bern, November 2005. Accessed November 20, 2014. <http://www.occc.ch/products/heatwave03/heatwave03_bericht.html>.

Prognos. "Investitionen durch den Ausbau erneuerbarer Energien in Deutschland." Accessed April 23, 2013. <http://www.bee-ev.de/_downloads/publikationen/studien/2010/1005_Prognos-Studie_Investitionen_BEE-Ausbauprognose_lang.pdf>.

Prognos. "Konsequenzen eines Ausstiegs aus der Kernenergie bis 2022 für Deutschland und Bayern." Accessed April 23, 2013. <http://www.prognos.com/fileadmin/pdf/publikationsdatenbank/Prognos_vbw_Studie__Ausstieg_Kernenergie_bis_2022.pdf>.

Prognos. "Ermittlung der Wachstumswirkungen der KfW-Programme zum Energieeffizienten Bauen und Sanieren." April 23, 2013. <https://www.kfw.de/migration/Weiterleitung-zur-Startseite/Startseite/KfW-Konzern/KfW-Research/Economic-Research/Evaluationen/PDF-Dokumente-Evaluationen/Prognos_Wachstums effekte_KfW_EBS_Endbericht.pdf>.

Propp, Vladimir Yakovlevich, *Morfologija skazki Морфология сказки* [Morphology of the Folk Tale], *Voprosy poėtiki* [Problems of Poetics] 21. Leningrad: Akademija, 1928.

Provincie Zuid-Holland, and Ministerie van Verkeer en Waterstaat. *De Zandmotor Delta Duin*. Published 2011. Accessed January 27, 2014. <http://www.wshvh.nl/downloads/21-4_factsheetsZandmotor.pdf>.

Provincie Zuid-Holland. *De Zandmotor: Van Zand naar Land. The Sand Engine: From Sand to Land*. Published 2009. Accessed January 27, 2014. <http://www.dezandmotor.nl/uploads/2011/03/090525-brochure-zandmotor-1.pdf>.

Przyborski, Aglaja, and Monika Wohlrab-Sahr. *Qualitative Sozialforschung: Ein Arbeitsbuch*. München: Oldenbourg 2008.

Pushkin, Aleksandr Sergeyevich. *Izbrannye proizvedenija v dvukh tomakh. Tom pervy: stikhotvoreniya, skazki, poëmy* [Selected Works in Two Volumes: Poems, Fairy Tales, Verse Novels]. Moscow: Chudojestvennaja Literatura, 1968.

Pustil'nik, Lev, and Gregory Yom Din. "On Possible Influence of Space Weather on Agricultural Markets: Necessary Conditions and Probable Scenarios." *Sun and Geosphere* 4 (2009): 35-43.

Quesada, Benjamin, Robert Vautard, Pascal Yiou, Martin Hirschi, and Sonia I. Seneviratne. "Asymmetric European Summer Heat Predictability from Wet and Dry Southern Winters and Springs." *Nature Climatic Change* 2 (2012): 736-741. doi:10.1038/nclimate1536.

Radkau, Joachim. *Natur und Macht: Eine Weltgeschichte der Umwelt*. München: Beck, 2000.

Radkau, Joachim, and Lothar Hahn. *Aufstieg und Fall der deutschen Atomwirtschaft*. München: Oekom, 2013.

Rajamani, Lavanya. "The Durban Platform for Enhanced Action and the Future for Climate." *International and Comparative Law Quarterly* 61.2 (2012): 501-518.

Ramel, Frédéric. "Marcel Mauss et l'étude des relations internationales: un héritage oublié." *Sociologie et sociétés* 36.2 (2004): 227-245.

Rapp, Tobias, Christian Schwägerl, and Gerald Traufetter. "The Copenhagen Protocol: How China and India Sabotaged the UN Climate Summit." *Spiegel Online International*, May 5, 2010. Accessed October 21, 2013. <http://www.spiegel.de/international/world/the-copenhagen-protocol-how-china-and-india-sabotaged-the-un-climate-summit-a-692861.html>.

Rappel, Simone. *'Macht euch die Erde untertan': Die ökologische Krise als Folge des Christentums?* Paderborn: Schöningh, 1996.

Rayner, Tim, and Andrew Jordan. "The European Union: The Polycentric Climate Policy Leader?" *Wiley Interdisciplinary Reviews* 4.2 (2013): 75-90.

RECLIDO. "REd para la reconstrucción del CLIma a partir de fuentes DOcumentales." Accessed September 13, 2014. <http://pendientedemigracion.ucm.es/info/reclido/>.

Redman, Charles L. "Resilience in Archaeology." *American Anthropology* 107.1 (2005): 70-77.

Regiegroep Natura2000. Accessed January 27, 2014. <http://natura2000.nl>.

Regierungsblatt für das Königreich Württemberg. May 10, 1847.

Reichholf, Josef. *Stabile Ungleichgewichte: Die Ökologie der Zukunft*. Frankfurt a.M.: Suhrkamp, 2008.

Reis, Oliver. *Nachhaltigkeit – Ethik – Theologie: Eine theologische Beobachtung der Nachhaltigkeitsdebatte*. Forum Religion & Sozialkultur, vol. 18. Münster: LIT, 2003.

Relazione dell'origine et successi della terra di Varese [Varese Ligure] *descritta dal R. P. Antonio Cesena l'anno 1558. Studi e Documenti di Lunigiana*, vol. 6. La Spezia: Academia lunigianese di Scienze Giovanni Cappellini, 1982.

Reuver, Marc, Friedhelm Solms, and Gerrit Huizer, ed. *The Ecumenical Movement Tomorrow: Suggestions for Approaches and Alternatives*. Kampen, Geneva: Kok Publishing House and WCC Publications, 1993.

Reynman, Leonhard. [Wetterbüchlein.] Von warer erkantnuß des wetters. Also das ain ÿeder er seÿ gelert oder ungelert durch alle natürliche anzaigung die änndrung des wetters aÿgentlich vnd augenscheinlich erkennen mag etc. (München: Hannss Schobsser, 1510) [(Neudrucke von Schriften und Karten über Meteorologie und Erdmagnetismus No.1) (Berlin: Asher 1893)], accessed March 27, 2017. <http://nbn-resolving.de/urn:nbn:de:bvb:12-bsb00004843-3>.

Ricci, Franco Maria, ed. *Arcimboldo*. Genf: Weber, 1978.

Riemann, Dirk. "Methoden zur Klimarekonstruktion aus historischen Quellen am Beispiel Mitteleuropas." In *Freiburger Geographische Hefte* Nr. 69 (2012).

Riemann, Dirk, Rüdiger Glaser, Michael Kahle, and Steffen Vogt. "The CRE tambora.org – New Data and Tools for Collaborative Research in Climate and Environmental History." *Geoscience Data Journal* (2016). <http://dx.doi.org/10.1002/gdj3.30.1-15>.

Rijswick, Marleen van, and Herman Havekes. *European and Dutch Water Law*. Groningen: Europa Law Publishing, 2012.

Rink, Dieter, Monika Wächter, and Thomas Potthast. "Naturverständnisse in der Nachhaltigkeitsdebatte: Grundlagen, Ambivalenzen und normative Implikationen." In *Naturverständnisse in der Nachhaltigkeitsforschung*, edited by Dieter Rink, and Monika Wächter, 11-34. Frankfurt: Campus, 2004.

Roberts, Neil, Ana Moreno, Blas L. Valero-Garcés, Juan Pablo Corella, Matthew Jones, Samantha Allcock, Jessie Woodbridge, Mario Morellón, Juerg Luterbacher, Elena Xoplaki, and Murat Türkeş. "Palaeolimnological Evidence for an East–West Climate See-Saw in the Mediterranean Since AD 900." *Global and Planetary Change* 84-85 (2012): 23-34.

Robertson, Roland. "Glocalization: Time-Space and Homogeneity-Heterogeneity." In *Global Modernities*, edited by Mike Featherstone, 25-44. London: Sage Publications, 1995.

Rockström, Johan Will Steffen, Kevin Noone, Åsa Persson, F. Stuart III Chapin, Eric F. Lambin, Timothy M. Lenton, Marten Scheffer, Carl Folke, Hans J. Schellnhuber, Björn Nykvist, Cynthia A. de Wit, Terry Hughes, Sander van der Leeuw, Henning Rodhe, Sverker Sörlin, Peter K. Snyder, Robert Costanza, Uno Svedin, Malin Falkenmark, Louise Karlberg, Robert W. Corell, Victoria J. Fabry, James Hansen, Brian Walker, Diana Liverman, Katherine Richardson, Paul Crutzen, and Jonathan A. Foley. "A Safe Operating Space for Humanity." *Nature* 461 (2009), 472-475. doi:10.1038/461472a.

Rockström, Johan Will Steffen. "Planetary Boundaries: Exploring the Safe Operating Space for Humanity." *Ecology and Society* 14.2 (2009): Art. 32. Accessed November 9, 2015. <http://www.ecologyandsociety.org/vol14/iss2/art32/>.

Rösche, Johannes. *Robert Fludd. Der Versuch einer hermetischen Alternative zur neuzeitlichen Naturwissenschaft.* Göttingen: V&R unipress 2008.

Rohland, Eleonora. *Sharing the Risk: Fire, Climate, and Disaster: Swiss Re 1864-1906.* Lancaster: Crucible Books, 2011.

Rosa, Hartmut. *Beschleunigung. Die Veränderung der Zeitstrukturen in der Moderne.* Frankfurt a.M.: Suhrkamp, 2005.

Rosa, Hartmut. *Alienation and Acceleration: Towards a Critical Theory of Late-Modern Temporality.* Aarhus: Aarhus Universitetsforlag, 2010.

Rosa, Hartmut. *Social Acceleration: A New Theory of Modernity.* New York: Columbia University Press, 2013.

Rosenberger, Michael. *Was dem Leben dient: Schöpfungsethische Weichenstellungen im konziliaren Prozeß der Jahre 1987-89.* Stuttgart: Kohlhammer, 2001.

Rubio, Mar, César Yáñez, Mauricio Folchi, and Albert Carreras. "Energy as an Indicator of Modernization in Latin America, 1890-1925." *The Economic History Review* 63.3 (2010).

Ruggieri, Giuseppe. "Die Zeichen der Zeit: Herkunft und Bedeutung einer hermeneutischen Chiffre der kirchlichen Praxis." In *Das Zweite Vatikanische Konzil und die Zeichen der Zeit heute,* edited by Peter Hünermann, 61-70. Freiburg: Herder, 2006.

Rummel, Walter, and Rita Voltmer. *Hexen und Hexenverfolgung in der Frühen Neuzeit.* Darmstadt: Wissenschaftliche Buchgesellschaft, 2008.

Rupnow, Dirk. "Laconicum Europae Speculum—Stereotype ohne Schimpf und Vorurteil?" In *Europäischer Völkerspiegel: Imagologisch-ethnographische Studien zu den Völkertafeln des frühen 18. Jahrhunderts,* edited by Franz K. Stanzel, 75-95. Heidelberg: Winter, 1999.

Ryff, Fridolin. "Die Chronik des Fridolin Ryff 1514-1541, mit der Fortsetzung des Peter Ryff 1543-1585: Die Chroniken des Karthäuser Klosters in Klein-Basel 1401-1532." In *Basler Chroniken,* vol. 1, edited by Historische und Antiquarische Gesellschaft in Basel, 18-192. Leipzig: S. Hirzel, 1872.

Sabbattini, Nicola. *Pratica di fabricar scene, e machine ne' teatri.* Pesaro: Concordia, 1637.

Sabbattini, Nicola. *Pratica Di Fabricar Scene, E Machine Ne'Teatri.* Ravenna: Pietro de' Paoli, Giovanni Battista Giouannelli, 1638.

Sachs, Wolfgang. *Nach uns die Zukunft: Der globale Konflikt um Gerechtigkeit und Ökologie.* Frankfurt a.M.: Brandes & Apsel, 2003.

Sandbothe, Mike. "Die Verzeitlichung der Zeit in der modernen Philosophie." In *Die Wiederentdeckung der Zeit: Reflexionen – Analysen – Konzepte,* edited by Antje Gimmler, Mike Sandbothe, and Walther Ch. Zimmerli. Darmstadt: PRIMUS Verlag und Wissenschaftliche Buchgesellschaft, 1997.

Sander, Hans-Joachim. “Theologischer Kommentar zur Pastoralkonstitution über die Kirche in der Welt von heute Gaudium et spes.” In *Herders Theologischer Kommentar zum Zweiten Vatikanischen Konzil*, vol. 4, edited by Peter Hünermann, and Bernd Jochen Hilberath, 581-886. Freiburg, Basel, Wien: Herder, 2005.

Satō Taketoshi 佐藤武敏, ed. *Chūgoku saigaishi nenpyō* 中国災害史年表 [Chronology of the History of Natural Disasters in China]. Tōkyō: Kokusho Kankokai, 1993.

Schaal, Gary P. “Responsivität – selbstzerstörerisches Ideal liberaler Demokratie?” In *Bedrohungen der Demokratie*, edited by André Brodocz, Marcus Llanque and Gary p. Schaal, 353-369. VS Verlag für Sozialwissenschaften, 2009.

Schaal, Gary P., and Claudia Ritzi. “Politische Führung in der ‘Postdemokratie’.” *Aus Politik und Zeitgeschichte* 2-3 (2010): 9-14.

Schär, Christoph, Pier Luigi Vidale, Daniel Lüthi, Christoph Frei, Christian Häberli, Mark A. Liniger, and Christof Appenzeller. “The Role of Increasing Temperature Variability in European Summer Heatwaves.” *Nature* 427 (2004): 335.

Schaller, Daniel. Herold: *Ausgesandt in allen Landen öffentlich zu verkündigen und auszurufen. Dass diese Welt mit ihrem Wesen bald vergehen werde, und der Jüngste Gerichtstag gar nahe vor der Tür sei.* Magdeburg, 1595.

Schelsky, Helmut. *Die skeptische Generation: Eine Soziologie der deutschen Jugend.* München: Diedrichs, 1957.

Schellnhuber, Hans Joachim. *Selbstverbrennung. Die fatale Dreiecksbeziehung zwischen Klima, Mensch und Kohlenstoff.* München: C. Bertelsmann, 2015.

Schiller, Friedrich. “Über das Erhabene.” In *Schillers Werke: Nationalausgabe*, vol. 21. Weimar: Böhlau, 1963.

Schiesser, Hans-Heinrich. *Hagelstürme in der Schweiz: Wiederkehrperioden von schadenbringenden Hagelkorngrössen— Eine Abschätzung*. Zürich, 2006.

Schiesser, Hans-Heinrich. *Klimatologie der Stürme und Sturmsysteme anhand von Radar- und Schadendaten*. Zürich: Vdf, 1997.

Schmitt, Lea. *Klimawandel, Raumwandel, Zeitwandel: Adaptationen und Konflikte auf der westfriesischen Insel Ameland.* PhD diss., University of Duisburg-Essen, 2013.

Schnell'sche Chronik: Manuscript. Stadtarchiv Lindau, Lit 29.

Schöllgen, Gregor. *Gerhard Schröder: Die Biographie*. München: DVA Sachbuch, 2015.

Scholtz, Gunter. “Musik.” In *Historisches Wörterbuch der Philosophie*, vol. 6., edited by Joachim Ritter, and Karlfried Gründer, 249-250. Darmstadt: Wissenschaftliche Buchgesellschaft, 1984.

Schütz, Alfred. “Tiresias oder unser Wissen von zukünftigen Ereignissen [1959].” In *Studien zur soziologischen Theorie*, edited by Alfred Schütz, 259-278. Den Hag: Nijhoff, 1972.

Schulze Altcappenberg, Hein-Thomas. “An der Wiege der Romantik: Die Jahreszeiten Caspar David Friedrichs von 1803.” In *An der Wiege der Romantik: Caspar David*

Friedrichs Jahreszeiten von 1803, edited by Klaus-Dieter Lehmann, 15-24. Berlin: Staatliche Museen, 2006.

Schunz, Simon. *Beyond Leadership by Example: Towards a Flexible European Union Foreign Climate Policy*. SWP *Working Paper* FG8, 2011/1. Berlin: Stiftung Wissenschaft und Politik (SWP), 2011.

Schuppert, Gunnar Folke, and Andreas Voßkuhle, ed. *Governance von und durch Wissen: Interdisziplinäre Tagung "Governance von und durch Wissen" (Berlin)*. Baden-Baden: Nomos, 2008.

Schusky, Renate. "Klänge durch das Jahr: Die vier Jahreszeiten in der Musik." In *Das Reich der Jahreszeiten*, edited by Rainer Gruenter, 137-138. Zürich: Offizin 1989.

Schweingruber, Fritz H. *Jahrringe und Umwelt – Dendroökologie*. Birmensdorf: Eidgenössische Forschungsanstalt für Wald, Schnee und Landschaft, 1993.

Sella, Domenico. *Crisis and Continuity: The Economy of Spanish Lombardy in the Seventeenth Century*. Cambridge: Harvard University Press, 1979.

Seneviratne, Sonia I., Thierry Corti, Edouard L. Davin, Martin Hirschi, Eric B. Jaeg, Irene Lehner, Boris Orlowsky, and Adriaan J. Teuling. "Investigating Soil Moisture–Climate Interactions in a Changing Climate: A Review." *Earth-Science Reviews* 99.3-4 (2010): 125-161. doi:10.1016/j.earscirev.2010.02.004.

Serres, Olivier de [Seigneur de Pradel]. *Le Theatre d'agriculture et mesnage des champs. Seconde edition reueuë et augmentee par l'auteur*. Paris: Abr[aham] Savgrain, 1603. Accessed March 27, 2017. <http://www.mdz-nbn-resolving.de/urn/resolver.pl?urn=urn:nbn:de:bvb:12-bsb10229138-5>.

Serres, Michel. *La naissance de la physique dans le texte de Lucrèce. Fleuves et turbulences*. Paris: Editions de Minuit 1977 [The Birth of Physics. Manchaster: Clinamen, 2000].

Sforza, Giolio, ed. *Memorie e documenti per servire alla Storia di Pontremoli*. Lucca: Giusti, 1887.

Shakespeare, William. *A Midsummer Night's Dream*. London, 1623.

Sieber, Jeannette. "Impacts of, and Adaptation Options to, Extreme Weather Events and Climate Change Concerning Thermal Power Plants." *Climatic Change* 121.1 (2013): 55-66. doi:10.1007/s10584-013-0915-0.

Sieferle, Rolf P. *The Subterranean Forest: Energy Systems and the Industrial Revolution*. Cambridge: White Horse Press, 2001.

Sieglerschmidt, Jörn. "Spielt die Natur mit dem Menschen? Zum Einfluß der Natur auf die soziale Umwelt des Menschen." In *Ad historiam humanam: Aufsätze für Hans-Christoph Rublack*, edited by Thomas Max Safley, 193-211. Epfendorf: Bibliotheca Academica, 2005.

Sieglerschmidt, Jörn, and Birgit Biehler. "Natur." In *Enzyklopädie der Neuzeit*, Vol 8, edited by Friedrich Jäger, 1133-1159.Stuttgart: Metzler 2008.

Sieglerschmidt, Jörn. "Zodiacus." In *Enzyklopädie der Neuzeit*, vol. 15, edited by Friedrich Jäger, 539-544. Stuttgart: Metzler, 2012.

Sigl, M., M. Winstrup, J.R. McConnell, K.C. Welten, G. Plunkett, F. Ludlow, U. Büntgen, M. Caffee, N. Chellman, D. Dahl-Jensen, H. Fischer, S. Kipfstuhl, C. Kostick, O.J. Maselli, F. Mekhaldi, R. Mulvaney, R. Muscheler, D.R. Pasteris, J.R. Pilcher, M. Salzer, S. Schüpbach, J.P. Steffensen, B.M. Vinther, and T.E. Woodruff. "Timing and Climate Forcing of Volcanic Eruptions for the Past 2,500 Years." *Nature* 533 (2015): 543-549.

Sirocko, Frank. *Wetter, Klima, Menschheitsentwicklung*. Darmstadt: Wissenschaftliche Buchgesellschaft, 2009.

Sjöstedt, Gunnar. "Leadership in Multilateral Negotiations: Crisis or Transition?" In *International Negotiation: Actors, Structure/Process & Values*, edited by Peter Berton, Hiroshi Kimura, and William I. Zartman, 223-253. Hampshire: Macmillan, 1999.

Skeat, Walter W. ed. *Fitzherberts Book of Husbandry 1534*. London: Trübner, 1882.

Slack, Paul. *The Impact of Plague in Tudor and Stuart England.* London: Routledge & Kegan Paul, 1985.

Sloterdijk, Peter. "Das 21. Jahrhundert beginnt mit dem Debakel vom 19. Dezember 2009." *Süddeutsche Zeitung,* December 21, 2009.

Sloterdijk, Peter. *Zeilen und Tage: Notizen 2008-2011.* Frankfurt a.M.: Suhrkamp, 2012.

Smil, Vaclav. *Energy in Nature and Society: General Energetics of Complex Systems*. Cambridge MA: The MIT Press, 2008.

Sloterdijk, Peter. *Making the Modern World: Materials and Dematerialization*. Chichester: Wiley, 2014.

Società italiana di demografia storica. *La popolazione italiana nel Seicento*. Bologna: CLUEB, 1999.

Solar, Peter M. "Poor Relief and English Economic Development before the Industrial Revolution." *Economic History Review* 48 (1995): 1-22.

Song Zhenghai 宋正海. *Zhongguo gudai zhongda ziran zaihai he yichang nianbiao zongji* 中国古代重大自然灾害和异常年表总集 [A Comprehensive Chronology of Significant Natural Disasters and Unusual Occurences in Ancient China] Guangzhou: Guangdong jiaoyu chubanshe, 1992.

Spencer, Thomas, Kristian Tangen, and Anna Korppoo. *The EU and the Global Climate Regime: Getting Back in the Game. Briefing Paper* 55. Helsinki: The Finnish Institute of International Affairs, 2010.

Spicker-Beck, Monika. *Räuber, Mordbrenner, Umschweifendes Gesind: Zur Kriminalität im 16. Jahrhundert.* Freiburg i.Br.: Rombach, 1995.

Spradley, James P. *The Ethnographic Interview*. Belmont: Wadsworth Thomson Learning, 2008.

Stahle, David W, Edward R. Cook, Malcolm K. Cleaveland, Matthew D. Therrell, David M. Meko, Henri D. Grissino-Mayer, Emma Watson, and Brian H. Luckman. "Tree-Ring Data Document 16th-Century Megadrought over North America." *EOS* 81.12 (2000): 121-125.

Stahle, David W., Falko K. Fye, Edward R. Cook, and R. Daniel Griffin. "Tree-Ring Reconstructed Megadroughts over North America since AD 1300." *Climatic change* 83 (2007): 133-49.

Stanzel, Franz K. ed. *Europäischer Völkerspiegel. Imagologisch-ethnographische Studien zu den Völkertafeln des frühen 18. Jahrhunderts*. Heidelberg: Winter, 1999.

Stefanon, Marc, Fabio D'Andrea, and Philippe Drobinski. "Heatwave Classification over Europe and the Mediterranean Region." *Environmental Research Letters* 7. (2012): 1-9. doi:10.1088/1748-9326/7/1/014023.

Steffen, Will, Jacques Grinevald, Paul Crutzen and John R. McNeill. "The Anthropocene: Conceptual and Historical Perspectives." *Philosophical Transactions*, Series A 369.1938 (2011): 842-867.

Steffen, Will, Wendy Broadgate, Lisa Deutsch, Owen Gaffney and Cornelia Ludwig. "The Trajectory of the Anthropocene: The Great Acceleration." *The Anthropocene Review* 2.1 (2015): 81-98.

Steffen, Will, Katherine Richardson, Johan Rockström, Sarah E. Cornell, Ingo Fetzer, Elena M. Bennett, Reinette Biggs, Stephen R. Carpenter, Wim de Vries, Cynthia A. de Wit, Carl Folke, Dieter Gerten, Jens Heinke, Georgina M. Mace, Linn M. Peterson, Veerabhadran Ramanathan, Belinda Reyers, Sverker Sörlin. "Planetary Boundaries: Guiding Human Development on a Changing Planet." *Science* (2015). doi:10.1126/science.1259855.

Stern, Nicholas Herbert. *A Blueprint for a Safer Planet: How to Manage Climate Change and Create a New Era of Progress and Prosperity*. London: Bodley Head, 2009.

Stieffel, Philip: *Witterungskunde. Mit Rücksicht auf vermuthliche Witterung überhaupt und des Jahres 1842 insbesondere.* Karlsruhe: Verlag der Chr. Fr. Müller'schen Hofbuchhandlung, 1842.

Stiner, Mary C., and Gillian Feeley-Harnik. "Energy and Ecosystems." In *Deep History: The Architecture of Past and Present*, edited by Andrew Shryock, Daniel Lord Smail, and Timothy K. Earle. Berkeley: University of California Press, 2011.

Stolz, Wolfram. *Die Hans Stolz'sche Gebweiler Chronik: Zeugenbericht über den Bauernkrieg am Oberrhein*. Freiburg i.Br.: Steinmann Druck und Verlag, 1979.

Strauss, Sarah. "Weather Wise: Speaking Folklore to Science in Leukerbad." In *Weather, Climate, Culture*, edited by Sarah Strauss, and Benjamin S. Orlove, 48-51. Oxford: Berg, 2003.

Strohm, Hans. "Einleitung." In *Aristoteles, Meteorologie. Über die Welt, Aristoteles, Werke in deutscher Übersetzung*, vol. 12, edited by Hans Strohm, 121-134. Berlin: Akademie, 1984.

Strype, John, ed. *Annals of the Reformation and Establishment of Religion and Other Various Occurrences in the Church of England, during Queen Elizabeth's Happy Reign*. 4 Vols. Oxford: Clarendon Press, 1824.

Stuart, Tristram. *Waste: Uncovering the Global Food Scandal*. London: Penguin, 2009.

Stückelberger, Johannes. *Wolkenbilder. Deutungen des Himmels in der Moderne*. München: Fink, 2010.

SwissRe (ed.). "Natur- und Man-made-Katastrophen im Jahr 2003: Zahlreiche Todesopfer, vergleichsweise moderate Versicherungsschäden." *Sigma* 1.1 (2004): 1-44.

Swope, Kenneth W. *A Dragon's Head and a Serpent's Tail: Ming China and the First Great East Asian War, 1592-1598*. Norman: University of Oklahoma Press, 2009.

Szulecki, Kacperand Fischer, Severin and Gullberg, Anne Therese and Sartor, Oliver, "Giving Shape to the Energy Union: Evolution, National Expectations and Implications for EU Energy and Climate Governance" (June 17, 2015). Available at SSRN: <http://ssrn.com/abstract=2623892 or http://dx.doi.org/10.2139/ssrn.2623892>.

Tännsjö, Torbjörn. "Future People, the All Affected Principle, and the Limits of the Aggregation: Model of Democracy." In *Hommage à Wodlek: 60 Philosophical Papers Dedicated to Wodlek Rebinowicz*, edited by Jonas Josefsson and Dan Egonsson. Published by Lund University, 2007. Accessed October 13, 2015. <http://www.fil.lu.se/hommageawlodek/site/papper/TannsjoTorbjorn.pdf>.

Taleb, Nassim N. *The Black Swan: The Impact of the Highly Improbable*. London: Allen Lane, 2007.

Tamanini, Jeremy, Andrea Bassi, Camila Hoffman, and Julieth Valenciano. "The Global Green Economy Index GGEI 2014." *Strategic Review 2014* (Washington, New York, 2014). Accessed July 17, 2016. <http://dualcitizeninc.com/GGEI-Report2014.pdf>.

Taub, Liba. *Ancient Meteorology*. London, New York: Routledge, 2003.

Taylor, Charles. *A Secular Age*. Cambridge: Harvard University Press, 2007.

Teodoreanu, Elena. "Preliminary Observations on the Little Ice Age in Romania." *Present Environment and Sustainable Development* 5 (2011): 187-194.

Termeer, Katrien, Art Dewulf, and Gerard Breeman. "Governance of Wicked Climate Adaptation Problems." In *Climate Change Governance*, edited by Jörg Knieling and Walter Leal Filho, 27-40. Berlin: Springer, 2013.

Thomas Aquinas. *Summa theological*, vol. 2. Parma: Petrus Fiaccadorus, 1852. Accessed March 27, 2017. <http://www.mdz-nbn-resolving.de/urn/resolver.pl?urn=urn:nbn:de:bvb:12-bsb10744125-5>.

Thompson, Dennis F. "Representing Future Generations: Political Presentism and Democratic Trusteeship." *Democracy, Equality, and Justice, Critical Review of International Social and Political Philosophy* 13.1 (2010): 17-37.

Thompson, Irving A.A. "Oposición política y juicio del gobierno en las Cortes de 1592-98." *Studia Historica, Historia Moderna*, 17 (1997): 37-62.

Thouret, J.-C., E. Juvigné, A. Gourgaud, P. Boivin, J. Dávila. "Reconstruction of the AD 1600 Huaynaputina Eruption Based on the Correlation of Geologic Evidence with Early Spanish Chronicles." *Journal of Volcanology and Geothermal Research* 115 (2002): 529-70.

"Tie Your Hands, Please: Is Germany's Fiscal Straitjacket an Example for Others?" *The Economist*, December 10, 2011. Accessed November 9, 2015. <http://www.economist.com/node/21541459>.

Tingley, M.P., and P Huybers. "Recent Temperature extremes at High Northern Latitudes Unprecedented in Past 600 Years." *Nature* 496 (2013), 201-5.

Tomorrow, France 2015 (Directors: Melanie Laurent, Cyril Dion; Producer: Bruno Levy; Director of photography: Alexandre Leglise, 118 minutes).

Touchan, Ramzi. "Preliminary Reconstructions of Spring Precipitation in South-Western Turkey from Tree-Ring Width." *International Journal of Climatology* 23 (2003): 157-171.

Touchan, Ramzi. "Reconstruction of Spring/Summer Precipitation for the Eastern Mediterranean from Tree-ring Widths and its Connection to Large-Scale AtmosphericCcirculation." *Climate Dynamics* 25 (2005): 75-98.

Touchan, Ramzi, Ünal Akkemik, Malcolm K. Hughes, and Nesat Erkan. "May–June Precipitation Reconstruction of Southwestern Anatolia, Turkey during the Last 900 Years from Tree Rings." *Quaternary Research* 68.2 (2007): 196-202.

Touchan, Ramzi, Kevin J. Anchukaitis, David M. Meko, Mohammad Sabir, Said Attalah, and Ali Aloui. "Spatiotemporal Drought Variability in Northwestern Africa Over the Last Nine Centuries." *Journal of Climate Dynamics* 37.1-2 (2010): 237-252.

Traber, Thure, Claudia Kemfert, and Jochen Diekmann. "Strompreise: Künftig nur noch geringe Erhöhung durch erneuerbare Energien." *DIW Wochenbericht* 6 (2011): 2-9.

Traber, Thure, and Claudia Kemfert. "Die Auswirkungen des Atomausstiegs in Deutschland auf Strompreise und Klimaschutz in Deutschland und Europa." Study conducted Greenpeace e.V. Published 2012. Accessed August 17, 2012. <https://www.greenpeace.de/sites/www.greenpeace.de/files/20120702-Auswirkungen-Atomausstieg-Gutachten.pdf>.

Trentmann, Frank, ed. *The Oxford Handbook of the History of Consumption*. Oxford, New York: Oxford University Press, 2012.

Trentmann, Frank. *Empire of Things: How We Became a World of Consumers, from the Fifteenth Century to the Twenty-First*. London: Allen Lane, 2016.

Trepl, Ludwig. "Ökologie als Heilslehre: Zum Naturbild der Umweltbewegung." *Politische Ökologie* 25 (1991): 30-45.

Trigo, Ricardo M., José M.C. Pereira, MáRio G. Pereira, Bernardo Mota, Teresa J. Calado, Carlos C. Dacamara, and Fátima E. Santo. "Atmospheric Conditions Associated with the Exceptional Fire Season of 2003 in Portugal." *International Journal of Climatology* 26.13 (2006): 1741-1757. doi:10.1002/joc.1333.

Trouet, Valérie, Jan Esper, Nicholas E. Graham, Andy Baker, James D. Scourse, and David C. Frank. "Persistent Positive North Atlantic Oscillation Mode Dominated the Medieval Climate Anomaly." *Science* 324.5923 (2009): 78-80.

Tschamser, Malachias. *Annales oder Jahrs-Geschichten der Baarfüsern oder Minderen-Brüdern S. Franc. ord. insgemein Conventualen genannt, zu Thann*. Colmar: no publisher, 1864.

Tsing, Anna Lowenhaupt. *Friction: An Ethnography of Global Connection*. Princeton: Princeton University Press, 2005.

Turgot, Anne-Robert-Jacques. *Turgot on Progress, Sociology and Economics: A Philosophical Review of the Successive Advances of the Human Mind, on Universal History* [and] *Reflections on the formation and the Distribution of Wealth*, translated, edited and with an introduction by Ronald L. Meek. Cambridge: Cambridge University Press, 1973.

Turner, Billie L., Roger E. Kasperson, Pamela A. Matson, James J. McCarthy, Robert W. Corell, Lindsey Christensen, Noelle Eckley, Jeanne X. Kasperson, Amy Luers, Marybeth L. Martello, Colin Polsky, Alexander Pulsipher, and Andrew Schiller. "A Framework for Vulnerability Analysis in Sustainability Science." *Proceedings of the National Academy of Sciences* 100.14 (2003): 8074-8079.

Übertragungsnetzbetreiber. "Netzentwicklungsplan 2012." Accessed August 17, 2012. <http://data.netzausbau.de/2022/NEP/NEP2022_UENB-Entwurf2a.pdf>.

Underdal, Arild. "Leadership Theory: Rediscovering the Arts of Management." In *International Multilateral Negotiation: Approaches to the Management of Complexity*, edited by William I. Zartman, 178-197. San Francisco: Jossey-Bass, 1994.

Universität Freiburg, Universitätsbibliothek und Physische Geographie. "Tambora.org: The Climate and Environmental History Collaborative Research Environment." Accessed September 13, 2014. <https://www.tambora.org/>.

Utterström, Gustav. "Climatic Fluctuations and Population Problems in Early Modern History." *Scandinavian Economic History Review* 3 (1955): 3-47.

van Schaik, Louise, and Simon Schunz. "Explaining EU Activism and Impact in Global Climate Politics: Is the Union a Norm- or Interest-Driven Actor?" *Journal of Common Market Studies* 50.1 (2012): 169-186.

van Schaik, Louise. *The EU and the Progressive Alliance Negotiating in Durban: Saving the Climate? ODI Working Paper* 354, London: Overseas Development Institute (ODI), 2012.

Vasak, Anouchka. *Météorologies. Discours sur le ciel et le climat, des Lumières au romantisme. Les dix-huitièmes siècles*, vol. 112. Paris: Champion, 2007.

Vergilius Maro, Publius. *Bucolica* [...], *Georgica* [...]. Florenz, 1487/1488. Accessed March 2017. <http://nbn-resolving.de/urn:nbn:de:bvb:12-bsb00065268-3>.

Verolme, Hans. *European Climate Leadership—Durban and Beyond. Discussion Paper*. Brussels: Heinrich-Böll-Stiftung, 2012.

Vieweg, Marion, Bill Hare, Niklas Höhne, Michiel Schaeffer, Marcia Rocha, Julia Larkin, Hanna Fekete, Johannes Gütschow, and Louise Jeffery. "Climate Shuffle: Climate Action Tracker Update." Published by Climate Analytics, Ecofys, PIK. Accessed

December 22, 2013. <http://www.ecofys.com/files/files/ecofys-ca-pik-2013-climate-action-tracker-bonn-update.pdf>.

Vinet, Freddy. "Climatology of Hail in France." *Atmospheric Research* 56 (2001).

Vogler, John, and Hannes Stephan. "The European Union in Global Environmental Governance: Leadership in the Making?" *International Environmental Agreements: Politics, Law and Economics* 7.4 (2007): 389-413.

Vogt, Markus. "Naturverständnis und christliche Ethik." In *Berichte der ANL: Wir und die Natur – Naturverständnis im Strom der Zeit*, vol. 25, edited by Bayerische Akademie für Naturschutz und Landschaftspflege, 103-111. Laufen: ANL, 2001.

Vogt, Markus. *Climate Justice*. München: Rachel Carson Center for Environment and Society, 2010.

Vogt, Markus. *Prinzip Nachhaltigkeit: Ein Entwurf aus theologisch-ethischer Perspektive*. München: oekom, 3. Auflage 2013.

Vogt, Markus. „Empirie in der Ethik." In *Interdisziplinarität der Christlichen Sozialethik*, edited by Peter Schallenberg, and Arnd Küppers, 405-424. Paderborn: Schöningh, 2013.

Vogt, Steffen, Rüdiger Glaser, Jürg Luterbacher, Dirk Riemann, Ghazi Al Dyab, Johannes Schönbein, and Elena Garcia Bustamante. "Assessing the Medieval Climate Anomaly in the Middle East: The Potential of Arabic Documentary Sources." *PAGES Newsletter* 19.1 (2011): 28-29.

Wack, Pierre. "Scenarios: Uncharted Waters Ahead." *Harvard Business Review* 63.5 (1985): 73-89.

Wagner, Hans ed. *Aristoteles, Physikvorlseung. Aristoteles Werke in deutscher Übersetzung*, vol. 11. Darmstadt: Wissenschaftliche Buchgesellschaft, 196).

Wahl, Laurent, Patrice Paul, Claude Pichard, Nicolas Mory, Gilles Drogue, and Laurent Pfister. "Les canicules de l'été 2003: Un événement météorologique exceptionnel dans le quart nord-est de la France." *Revue géographique de l'est* 45.2 (2005): 67-77.

Walser, Gabriel. *Neue Appenzeller Chronick oder Geschichte des Landes Appenzell der Innern und Äussern Rhoden*. St. Gallen: Selbstverlag, 1740.

Walter, John. "A 'Rising of the People'? The Oxfordshire Rising of 1596." *Past and Present* 107 (1985): 90-143.

Waquet, Henri, ed. Mémoires du Chanoine Jean Moreau sur les Guerres de la Ligue en Bretagne. Quimper: Archives historiques de Bretagne, 1960.

Ward, David M. "The Effect of Weather on Grid Systems and the Reliability of Electricity Supply." *Climatic Change* 121.1 (2013): 103-113. doi:10.1007/s10584-013-0916-z.

"Wasser – das wichtigste Futtermittel für Kühe." Published by Tiergesundheit und mehr. Last modified October 9, 2013. Accessed April 6, 2017. <http://www.tiergesundheitundmehr.de/wasser-fuer-kuehe.pdfx>.

Wattenbach, Wilhelm, Ernst Dümmler, and Franz Huf. Deutschlands Geschichtsquellen im Mittelalter: Frühzeit und Karolinger. 2 Vols. Kettwig: Phaidon, 1991.

Watts, Michael J., and Hans-Georg Bohle. "The Space of Vulnerability: The Causal Structure of Hunger and Famine." *Progress in Human Geography* 17.1 (1993): 43-67.

Weart, Spencer R. *The Discovery of Global Warming*. Cambridge, London: Harvard University Press, 2003.

Weart, Spencer R. *The Discovery of Global Warming*. Cambridge: Harvard University Press, 2008.

Weart, Spencer R. *The Discovery of Global Warming* (online), chapter "Money for Keeling: Monitoring CO_2 Levels." Accessed April 5, 2017. <http://history.aip.org/climate/Kfunds.htm>.

Weber, Erich Martin. *Untiefen, Flut und Flauten: Der Güterverkehr auf dem Rhein zwischen 1750 und 1850: Die Modernisierung der vorindustriellen Rheinschifffahrt aus einer wirtschafts-, sozial- und umweltgeschichtlichen Perspektive betrachtet*. Bern: Selbstverlag, 2005. Accessed November 20, 2013. <http://www.zb.unibe.ch/download/eldiss/05weber_e.pdf>.

Wehry, Werner, and Franz Ossing, ed. *Wolken – Malerei – Klima in Geschichte und Gegenwart*. Berlin: Deutsche Meteorologische Gesellschaft, 1997.

Weichenhan, Michael. *'Ergo perit coelum...' Die Supernova des Jahres 1572 und die Überwindung der aristotelischen Kosmologie*. Stuttgart: Steiner, 2004.

Weigand, Kurt ed. *Montesquieu* [Charles-Louis de Secondat Baron de Brède et de], *Vom Geist der Gesetze*. Stuttgart: Reclam, 1965.

Weikinn, Curt, Michael Börngen, and Gerd Tetzlaff ed. *Quellentexte zur Witterungsgeschichte Europas von der Zeitenwende bis zum Jahre 1850. Hydrographie.* 6 Vols. Berlin: Akademie, 1958-1963, 2000/2002.

Weiler, Ingomar. "Ethnographische Typisierungen im antiken und mittelalterlichen Vorfeld der' 'Völkertafel'." In *Europäischer Völkerspiegel: Imagologisch-ethnographische Studien zu den Völkertafeln des frühen 18. Jahrhunderts*, edited by Franz K. Stanzel, 97-118. Heidelberg: Winter, 1999.

Wiemann, Elsbeth ed. *Die Entdeckung der Landschaft. Meisterwerke der niederländischen Kunst des 16. und 17. Jahrhunderts*. Stuttgart: Staatsgalerie; Köln: Dumont, 2005.

Welker, Michael. *Schöpfung und Wirklichkeit*. Neukirchen-Vluyn: Neukirchener, 1995.

Wells, H.G. *The Time Machine*.1895.

Welzer, Harald. "Vergangene Zukünfte und zukünftige Vergangenheiten." In *Erzählte Zukunft: Zur inter- und intragenerationellen Aushandlung von Erwartungen*, edited by Jens Kroh and Sophie Neuenkirch, 19-26. Göttingen: Wallstein, 2011.

Werner, Hans-Joachim. *Eins mit der Natur: Mensch und Natur bei Franz von Assisi, Jakob Boehme, Albert Schweitzer, Teilhard de Chardin*. München: C.H. Beck, 1986.

Wetter, Oliver, and Christian Pfister. "Spring–Summer Temperatures Reconstructed for Northern Switzerland and Southwestern Germany from Winter Rye Harvest Dates, 1454-1970." *Climate of the Past* 7.4 (2011): 1307-1326.

Wetter, Oliver, Christian Pfister, Rolf Weingartner, Jürg Luterbacher, Tom Reist, and Jürg Trösch. "The Largest Floods in the High Rhine Basin since 1268 Assessed from Documentary and Instrumental Evidence." *Hydrological Science Journal* 56.5 (2011): 733-758.

Wetter, Oliver, and Christian Pfister. "An Underestimated Record Breaking Event: Why Summer 1540 Was Likely Warmer than 2003." *Climate of the Past* 9 (2013): 1-16. doi:10.5194/cp-9-41-2013.

Wetter, Oliver, Christian Pfister, Johannes P. Werner, Eduardo Zorita, Sebastian Wagner, Sonia I. Seneviratne, Jürgen Herget, Uwe Grünewald, Jürg Luterbacher, Maria-Joao Alcoforado, Mariano Barriendos, Ursula Bieber, Rudolf Brázdil, Karl H. Burmeister, Chantal Camenisch, Antonio Contino, Petr Dobrovolný, Rüdiger Glaser, Iso Himmelsbach, Andrea Kiss, Oldřich Kotyza, Thomas Labbé, Danuta Limanówka, Laurent Litzenburger, Øyvind Nordli, Kathleen Pribyl, Dag Retsö, Dirk Riemann, Christian Rohr, Werner Siegfried, Johan Söderberg, and Jean-Laurent Spring. "The Year-long Unprecedented European Heat and Drought of 1540—A Worst Case." *Climatic Change* 125 (2014): 349-363. doi: 0.1007/s10584-014-1184-2.

White, Jonathan. "Thinking Generations." *British Journal of Sociology* 64.2 (2013): 216-247.

White, Leslie A. "Energy and the Evolution of Culture." *American Anthropologist* 45.3 (1943): 335-356.

White, Sam. *The Climate of Rebellion in the Early Modern Ottoman Empire*. Cambridge: Cambridge University Press, 2011.

White, Sam. "Cold, Drought, and Disaster: The Little Ice Age and the Spanish Conquest of New Mexico." *New Mexico Historical Review* 89 (2014): 425-58.

White, Sam. *A Cold Welcome. The Little Ice Age and Europe's Encounter with North America* (forthcoming 2017).

Wielenga, Friso. "Konsens im Polder: Politik und politische Kultur in den Niederlanden nach 1945." In *Länderbericht Niederlande: Geschichte – Wirtschaft – Gesellschaft*, edited by Friso Wielenga und Ilona Taute, 13-130. Bonn: Bundeszentrale für Politische Bildung, 2004.

Wielenga, Friso. "Grundzüge niederländischer Politik im 20. Jahrhundert." In *Nachbar Niederlande: Eine landeskundliche Einführung*, edited by Friso Wielenga, 13-60. Münster: Aschendorff, 2007.

"Wind Water en Zand Spelen hun Spel bij Strand Ballum." *Leeuwarder Courant*, August 11, 2000, 9.

Wilkinson, Angela and Roland Kupers. "Living in the Futures: How Scenario Planning Changed Corporate Strategy." *Harvard Business Review* 91.5 (2013): 3-11.

Wilkinson, Angela and Roland Kupers. *The Essence of Scenarios: Learning from Shell Experience*. Amsterdam: Amsterdam University Press, 2014.

Wimmer, Andreas, and Nina Glick-Schiller. "Methodological Nationalism and Beyond: Nation-State Building, Migration and the Social Sciences." *Global Networks* 2.4 (2002): 301-334.

World Bank. "The Turn Down Heat: Why a 4°C Warmer World Must be Avoided." Report for the World Bank by the Potsdam Institute for Climate Impact Research and Climate Analytics. Published November 2012. Accessed November 20, 2014. <http://www.worldbank.org/en/topic/climatechange/publication/turn-down-the-heat-climate-extremes-regional-impacts-resilience>.

World Bank. *Turn Down the Heat*. New York: International Bank for Reconstruction and Development/The World Bank, 2012.

Wrightson, Keith, and John Walter. "Dearth and the Social Order in Early Modern England." *Past and Present* 71 (1976): 22-42.

Wriothesley, Charles. *A Chronicle of England during the Reigns of the Tudors, from AD 1485 to 1559*. Edited by William Douglas Hamilton. Westminster: J.B. Nichols and Sons, 1875.

Wulf, Andrea. *The Invention of Nature. The Adventures of Alexander von Humboldt—The Last Hero of Science*. London: John Murray, 2015.

Wurstisen, Christian. *Basler Chronik*. Basel: Sebastian Henricpetri, 1580.

Yılmazer, Ziya., ed. *Topçular Kâtibi 'Abdülkādir (Kadrî) Efendi Tarihi* [The Chronicle of Abdulkadir Efendi, Secretary of the Artillery Corps]. 2 Vols. Ankara: Türk Tarih Kurumu, 2003.

Yong, Xue. "Agrarian Urbanization: Social and Economic Changes in Late Imperial Jiangnan." PhD diss., Yale University, 2006.

Young, Oran R. "Political Leadership and Regime Formation: On the Development of Institutions in International Society." *International Organization* 45.3 (1991): 281-308.

Zumbühl, Heinz J. *Die Schwankungen der Grindelwaldgletscher in den historischen Bild- und Schriftquellen des 12. bis 19. Jahrhunderts*. Basel: Birkhäuser Verlag AG, 1980.

Züricher, Gertrud, ed. *Karl Friedrich Ludwig Lohners Chronik der Stadt Thun*. Bern: Haupt, 1935.

Zweifel, Peter, and Roland Eisen. *Versicherungsökonomie*. Berlin: Springer, 2003.

Zwierlein, Cornel. *Der gezähmte Prometheus. Feuer und Sicherheit zwischen früher Neuzeit und Moderne*. Göttingen: Vandenhoeck & Ruprecht, 2011.

Wisner, Ben, Piers Blaikie, Terry Cannon, and Ian Davis. *At Risk: Natural Hazards, People's Vulnerability and Disasters*. London: Routledge, 2004.

Württembergische Jahrbücher, vol. 1. Stuttgart, 1818.

Wikipedia. "Hitzewelle 2003." Last modified January 21, 2014. Accessed April 6, 2017. <http://de.wikipedia.org/wiki/Hitzewelle_2003>.

Wikipedia. "Palmer Drought Index." Accessed December 23, 2014. <http://en.wikipedia.org/wiki/Palmer_Drought_Index>.

Zacharasiewicz, Waldemar. "Klimatheorie und Nationalcharakter auf der 'Völkertafel'." in *Europäischer Völkerspiegel: Imagologisch-ethnographische Studien zu den Völkertafeln des frühen 18. Jahrhunderts*, edited by Franz K. Stanzel, 119-137. Heidelberg: Winter, 1999.

Zacharasiewicz, Waldemar. *Die Klimatheorie in der englischen Literatur und Literaturkritik von der Mitte des 16. Jahrhunderts bis zum frühen 18. Jahrhundert. Wiener Beiträge zur englischen Philologie*, vol. 77. Stuttgart: Braumüller, 1977.

Zander, Helmut. *Anthroposophie in Deutschland. Theosophische Weltanschauung und gesellschaftliche Praxis 1884 –1945*. Göttingen: Vandenhoeck, 2007/2008.

Zittel, Claus, ed, *René Descartes, Les météores / Die Meteore*. Frankfurt a.M.: Klostermann 2006.

Zittel, Claus. *Theatrum philosophicum. Descartes und die Rolle ästhetischer Formen in der Wissenschaft.* Berlin: Akademie 2009.

Index

www.ingramcontent.com/pod-product-compliance
Lightning Source LLC
LaVergne TN
LVHW021724021026
849161LV00004BA/670